Calcium Entry Channels in Non-Excitable Cells

METHODS IN SIGNAL TRANSDUCTION SERIES

Joseph Eichberg, Jr. and Michael X. Zhu

Series Editors

Published Titles

Calcium Entry Channels in Non-Excitable Cells, Juliusz Ashot Kozak and James W. Putney, Jr.

Lipid-Mediated Signaling Transduction, Second Edition, Eric Murphy, Thad Rosenberger, and Mikhail Golovko

Signaling Mechanisms Regulating T Cell Diversity and Function, Jonathan Soboloff and Dietmar J. Kappes

Gap Junction Channels and Hemichannels, Donglin Bai and Juan C. Sáez

Cyclic Nucleotide Signaling, Xiaodong Cheng

TRP Channels, Michael Xi Zhu

Lipid-Mediated Signaling, Eric J. Murphy and Thad A. Rosenberger

Signaling by Toll-Like Receptors, Gregory W. Konat

Signal Transduction in the Retina, Steven J. Fliesler and Oleg G. Kisselev

Analysis of Growth Factor Signaling in Embryos, Malcolm Whitman and Amy K. Sater

Calcium Signaling, Second Edition, James W. Putney, Jr.

G Protein-Coupled Receptors: Structure, Function, and Ligand Screening, Tatsuya Haga and Shigeki Takeda

G Protein-Coupled Receptors, Tatsuya Haga and Gabriel Berstein

Signaling Through Cell Adhesion Molecules, Jun-Lin Guan

G Proteins: Techniques of Analysis, David R. Manning

Lipid Second Messengers, Suzanne G. Laychock and Ronald P. Rubin

Calcium Entry Channels in Non-Excitable Cells

Edited by

Juliusz Ashot Kozak and James W. Putney, Jr.

CRC Press
Taylor & Francis Group
Boca Raton London New York

CRC Press is an imprint of the
Taylor & Francis Group, an **informa** business

CRC Press
Taylor & Francis Group
6000 Broken Sound Parkway NW, Suite 300
Boca Raton, FL 33487-2742

Printed on acid-free paper

International Standard Book Number-13: 978-1-4987-5272-5 (Hardback)

Library of Congress Cataloging-in-Publication Data

Names: Kozak, Juliusz Ashot, editor. | Putney, James W., Jr., editor.
Title: Calcium entry channels in non-excitable cells / [edited by] Juliusz Ashot Kozak and James W. Putney, Jr.
Other titles: Methods in signal transduction.
Description: Boca Raton : Taylor & Francis, 2017. | Series: Methods in signal transduction series | Includes bibliographical references and index.
Identifiers: LCCN 2016053681 | ISBN 9781498752725 (hardback : alk. paper)
Subjects: | MESH: Calcium Channels--physiology | Calcium Signaling--physiology
Classification: LCC QP552.C24 | NLM QU 55.7 | DDC 572/.696--dc23
LC record available at https://lccn.loc.gov/2016053681

Visit the Taylor & Francis Web site at
http://www.taylorandfrancis.com

and the CRC Press Web site at
http://www.crcpress.com

Contents

Series Preface

The concept of signal transduction is now long established as a central tenet of biological sciences. Since the inception of the field close to 50 years ago, the number and variety of signal transduction pathways, cascades, and networks have steadily increased and now constitute what is often regarded as a bewildering array of mechanisms by which cells sense and respond to extracellular and intracellular environmental stimuli. It is not an exaggeration to state that virtually every cell function is dependent on the detection, amplification, and integration of these signals. Moreover, there is increasing appreciation that in many disease states, aspects of signal transduction are critically perturbed.

Our knowledge of how information is conveyed and processed through these cellular molecular circuits and biochemical switches has increased enormously in scope and complexity since this series was initiated 15 years ago. Such advances would not have been possible without the supplementation of older technologies, drawn chiefly from cell and molecular biology, biochemistry, physiology, pharmacology, with newer methods that make use of sophisticated genetic approaches, as well as structural biology, imaging, bioinformatics, and systems biology analysis.

The overall theme of this series continues to be the presentation of the wealth of up-to-date research methods applied to the many facets of signal transduction. Each volume is assembled by one or more editors who are preeminent in their specialty. In turn, the guiding principle for editors was to recruit chapter authors who can describe procedures and protocols with which they are intimately familiar in a reader-friendly format. The intent is to assure that each volume is of maximum practical value to a broad audience, including students and researchers just entering an area, as well as seasoned investigators.

It is hoped that the information contained in the books of this series will constitute a useful resource to the life sciences research community well into the future.

Joseph Eichberg
Michael Xi Zhu
Series Editors

Preface

Virtually every eukaryotic cell expresses at least some type of calcium channel, in the plasma membrane, in intracellular organelles, or typically both. Calcium channels in the plasma membrane have long been of interest to biologists because of their vulnerability to modulation by extracellular factors, with either pathological or therapeutic consequences. Plasma membrane calcium channels are regulated by membrane voltage or by ligands, and in some cases by both. Not surprisingly, voltage-activated channels are generally encountered in cells that depend largely on excitable behavior, for example, muscle and nerve. Calcium channels that are activated by ligands are more broadly distributed but are the exclusive mediators of transmembrane calcium flux in non-excitable cells, for example, blood cells and epithelial cells.

The history of voltage-activated calcium channels is nicely summarized in a review by Richard Tsien [1], dating back to the seminal work of Sidney Ringer at the end of the nineteenth century. In non-excitable cells, the history is more recent, and much of the early work was carried out in smooth muscles, a tissue that expresses both voltage-activated and ligand-gated calcium channels. It might be said that the first mechanistic insights into calcium channels in non-excitable cells came not from studies of the channels themselves but of the initial receptor mechanisms upstream. By far, the most commonly encountered receptor signaling mechanism leading to calcium regulation in non-excitable cells is the receptor-activated phosphatidylinositol 4,5-*bis*phosphate–directed phospholipase C. Through distinct downstream mechanisms, this initial enzymatic event gives rise to the activation of the two major classes of non-voltage-activated calcium channels, the store-operated channels and TRPC channels, a subgroup of the larger TRP ion channel superfamily. The store-operated channels are activated in response to the depletion of endoplasmic reticulum stores by the phospholipase C product, inositol 1,4,5-trisphosphate, while the TRPCs appear to be activated by phospholipase C–mediated changes in membrane lipid composition. And not surprisingly, there are instances in which the two channels interact and regulate one another. The history of the phospholipase C signaling field is well described by Robert Michell [2], and the histories of the TRP channels and store-operated channels are well described by Craig Montell and Roger Hardie [3,4] and Jim Putney [5], respectively. This volume focuses on physiologically and potentially clinically important channels in non-excitable cells, primarily the store-operated channels and TRP channels, including TRPC channels and the highly Ca^{2+}-selective TRPV5 and TRPV6. In keeping with the theme of this series, there is considerable emphasis on the specific methods required for studying them. This topic was addressed in part in an earlier volume in this series published in 1999 focusing more broadly on calcium signaling [6]. An underlying impetus to produce the current volume comes from the remarkable advances in this field in the past few years. Following the discovery of the first mammalian TRP channels by Zhu and Birnbaumer, the superfamily of TRP channels virtually exploded. After years of false leads, the molecular components of store-operated channels were revealed by the use of modern high-throughput genetic screens, first the calcium sensor STIM, and subsequently the pore-forming store-operated channel subunit, Orai.

In addition to the in-depth analysis of the molecular and physiological aspects of these channels, the reader will no doubt be struck by the breadth of methodological approaches involved in studying them. Enough cannot be said of the contributions of the late Roger Tsien in developing readily available chemical as well as genetically encoded calcium indicators, making possible the measurement of calcium concentration in the cytoplasm and in organelles in real time. Complementing this approach is the use of the techniques of electrophysiology, FRET, optogenetics, and x-ray crystallography, to name a few. Hopefully, this collection will be of use to scientists investigating these important channels, from professor to graduate student, and in academia, government, or industry.

MATLAB® is a registered trademark of The MathWorks, Inc. For product information, please contact:

The MathWorks, Inc.
3 Apple Hill Drive
Natick, MA 01760-2098 USA
Tel: 508-647-7000
Fax: 508-647-7001
E-mail: info@mathworks.com
Web: www.mathworks.com

REFERENCES

1. Tsien, R.W. and Barrett, C.F. 2005. A brief history of calcium channel discovery. In *Voltage Gated Calcium Channels*. G. Zamponi, ed. Plenum Press, New York, pp. 27–47.
2. Michell, R.H. 1986. Inositol lipids and their role in receptor function: History and general principles. In *Phosphoinositides and Receptor Mechanisms*. J.W. Putney, ed. Alan R. Liss, Inc., New York, pp. 1–24.
3. Hardie, R.C. 2011. A brief history of TRP: Commentary and personal perspective. *Pflügers Archiv: European Journal of Physiology* 461:493–498.
4. Montell, C. 2011. The history of TRP channels, a commentary and reflection. *Pflügers Archiv: European Journal of Physiology* 461:499–506.
5. Putney, J.W. 2007. Recent breakthroughs in the molecular mechanism of capacitative calcium entry (with thoughts on how we got here). *Cell Calcium* 42:103–110.
6. Putney, J.W. 1999. *Calcium Signaling*. CRC Press, Boca Raton, FL.

Contributors

Ethel Adap
Department of Physiology and Biophysics
Weill Cornell Medicine Qatar
Doha, Qatar

Alexander Asanov
TIRF Labs Inc.
Cary, North Carolina

Gary S. Bird
Signal Transduction Laboratory
National Institute of Environmental Health Sciences
National Institutes of Health
Research Triangle Park, North Carolina

Laura G. Ceballos
Department of Cell and Developmental Biology
Institute of Cellular Physiology
National Autonomous University of Mexico
Mexico City, Mexico

Raphael Courjaret
Department of Physiology and Biophysics
Weill-Cornell Medical College
Doha, Qatar

Maya Dib
Department of Physiology and Biophysics
Weill Cornell Medicine Qatar
Doha, Qatar

Marc Fahrner
Life Science Center
Institute of Biophysics
Johannes Kepler University
Linz, Austria

Nicola Fameli
Institute of Biophysics
Medical University of Graz
Graz, Austria

Jianlin Feng
Department of Cell Biology
University of Connecticut School of Medicine
Farmington, Connecticut

Donald L. Gill
Department of Cellular and Molecular Physiology
Pennsylvania State University College of Medicine
Hershey, Pennsylvania

Klaus Groschner
Institute of Biophysics
Medical University of Graz
Graz, Austria

Aparna Gudlur
Division of Signaling and Gene Expression
La Jolla Institute for Allergy and Immunology
La Jolla, California

Maxime Gueguinou
Department of Cellular and Molecular Physiology
Pennsylvania State University College of Medicine
Hershey, Pennsylvania

Yousang Gwack
Department of Physiology
University of California
Los Angeles, California

Gergely Gyimesi
Institute of Biochemistry and Molecular Medicine
and
Swiss National Center of Competence in Research
TransCure
University of Bern
Bern, Switzerland

Lian He
Institute of Biosciences and Technology
Texas A&M University Health Science Center
Houston, Texas

Matthias A. Hediger
Institute of Biochemistry and Molecular Medicine
and
Swiss National Center of Competence in Research
TransCure
University of Bern
Bern, Switzerland

Rawad Hodeify
Department of Physiology and Biophysics
Weill Cornell Medicine Qatar
Doha, Qatar

Patrick Hogan
Division of Signaling and Gene Expression
La Jolla Institute for Allergy and Immunology
La Jolla, California

Yun Huang
Center for Epigenetics and Disease Prevention
Texas A&M University Health Science Center
Houston, Texas

Satanay Hubrack
Department of Physiology and Biophysics
Weill Cornell Medicine Qatar
Doha, Qatar

J. Ashot Kozak
Department of Neuroscience, Cell Biology and Physiology
Boonshoft School of Medicine
Wright State University
Dayton, Ohio

Xiuyuan Lu
Division of Molecular Immunology
Medical Institute of Bioregulation
Kyushu University
Fukuoka, Japan

Khaled Machaca
Department of Physiology and Biophysics
Weill Cornell Medicine Qatar
Doha, Qatar

Nancy Nader
Department of Physiology and Biophysics
Weill Cornell Medicine Qatar
Doha, Qatar

Masatsugu Oh-Hora
Division of Molecular Immunology
Medical Institute of Bioregulation
Kyushu University
Fukuoka, Japan

Anant B. Parekh
Department of Physiology, Anatomy and Genetics
University of Oxford
Oxford, United Kingdom

Ji-Bin Peng
Division of Nephrology
Nephrology Research and Training Center
and
Department of Urology
University of Alabama at Birmingham
Birmingham, Alabama

James W. Putney, Jr.
Signal Transduction Laboratory
National Institute of Environmental Health Sciences
National Institutes of Health
Research Triangle Park, North Carolina

Christoph Romanin
Life Science Center
Institute of Biophysics
Johannes Kepler University
Linz, Austria

Grigori Rychkov
School of Medicine
University of Adelaide
and
South Australian Health and Medical Research Institute
Adelaide, South Australia, Australia

Kei Saotome
Department of Neuroscience
Howard Hughes Medical Institute
Dorris Neuroscience Center
The Scripps Research Institute
La Jolla, California

and

Department of Integrative Structural and Computational Biology
The Scripps Research Institute
La Jolla, California

Rainer Schindl
Life Science Center
Institute of Biophysics
Johannes Kepler University
Linz, Austria

Niroj Shrestha
Institute of Biophysics
Medical University of Graz
Graz, Austria

Appu K. Singh
Department of Biochemistry and Molecular Biophysics
Columbia University
New York, New York

Alexander I. Sobolevsky
Department of Biochemistry and Molecular Biophysics
Columbia University
New York, New York

Sonal Srikanth
Department of Physiology
University of California
Los Angeles, California

Lu Sun
Department of Physiology and Biophysics
Weill Cornell Medicine Qatar
Doha, Qatar

Yoshiro Suzuki
Division of Cell Signaling
National Institutes of Natural Sciences
and
Department of Physiological Sciences
SOKENDAI (The Graduate University for Advanced Studies)
Okazaki, Japan

Mohamed Trebak
Department of Cellular and Molecular Physiology
Pennsylvania State University College of Medicine
Hershey, Pennsylvania

Luis Vaca
Department of Cell and Developmental Biology
Institute of Cellular Physiology
National Autonomous University of Mexico
Mexico City, Mexico

Youjun Wang
Beijing Key Laboratory of Gene Resources and Molecular Development
Beijing Normal University
Haidian, Beijing, People's Republic of China

Jin Seok Woo
Department of Physiology
University of California
Los Angeles, California

Yi-Chun Yeh
Department of Physiology, Anatomy and Genetics
University of Oxford
Oxford, United Kingdom

Albert S. Yu
Department of Cell Biology
University of Connecticut School of Medicine
Farmington, Connecticut

Fang Yu
Department of Physiology and Biophysics
Weill Cornell Medicine Qatar
Doha, Qatar

Lixia Yue
Department of Cell Biology
University of Connecticut School of Medicine
Farmington, Connecticut

Zhichao Yue
Department of Cell Biology
University of Connecticut School of Medicine
Farmington, Connecticut

Qian Zhang
Institute of Biosciences and Technology
Texas A&M University Health Science Center
Houston, Texas

Xuexin Zhang
Department of Cellular and Molecular Physiology
Pennsylvania State University College of Medicine
Hershey, Pennsylvania

Yandong Zhou
Department of Cellular and Molecular Physiology
Pennsylvania State University College of Medicine
Hershey, Pennsylvania

Yubin Zhou
Institute of Biosciences and Technology
Texas A&M University Health Science Center
Houston, Texas

and

Department of Medical Physiology
Texas A&M University Health Science Center
Temple, Texas

1 Electrophysiological Methods for Recording CRAC and TRPV5/6 Channels

J. Ashot Kozak and Grigori Rychkov

CONTENTS

1.1 INTRODUCTION

During the past two decades, great advances have been made in the electrophysiological and molecular identification of calcium entry pathways in non-excitable cells. The term "non-excitable" refers to a variety of cell types that are not capable of firing action potentials. Essentially, except for neurons, muscle cells, and some endocrine cells, all other cells in the body are non-excitable. For the most part, they lack the necessary levels of expression of voltage-gated Na^+ channels (Na_V family) and also voltage-gated Ca^{2+} channels (Ca_V family). Ca^{2+} influx in these cell types is therefore thought to rely on unrelated, voltage-independent channels, such as Orai (CRACM) and TRP family members. In this chapter, we will discuss direct electrophysiological methods used to record the electrical activity of these proteins, focusing on calcium-selective Orai/STIM and TRPV5/TRPV6 channels.

1.2 CHARACTERISTICS OF CALCIUM ENTRY IN NON-EXCITABLE CELLS

One of the first non-excitable cell types used to investigate the function of non-voltage-gated Ca^{2+} channels was cells of the immune system [1]. In T lymphocytes and mast cells, store-operated calcium entry is the main pathway providing cytoplasmic calcium elevations necessary for key cellular functions, such as antigenic activation, proliferation, and degranulation. Calcium stores inside the cell that were shown to be important for CRAC channel activation and functions are the endoplasmic reticulum (ER) and mitochondria [2–6].

Direct evidence for calcium entry following store depletion was demonstrated using the now classical calcium readdition protocol in cells loaded with calcium indicator dyes, such as Fura-2. Figure 1.1 shows a calcium imaging experiment in Jurkat T lymphocytes loaded with Fura-2 ratiometric calcium dye. Upon the removal of calcium and the simultaneous addition of sarcoplasmic/endoplasmic reticulum calcium ATPase (SERCA) pump inhibitor cyclopiazonic acid (CPA) [7], a transient calcium elevation was observed. This is believed to represent the release of ionized calcium from the ER into the cytoplasm (Fura-2 indicator is in the cytoplasm). Calcium release in this case is mediated through a "Ca^{2+} leak" pathway operating in the ER membrane. Normally, SERCA acts to sequester cytoplasmic calcium into the ER, counteracting this outwardly directed calcium leak. When SERCA is blocked by

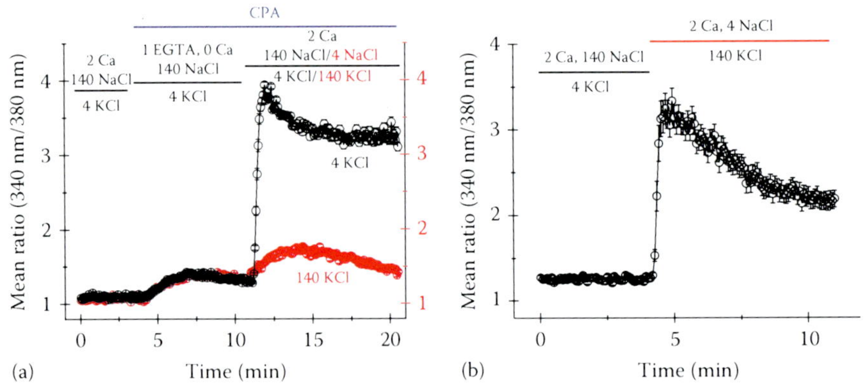

FIGURE 1.1 Fura-2 measurement of cytosolic calcium in intact human Jurkat T lymphocytes (a) and murine MIN6 β cells (b). (a) Calcium entry through endogenous Orai channels was measured with the calcium readdition protocol after ER store depletion with 20 μM CPA. External $[K^+]$ was either 4 mM (shown in black) or elevated to 140 mM (in red) after calcium was reintroduced to the bath. $[Na^+]$ was 140 and 4 mM for (a) and (b), respectively. What, if any, influence the concomitant changes in bathing $[Na^+]$ have on SOCE is not known. Each black and red symbol represents the mean response from 49 and 36 cells. (b) Calcium entry through endogenous Ca_V channels [142] was evoked by $[K^+]$ elevation in the bathing solution (as in (a)). Cl^- concentration was kept constant in (a) and (b). Each symbol represents the mean response from 14 cells. Error bars in (a) and (b) represent SEM.

CPA, however, the calcium flux into the cytoplasm due to the leak pathway is no longer counteracted and this is manifested as a calcium elevation. The transient nature of this elevation is likely due to plasma membrane Ca^{2+} ATPase (PMCA), which is not sensitive to CPA and can still expel calcium ions from the cytoplasm.

The basal ER "Ca^{2+} leak" pathway remains an enigma both in terms of its molecular identity and the factors that regulate it. It is thought to participate in determining the ER calcium content (also termed "calcium load"). Over the years several ion channels have been proposed to underlie the ER leak such as the translocon complex (translocation channel), pannexins, TRPP2 (polycystin), Bcl2 anti-apoptotic proteins, and even IP_3R channels functioning in the absence of ligand binding [8–12]. It is presently unclear if Ca^{2+} leak channels are regulated by ER calcium content and cytoplasmic Ca^{2+} or if they are constitutively open. In whole-cell patch-clamp experiments, the leak pathway appears to function throughout the duration of the recording, enabling the prolonged continuous detection of CRAC channel activity with passive store depletion. Almost no information is available on their pharmacology. Basal leak pathways have also been described in the cardiac sarcoplasmic reticulum, and ryanodine receptors have been suggested to participate in basal leak under specific circumstances [10].

The reintroduction of calcium to the bathing solution (in the presence of CPA) results in a large calcium elevation, which represents store-operated calcium entry (SOCE). This process is a direct consequence of emptying the ER calcium store, as this pathway is not active without CPA or thapsigargin application. In lymphocytes, SOCE is usually larger than the ER release transient and also decays more slowly. When extracellular [K^+] is increased from 4 to 140 mM (Figure 1.1b) calcium increases are drastically diminished. K^+ elevation moves the potassium Nernst potential from −90 mV to approximately 0 mV, a 90 mV depolarization. This is expected to depolarize the membrane potential of the Jurkat T cell, which is set by the voltage-gated $K_V1.3$ and calcium-activated $K_{Ca}3.1$ channels, as well as the two-pore voltage-independent K^+ channels [13,14]. Depolarized potentials result in a reduced driving force for Ca^{2+}, thereby decreasing Ca^{2+} influx. This in turn reduces $K_{Ca}3.1$ currents in a positive feedback loop, causing Ca^{2+} transients to decay faster [1]. Importantly, the shape (time course) of SOCE is set by PMCA activity [6,15–18]. In short, membrane depolarization reduces rather than increases SOCE, demonstrating that in lymphocytes, voltage-gated Ca_V channels are not the underlying calcium influx pathway. Accordingly, the blockade of $K_V1.3$ in lymphocytes invariably inhibits SOCE and suppresses proliferation, which requires SOCE [19,20]. Similar SOCE-K^+ channel systems exist in other cell types [21]. Note that the CPA-induced store calcium transient is not affected by the rise in the extracellular K^+ concentration since it does not depend on the plasma membrane potential. SOCE is not entirely abolished in high K^+, presumably because the calcium equilibrium potential is well above 0 mV [22]. By contrast, in excitable cells, such as pancreatic β cells, depolarizations caused by increasing [K^+] result in a substantial Ca^{2+} influx through voltage-gated Ca^{2+} channels of the Ca_V family (Figure 1.1b). In β cells the major Ca_V subtype is the dihydropyridine-sensitive L-type channel, which opens at membrane potentials above −10 mV [23,24].

1.3 CRAC CHANNELS

1.3.1 CRAC Channels in the Native Environment

Early electrophysiological evidence for the existence of a non-voltage-gated calcium channel was published in the 1990s (reviewed in [25]). Two groups demonstrated that the intracellular application of inositol trisphosphate (IP_3) and high concentrations of Ca^{2+} buffer EGTA resulted in the gradual development of an inwardly rectifying current that did not exhibit any dependence of gating on voltage. This Ca^{2+} current was characterized in detail in Jurkat T and rat basophilic leukemia (RBL) cell lines. The channels responsible were named CRAC, for calcium release–activated Ca^{2+}, even though they are in fact activated by store depletion and not calcium release *per se* [26,27]. It was debated for some time whether calcium release causes calcium influx [28] or if IP_3 directly activates calcium influx channels in the plasma membrane [29–33].

In the past decade, the molecular identity of CRAC channels has been discovered to consist of two key components, STIM and Orai (CRACM). STIM1, stromal interaction molecule, is a single-pass transmembrane protein residing in the ER (but also in the plasma membrane) that via its EF hand domains senses Ca^{2+} concentration in the ER lumen [34]. Upon the emptying of the Ca^{2+} stores, STIM1 concentrates in junctional ER in close apposition to the plasma membrane, being able now to activate the pore subunit of the CRAC channel, Orai [35] (see Chapter 3). Orai1–3 form a three-member family of four transmembrane domain proteins, which can be activated by store depletion in overexpression systems [35–37]. STIM2, a homologue of STIM1, has been shown to have a role in setting the resting cytoplasmic calcium levels by virtue of its sensing smaller reductions of ER [Ca^{2+}] [38] (see also Chapter 6).

In the following, we discuss in detail the steps required to record CRAC currents in whole-cell patch clamp. In order to deplete ER calcium stores, it is sufficient to simply include high concentrations of calcium chelators EGTA or BAPTA in internal recording solutions. Normally, CRAC channel activity is not detectable immediately after establishing the whole-cell configuration (i.e., break-in), because the calcium stores are full. As the chelator diffuses into the cell, cytoplasmic calcium is drastically lowered. The ER stores are then gradually emptied through ER leak channels [39,40], transporting calcium down its concentration gradient into the cytoplasm, where it is captured by the chelator and prevented from being pumped into ER by SERCA [41]. CRAC channel activation is proportional to the degree of store emptying, that is, inversely proportional to ER [Ca^{2+}] [42,43].

Under physiological conditions, the ER calcium stores are emptied by the second messenger inositol trisphosphate (IP_3) [2,44]. IP_3 is generated from the hydrolysis of the plasma membrane phospholipid $PI(4,5)P_2$ by phospholipase C enzymes [5,45]. Various types of PLC are stimulated by G proteins or tyrosine kinase–linked receptors, depending on the tissue and ligand [46]. IP_3 can bind to its receptors (IP_3R) in the ER membrane, which are calcium-permeable channels and provide a pathway for the diffusion of calcium into the cytoplasm. Compared to passive store depletion with chelators, the inclusion of IP_3 in internal solutions results in a faster activation of CRAC channels due to rapid store depletion [47].

1.3.2 CRAC Current–Voltage Relation

The most common protocol used to record CRAC channels is the application of command voltage ramps spanning −100 to +100 mV. Ramp durations can be anywhere between 50 and 500 ms in order to generate *bona fide* current–voltage (*I*–*V*) curves. Ramp durations shorter than 50 ms may result in deformed *I*–*V* curves. The advantage of a ramp protocol is that instantaneous *I*–*V* relations can be obtained every 1–2 s. This is important, particularly for monitoring the development of other, unrelated conductances, such as Mg^{2+}-inhibited cation (MIC/TRPM7) channels discussed in the following section. The drawback of ramp protocols is lack of detailed kinetics information. Thus, CRAC channels inactivate when calcium influx is increased by raising the bathing Ca^{2+} concentration [48–51]. In cell types traditionally used for recording native CRAC channels (e.g., lymphocytes, mast cells), even when measured at −100 mV, a nonphysiological membrane potential, the current magnitude is usually quite small, in the vicinity of 5–10 pA (corresponding to current densities of 0.5–3 pA/pF). The time course of current development is monitored by plotting the inward current amplitude at negative potentials (usually at −100 or −80 mV) where current is maximal. Outward whole-cell current at a positive membrane potential can also be plotted to ascertain that the leak does not increase during the recording. Because of the steep inward rectification and calcium selectivity of CRAC channels, an outward ionic current at +50 mV or above should be minimal even when calcium stores are completely empty.

1.3.3 Current Separation

All cell types have numerous ion channels in their membranes that contribute to the overall ionic conductance of the cells [52]. The particular channel complement is dependent on the cell type. In order to record CRAC channels in isolation, appropriate ion substitutions and membrane potentials are used. To minimize contribution from endogenous K^+ channels, pipette K^+ is substituted with Cs^+, which permeates poorly through most K^+-selective channels such as $K_V1.3$ highly expressed in lymphocytes [13]. High concentrations of tetraethyl ammonium (TEA) or tetramethyl ammonium (TMA) can be included in recording solutions for this purpose. The addition of external Cs^+ (1–10 mM) is often used to block inwardly rectifying K^+ channels, such as Kir2.1 present in RBL cells and macrophages [53,54]. Alternatively, external K^+ is entirely substituted with Cs^+. The holding membrane potential between ramps is near 0 mV, which minimizes CRAC channel amplitude but is also depolarized enough to inactivate contaminating voltage-dependent channels, such as $K_V1.3$. The frequency of voltage ramps is kept at 0.5–1 Hz, which favors K_V channel inactivation.

Contributions from intermediate or small conductance calcium-activated K^+ channels are reduced by Cs^+ substitution and inclusion of large amounts of EGTA or BAPTA in the pipette solution. Most Ca^{2+}-activated K^+ channels require free calcium of 200 nM or higher to open [55]. Also, these channels can be blocked with relatively specific blockers such as charybdotoxin or TRAM-34 [56–58].

Chloride currents are reduced by substituting chloride both in the pipette and in the bathing solution with larger anions, such as aspartate, glutamate, gluconate, methanesulfonate, or isethionate. It should be kept in mind, however, that most organic anions bind Ca^{2+} and Mg^{2+} to some extent and can significantly reduce the concentrations of free divalent cations in the solution. This chelating effect is especially pronounced with citrate. In most cell types, large volume-sensitive chloride channels are expressed and an effort should be made to prevent cell swelling, which transiently activates these channels [59]. It needs to be noted that even the said large anions can pass through chloride channels to some degree [60]. Assuming that the background chloride conductance is constant throughout the recording, it can be subtracted out in most cases [61]. Maintaining correct and consistent osmolality of the internal and external recording solutions is also important to minimize volume-activated chloride channel activation. Osmolality can be adjusted with mannitol, which does not affect CRAC channels.

Magnesium-inhibited cation (MIC/TRPM7) channels are ubiquitously expressed and have in the past contaminated many recordings of CRAC channels. Like CRAC channels, MIC channels were first identified by patch-clamp electrophysiology in RBL and Jurkat cells [54,62,63]. All commonly used mammalian cell lines such as HEK293, CHO-K1, HeLa, COS-7, Jurkat, RBL1 and 2H3, Neuro-2A, and others express large Mg^{2+}-inhibited cation currents encoded by TRPM7 (or possibly by TRPM6 which is also inhibited by Mg^{2+}). TRPM7 channels are inhibited by micromolar to millimolar cytosolic Mg^{2+} concentrations in a non-voltage-dependent manner [54,64,65]. This current was initially confused with CRAC channels because the conditions for activation of both channels were similar; in both cases, high internal EGTA concentrations are used. While EGTA is used to passively deplete ER Ca^{2+} stores (see preceding text), it also chelates cellular Mg^{2+}, albeit more weakly. In the absence of Mg^{2+}, it can lead to the slow development of TRPM7 channels by virtue of gradual channel disinhibition. Moreover, the slow time course of current development is reminiscent of CRAC current development. However, major differences exist between these two conductances: the current voltage relation of TRPM7 is steeply outwardly rectifying although this is only apparent at membrane potentials above +40 mV. Upon removal of external divalent cations, TRPM7 *I–V* is drastically modified, becoming semilinear, as TRPM7 conducts Na^+ and Cs^+ equally well. By contrast, CRAC channels maintain their inward rectification both in the presence of Ca^{2+} and other divalent cations and in their absence. Na^+ is permeant, whereas Cs^+ is almost impermeant through CRAC channels [54,63,66,67]. TRPM7 inward unitary conductance is approximately 39 pS in the absence of divalent cations [64]. By contrast, CRAC channel unitary conductance is much smaller and cannot be measured directly. From noise analysis, it is inferred to be in the range of 100 fS (see succeeding text). CRAC channel inward rectification is not Mg^{2+} dependent [54,68] unlike TRPV6 or inwardly rectifying K^+ channels (KIR) and appears to be intrinsic to the protein. It should be noted that the voltage-gated Ca_V open channel *I–V* plot is also inwardly rectifying (e.g., [69]). In addition to ion substitution, the pharmacological properties of CRAC channels can also be used for current separation and are discussed in some detail in Chapter 16.

1.3.4 Perforated-Patch Recording

The perforated-patch variety of the whole-cell patch clamp has been successfully used to record CRAC channels [70–73]. It was originally used to prevent a rundown of voltage-gated calcium channels [74]. In this method, the patch pipette is back-filled with internal solution supplemented with dissolved nystatin or amphotericin B. These are polyene antibiotic ionophores that form ion channels in mammalian cell membranes. Nystatin and amphotericin channels are permeable to monovalent cations and anions but not to divalent ions, such as Ca^{2+}, Mg^{2+}, or SO_4^{2-} [75]. Such ion selectivity allows the maintenance of physiologically relevant buffering of these ions for the duration of the recording. The pipette tip is filled with an antibiotic-free internal solution. After seal formation, it takes roughly 10 min for the ionophore to diffuse to the pipette tip and the cell surface and begin forming channels. This is detectable by the increased slow membrane capacitance transients, a kind of whole-cell break-in "in slow motion." Perforated-patch recording requires more skill than the simple whole-cell configuration, however, mostly because at high concentrations these ionophores prevent gigaohm seal formation. The shape of the pipette is crucial for facilitating the diffusion of pore-forming antibiotics to the tip. Both nystatin and amphotericin are light sensitive, particularly the former, and lose their activity in 4–5 h when diluted in the internal solution. In our experience, amphotericin B is easier to use because it forms pores in the membrane more readily, and access resistance falls to lower values (~10 MΩ) than for nystatin [76]. If physiological anion concentrations are desired, then gramicidin can be used instead of nystatin or amphotericin, since gramicidin channels are believed not to conduct chloride, and the steady-state cellular anion concentrations will be undisturbed [24,75,77].

IP_3, EGTA, and BAPTA cannot pass through the antibiotic-formed channels; therefore, CRAC channels can be elicited by the bath application of antigens, mitogens, hormones, or SERCA inhibitors thapsigargin and CPA, which are uncharged and membrane permeant [43,70,78]. The perforated patch should also prevent Mg^{2+} depletion and spontaneous TRPM7 activation [79]. In perforated patch, care must be taken to rule out the accidental rupture of the patch and formation of a conventional whole-cell configuration. This can be done by including high (millimolar) concentrations of Ca^{2+} in the internal solution expected to kill the cell after entering the cytoplasm. In order to minimize Ca^{2+}-dependent inactivation [26], it is especially important to use a depolarized holding potential to avoid Ca^{2+} loading in the absence of EGTA or BAPTA.

1.3.5 CRAC Channel Activity with Various Permeating Cations

As with Ca_V channels, in order to maximize the CRAC channel current, the external calcium concentration can be elevated to 50 or ~110 mM [70,80]. (This concentration is close to the upper limit imposed by osmotic pressure.) This results in increased current amplitude, because the concentration of the conducting ion and driving force is increased. The increased calcium influx, however, also triggers calcium-dependent inactivation, both fast and slow, which eventually reduces the current amplitude [48,81]. Interestingly, the calcium-dependent inactivation appears to involve Orai and STIM proteins themselves but not calmodulin, unlike the case for TRPV5/6 channels (see succeeding text) [49–51]. A good trade-off between these two Ca^{2+} effects is a bathing concentration of 6–10 mM Ca^{2+}, which is sufficiently low to prevent substantial inactivation [37,82]. Command voltage ramps can be preceded by a conditioning hyperpolarizing step to bring inactivation to a quasi-steady state and avoid the distortion of the instantaneous *I–V*s due to fast Ca^{2+}-dependent inactivation [78].

It is now firmly established that in the virtual absence of divalent cations in the external medium, Ca^{2+}-selective channels become permeable to small monovalent cations. The amplitude of the monovalent cation current through Ca^{2+} channels in the absence of divalents is usually 6–10 times larger than when carried by Ca^{2+}. This was originally demonstrated for L-type voltage-gated (Ca_V) Ca^{2+} channels [83–85]. Since then, the removal of divalent cations has been tested for other Ca^{2+} channels and shown to generally increase current amplitude. CRAC channels also belong to this group [26]. The removal of divalent cations results in a significantly larger current amplitude but with a twist: the increase is transient, unlike the Ca^{2+}-mediated CRAC current recorded when internal solution contains high concentrations of chelators [86]. It is thought that this time-dependent reduction of monovalent CRAC current (termed inactivation) is a result of depotentiation caused by the removal of external Ca^{2+}. Other divalent metal cations such as Ni^{2+} and Zn^{2+} could not substitute for the potentiating effect of Ca^{2+} in RBL cells [80], but Ni^{2+} was found to be effective in Jurkat T cells [87–89]. In an overexpression system, it was demonstrated that under certain conditions the monovalent Orai current can be stable [37]. It is now thought that this is a consequence of changes in Orai to STIM1 ratios during transfection [90]. The CRAC *I–V* shape in divalent-free (DVF) solutions is identical to its shape in Ca^{2+}, preserving its inward rectification, unlike the case for TRPM7 [67].

The overwhelming majority of CRAC channel recordings have been performed at room temperature. Increasing the bath temperature to 37°C significantly increased CRAC current amplitudes in RBL cells [71,78]. Additionally, in perforated-patch recording, CRAC channels could only be activated by antigen application at 37°C but not at room temperature [78], unlike the case for SERCA inhibitors.

1.3.6 CRAC Single-Channel Conductance

One of the fundamental characteristics of an ion channel is its conductance. The first recordings of I_{CRAC} in mast cells and lymphocytes suggested that the conductance of a single CRAC channel is well below the levels that can be resolved by patch clamping as single-channel currents [26,70,91]. The standard approach in cases when single-channel conductance is too small for patch-clamp recording is to use the so-called nonstationary noise analysis (analysis of variance) [92]. Due to the stochastic nature of single-channel opening, a steady-state macroscopic current mediated by a constant number of channels fluctuates with time around its mean value. The nonstationary noise analysis is based on the fact that the magnitude of these fluctuations depends on the amplitude of the single-channel current, the number of channels, and their open probability [92,93]. The relationship between current variance

(σ^2; averaged squared current fluctuations from its mean) and single-channel current is given by the following equation:

$$\sigma^2 = iI - \frac{I^2}{N} \tag{1.1}$$

where

I is the mean amplitude of the macroscopic current
i is the single-channel current
N is the number of channels

I, i, and N parameters can be used to determine the open probability P_o:

$$P_o = \frac{I}{iN} \tag{1.2}$$

Originally, nonstationary noise analysis was used for voltage-gated channels, where current is recorded in response to a voltage step that promotes channels to either open or close, so that the P_o of the channels under investigation would change significantly during that voltage step, preferably by more than 50% [93]. The recorded current traces (at least 100 sweeps are normally required) are averaged to obtain a mean current trace and variance. The calculated variance is then plotted against the mean current and the data points fitted with a parabola (Equation 1.1), which gives unitary channel current i and the number of channels N [92,93].

Although CRAC channels show some voltage dependence due to fast Ca^{2+}-dependent inactivation at negative potentials [26,94], it is virtually impossible to record a large enough number of current sweeps in response to voltage steps below −100 mV in one cell due to an I_{CRAC} drift and rundown. The first attempts of CRAC unitary conductance estimation used traces of I_{CRAC} development in Jurkat T cells at a constant voltage of −80 mV in a bath solution containing 2 mM Ca^{2+}, which was then switched to 110 mM Ca^{2+} [70]. This approach was based on the assumption that the changes in I_{CRAC} amplitude during current development and its slow inactivation in a solution containing 110 mM Ca^{2+} were due to the changes in CRAC channel P_o. Current variance was calculated from 200 ms episodes taken from every 2 s of continuous recording and plotted against corresponding mean current amplitudes [70]. In these plots, variance showed linear dependence on the current amplitude, which suggested a low P_o of CRAC channels [70] (Figure 1.2). The calculated single-channel current amplitudes were −1.4 fA in 2 mM Ca^{2+} and −3.6 fA in 110 mM Ca^{2+} at −80 mV. The estimated single-channel conductances were 9 and 24 fS, respectively [70].

As discussed earlier, in the absence of Ca^{2+} and Mg^{2+}, CRAC channels become permeable to monovalent cations. Replacing physiological (Ca^{2+}-containing) bath solutions with divalent cation-free solutions causes a 7–8-fold increase in I_{CRAC} amplitude followed by current deactivation to about 20%–30% of its peak value within 30–60 s, which suggests a significant change in CRAC channels P_o in that time [86]. This property of I_{CRAC} was utilized to estimate monovalent CRAC single-channel currents

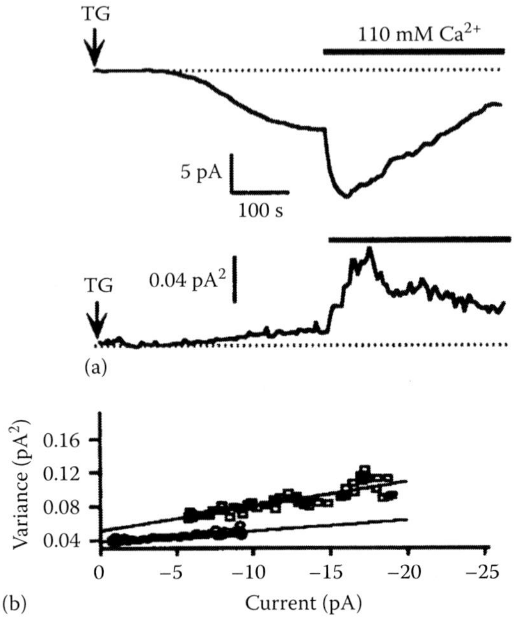

FIGURE 1.2 Fluctuation analysis of currents stimulated by thapsigargin in Jurkat T cells. (a) Macroscopic Ca^{2+} current (upper) and variance (lower) elicited by 1 μM thapsigargin. (b) Variance plotted against mean current for the data from (a). Linear regression fits are superimposed on the data collected in 2 and 110 mM $(Ca^{2+})_o$. Unitary current amplitudes are −1.3 fA (2 mM Ca^{2+}) and −2.9 fA (110 mM Ca^{2+}), as indicated by the slopes. (Reproduced from Zweifach, A. and Lewis, R.S., *Proc. Natl. Acad. Sci. USA*, 90, 6295, 1993, Figure 3.)

using an approach similar to that of Zweifach and Lewis [87]. In these experiments, the variance of Na^+ current through I_{CRAC} also exhibited linear dependence on the current amplitude, supporting the notion of low P_o of CRAC channels [70]. However, the calculated unitary CRAC channel conductance of 2.6 pS in the absence of divalent cations was significantly higher than what would be expected from the estimates of CRAC unitary conductance in 110 mM Ca^{2+} [70], considering that the removal of divalent cations causes only about a 7–8-fold increase in I_{CRAC} amplitude. The discrepancy was later explained by a possible contamination of these recordings with endogenous MIC/TRPM7 channels that were incompletely blocked by 3 mM $MgCl_2$ in the pipette solution [63]. Using a higher internal concentration of Mg^{2+} (8 mM), Prakriya and Lewis reported a unitary monovalent cation (Na^+) CRAC conductance of ~0.2 pS, which agreed well with Ca^{2+} unitary conductance [63].

All measurements described earlier were based on the assumption that the observed changes in I_{CRAC} amplitude resulted from changes in P_o and not the number of functional CRAC channels on the plasma membrane. The linear dependence of variance on current amplitude was explained by low P_o of CRAC channels [70]. The linear variance–current plot, however, could result from the changes in the number of active CRAC channels during I_{CRAC} development and depotentiation in divalent cation-free solutions, with constant and high P_o [89]. To investigate this possibility,

Prakriya and Lewis used nonstationary noise analysis of Na^+ current through CRAC channels in the presence of a range of micromolar Ca^{2+} concentrations in the bath solution [89]. An application of up to 20 μM of Ca^{2+} to the bath solution increased the Na^+ current noise, although the amplitude of the current declined [89], which suggested that, indeed, the P_o of CRAC channels was above 0.5 and the previous assumption of low P_o of CRAC channels was incorrect. The variance–current plot of the data obtained using a Ca^{2+} block of Na^+ CRAC conductance followed a parabolic function (Equation 1.1), giving a CRAC Na^+ conductance of ~0.7 pS and P_o of about 0.8 [89]. Similar results were later obtained for overexpressed Orai1/STIM1 and Orai3/STIM1 channels [95,96]. In conclusion, a higher than previously thought single-channel conductance and high P_o of CRAC channels opens a possibility of direct single-channel current measurements. The single-channel conductance of 0.7 pS corresponds to a 70 fA unitary current per 100 mV of driving force. Assuming that the equilibrium potential for Na^+ is about +50 mV, the amplitude of the monovalent cation unitary CRAC current at a membrane potential of −100 mV should be in the vicinity of 0.1 pA. A single-channel current of such amplitude can be recorded using low-noise patch-clamp amplifiers and appropriate filtering, provided that the right conditions for the current activation are met [97]. It remains to be determined, however, if this can be achieved in practice for CRAC channels.

1.3.7 Heterologously Expressed Orai/STIM Channels

The molecular components of CRAC channels were discovered more than a decade ago and include Orai1–3 (CRACM1–3) pore subunits and STIM1–2, ER resident luminal calcium sensors (reviewed in [33,35] and Chapters 2 and 3). Thus, mammalian cells express three different pore subunits, which was not predicted before they were cloned. Orai1 and Orai2 amino acid sequences are ~60% homologous. Orai3 also has a high percentage of homology with the remaining members: ~63% with Orai1 and ~66% with Orai2. In the transmembrane domains the homology percentages are even higher, reaching 90% [98,99]. All three isoforms are expressed in human T lymphocytes [82]; however, a familial defect in Orai1 results in severe combined immunodeficiency even in the presence of normal Orai2 and Orai3 expressed in these cells. This strongly suggests that Orai2 and 3 cannot substitute for Orai1 function, at least in that case [100]. It should be noted that in cells where Orai1 expression has been knocked down with siRNA, small SOCE signals mediated by Orai2 or Orai3 are still evident [101,102].

When coexpressed with STIM1, all three Orai proteins give rise to robust inwardly rectifying currents (Figure 1.3). Orai1–3 isoforms are permeable to Ca^{2+} and some other divalent cations (Ba^{2+}, Sr^{2+}) [90,103]. For Orai2 and Orai3, inward Ba^{2+} currents were larger than for Orai1, although they could only be detected in the presence of Na^+ but not TEA, suggesting that Na^+ may carry part of the current [103]. In the complete absence of divalent cations, Orai1–3 readily conduct monovalents, such as Na^+, Li^+, and Rb^+ [103,104]. Interestingly, Orai isoforms can be distinguished in heterologous expression systems by their sensitivity to 2-aminoethyl diphenyl borinate (2-APB), a compound originally identified as an IP_3 receptor blocker [105] (see Chapter 16). Orai1 and 2 are potentiated by low (5–10 μM) but

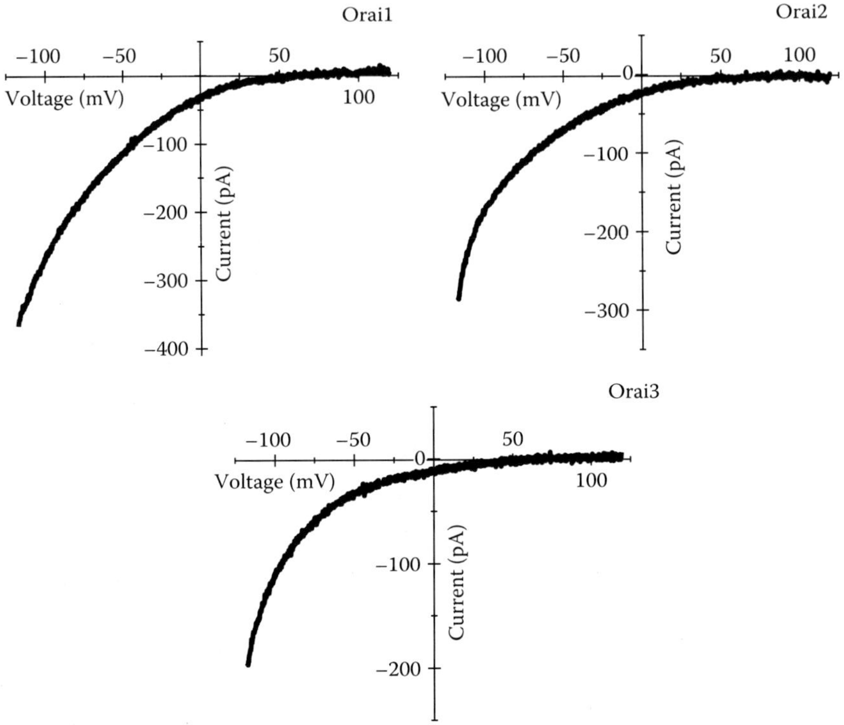

FIGURE 1.3 Instantaneous *I–V* plots of Orai1–3/STIM1-mediated I_{CRAC}. Traces were recorded in response to 100 ms voltage ramps ranging from −120 to 120 mV in HEK293T cells transfected with STIM1 and Orai1, Orai2, or Orai3 cDNA as previously described [90]. External [Ca^{2+}] = 10 mM.

inhibited by high 2-APB (above 50 μM) concentrations [106]. By contrast, 2-APB activates Orai3/STIM1, altering its *I–V* relation, which is normally inwardly rectifying (Figures 1.3 and 1.4). 2-APB-mediated activation results in an outwardly rectifying current, which can be conducted by both Cs^+ and Na^+. Similar to native CRAC channels, overexpressed Orai1 and Orai2 conduct Cs^+ poorly (~10% of Na^+ conductance) in the absence of divalent cations. Consequently, it has been suggested that 2-APB widens the Orai3 pore and allows it to conduct monovalent cations even in the presence of Ca^{2+} [107]. In contrast to Orai3, 2-APB does not change Orai1 and Orai2 selectivity [106]. Whether this divergence of 2-APB effects for overexpressed Orai1–3 can be used to distinguish endogenous Orai isoform activity is still unclear. It remains to be discovered if the Orai3 nonselective channel state can be achieved under physiological conditions without 2-APB and what that means in terms of cellular calcium metabolism.

Orai1 to STIM1 transfection ratios significantly influence the electrophysiological properties of these channels. Specifically, the kinetics of the channel activation and inactivation were changed in addition to selectivity to divalent metals (Ba^{2+} and Ca^{2+}). Increasing STIM1 to Orai1 expression ratios results in an inactivating current

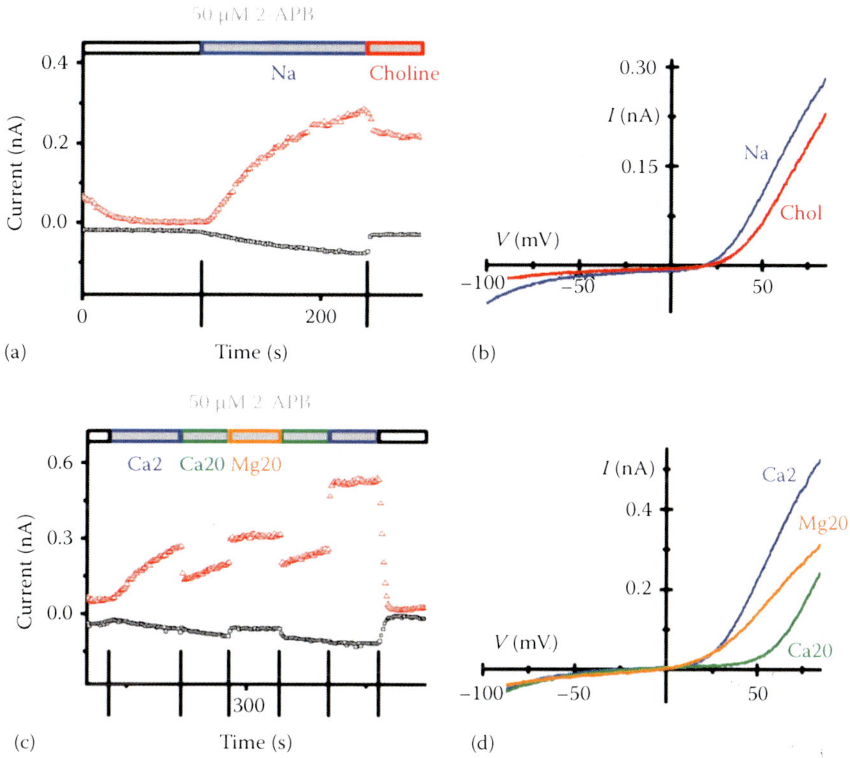

FIGURE 1.4 2-APB modified Orai3 conducts both Na^+ and Ca^{2+}. Whole-cell patch-clamp recordings of HEK293 cells transfected with Orai3 and STIM1. (a, c) Time courses of inward (black) and outward (red) current development in the presence of 50 μM 2-APB. (b, d) Individual *I–V* plots from the same cell as on the left. Effects of Na^+ (a, b) and Ca^{2+} (c, d) substitution and replacement in the external medium. (Reproduced from Zhang, S.L. et al., *J. Biol. Chem.*, 283, 17662, 2008, Figure 7.)

whereas Orai1 excess to STIM1 leads to non-inactivating currents [90]. Orai currents can be evoked by coexpressing full length STIM1 but also by overexpressing only the SOAR region [37] (discussed in Chapter 2).

Orai/STIM channels have primarily been studied in human embryonic kidney (HEK) cells transfected with their various isoforms. The recording conditions are essentially the same as for recording the native CRAC channels. The internal solution should contain calcium chelators alone or with IP_3 for depleting ER Ca^{2+} stores [104,106]. However, in most cases, overexpressed Orai/STIM channels are to some extent constitutively active, unlike CRAC channels in Jurkat or RBL cells [37]. The overexpression of either Orai or STIM proteins in isolation does not usually result in large CRAC currents.

The murine and human Orai2 exist in two forms, long and short [108–110]. The Orai2 expression pattern is somewhat different from that of Orai1, although in many cases they are coexpressed [109]. For example, neurons and chondrocytes show high expression of Orai2, even though its function in these cells is unknown [111]. When coexpressed with STIM1, human Orai2 form inwardly rectifying channels (Figure 1.3) sharing many biophysical properties with Orai1 [103]. In an early detailed study of overexpressed murine Orai2, Gross and colleagues recorded currents from HEK cells overexpressing both long and short splice variants [108]. Orai2 and STIM1 were transfected at a 2:1 ratio. Infusion of cells with high EGTA and IP_3-containing buffers evoked currents reminiscent of native CRAC channels. Interestingly, the levels of expression depended on the cell type chosen for transfection, HEK vs. RBL. One striking difference of Orai2 from Orai1 was the apparently increased (slow) calcium-dependent inactivation. It was concluded that calcium influx through Orai2 channels causes the inactivation that can be relieved by the fast removal of calcium from the vicinity of the channels. Inactivation was proportional to the peak current amplitude. The dependence of channel expression on the type of the background cell line was explained by the fact that HEK cells express very little Orai1 and Orai2 protein detected in Western blots, whereas RBL cells express both channels at high levels. Accordingly, the overexpression of STIM1 alone in HEK cells did not increase CRAC current density but doubled it in RBL cells. When coexpressed with Orai1, Orai2 was shown to reduce current expression, suggesting that it can form heteromers with other Orai isoforms [108]. Similar conclusions were reached by Inayama and colleagues by coexpressing Orai1 E106Q point mutant with the wild-type Orai2 in chondrocyte cell lines [111].

1.4 TRPV5 AND TRPV6 CHANNELS

1.4.1 Heterologously Expressed TRPV5/6

TRPV5 and TRPV6 channels are present in kidney and gastrointestinal epithelial cells at high levels [112,113]. TRPV5 (ECaC1) is thought to be expressed mostly in the kidney, whereas TRPV6 (CaT1, ECaC2) expression is higher in the gastrointestinal tract, placenta, and epididymal epithelia [114–118] (see Chapter 13). Both are absent in lymphocytes [119]. TRPV5/6 channels were first identified by patch-clamp electrophysiology only after their cDNA sequence was discovered. In an overexpression system (primarily plasmid-transfected HEK or CHO cells but also mRNA-injected *Xenopus* oocytes; Chapter 13), they are either constitutively active or activated (i.e., revealed) by calcium removal from the cytosol when they become permeable to monovalent cations [120–126]. Increasing bathing calcium concentration potentiates the currents as expected for a Ca^{2+}-selective channel. This appears to be a simple increase in ionic current due to increased driving force for calcium. The extracellular calcium-dependent potentiation effect described for CRAC channels has not been reported for these channels. Notably, the holding potential essentially determines the current magnitude in a given cell [121,127]. Current increase resulting from increased $[Ca^{2+}]_o$ is usually short-lived and depends on the holding membrane potential [120]. This is thought to be the consequence of calcium loading into the cell (in the vicinity of the channels) via constitutively open TRPV5/6 channels [122]. Accordingly, the dependence on the holding potential disappears when external DVF solutions are used. Internal Ca^{2+} inhibits TRPV5/6 channels with an IC_{50} of ~130 nM, explaining the current inactivation seen when Ca^{2+} is present in the bath solution [128]. As for Orai/STIM channels, internal recording solutions should contain high concentrations of EGTA or BAPTA, in this case to avoid Ca^{2+}-dependent inactivation.

In addition to Ca^{2+}, TRPV5 and 6 also conduct Sr^{2+}, Ba^{2+}, and Mn^{2+} [128]. When Ba^{2+} is used as the charge carrier instead of Ca^{2+}, inactivation is drastically reduced for both TRPV5 and 6 [128]. Mn^{2+} permeation through TRPV5/6 becomes relevant when Mn^{2+} quenching of Fura-2 signals is used to evaluate calcium entry [129,130].

Like Orai channels, TRPV5 and TRPV6 are steeply inwardly rectifying. And as with Orai channels, the removal of divalent cations from the bathing medium results in vastly magnified TRPV5 and TRPV6 currents. This simple maneuver has been used to record channel activity without the fast inactivation caused by Ca^{2+} influx through these channels (see preceding text) [127,131]. Primarily Na^+ has been used, but K^+ and especially Li^+ are also highly permeant [128]. Cs^+ is less permeant (~80% of Na^+), as is the case for voltage-gated Ca^{2+} channels. It appears that channel sensitivity to blockers, ruthenium red or econazole, is not affected by the permeating ion [131].

It was discovered early on that TRPV5 and 6 channels exhibit a rundown of activity during prolonged electrophysiological recordings [120]. Intracellular MgATP is required to maintain TRPV5 and TRPV6 channel activity and prevent rundown, whereas NaATP cannot substitute for this effect of MgATP [128]. Most likely the MgATP effect reflects maintaining $PI(4,5)P_2$ levels through lipid kinases, although other mechanisms have also been proposed [128,132,133]. Cytosolic Mg^{2+} alone induces slow inhibition which can be relieved by $PI(4,5)P_2$ binding [134]. The strong inward rectification of TRPV6 also depends on intracellular Mg^{2+} and is mediated by an aspartate residue [135].

1.4.2 Endogenous TRPV5/6 Channels

Endogenous TRPV5/6 channel activity has been reported mainly in two studies: TRPV5 channel currents were measured in rat renal cells [136] and TRPV6 in a human colonic cell line [137] using whole-cell patch clamp. In epithelial cells, native TRPV5/6 channels are not normally active/detectable in the presence of physiological concentrations of Ca^{2+} (1–2 mM). In order to detect measurable currents, bathing calcium was increased to 20 or 100 mM combined with EGTA-containing internal solutions [136,137]. The addition of the steroid hormone 17β-estradiol markedly increased TRPV5 and TRPV6 currents carried by Ca^{2+}. It remains to be discovered under what metabolic conditions native TRPV5/6 channels conduct measurable current at more physiological calcium and intracellular buffering. To date, however, most patch-clamp studies of TRPV5/6 have been performed in heterologous expression systems. It should be kept in mind that in the conventional whole-cell recording configuration used for TRPV5/6 recording, the ER stores will also most likely be depleted due to low calcium and the presence of chelators in the internal solution. A way around this would be to use a perforated patch, which however will prevent the depletion of cytoplasmic calcium and reduce current amplitudes by inactivation [121,122].

27. Dolmetsch, R.E. and Lewis, R.S. 1994. Signaling between intracellular Ca^{2+} stores and depletion-activated Ca^{2+} channels generates $[Ca^{2+}]_i$ oscillations in T lymphocytes. *J Gen Physiol* 103:365–388.
28. von Tscharner, V., Prod'hom, B., Baggiolini, M., and Reuter, H. 1986. Ion channels in human neutrophils activated by a rise in free cytosolic calcium concentration. *Nature* 324:369–372.
29. Gardner, P., Alcover, A., Kuno, M., Moingeon, P., Weyand, C.M., Goronzy, J., and Reinherz, E.L. 1989. Triggering of T-lymphocytes via either T3-Ti or T11 surface structures opens a voltage-insensitive plasma membrane calcium-permeable channel: Requirement for interleukin-2 gene function. *J Biol Chem* 264:1068–1076.
30. Ng, J., Gustavsson, J., Jondal, M., and Andersson, T. 1990. Regulation of calcium influx across the plasma membrane of the human T-leukemic cell line, JURKAT: Dependence on a rise in cytosolic free calcium can be dissociated from formation of inositol phosphates. *Biochim Biophys Acta* 1053:97–105.
31. van Rossum, D.B., Patterson, R.L., Kiselyov, K., Boehning, D., Barrow, R.K., Gill, D.L., and Snyder, S.H. 2004. Agonist-induced Ca^{2+} entry determined by inositol 1,4,5-trisphosphate recognition. *Proc Natl Acad Sci USA* 101:2323–2327.
32. Dellis, O., Dedos, S.G., Tovey, S.C., Taufiq Ur, R., Dubel, S.J., and Taylor, C.W. 2006. Ca^{2+} entry through plasma membrane IP^3 receptors. *Science* 313:229–233.
33. Hogan, P.G., Lewis, R.S., and Rao, A. 2010. Molecular basis of calcium signaling in lymphocytes: STIM and ORAI. *Annu Rev Immunol* 28:491–533.
34. Roos, J., DiGregorio, P.J., Yeromin, A.V., Ohlsen, K., Lioudyno, M., Zhang, S., Safrina, O. et al. 2005. STIM1, an essential and conserved component of store-operated Ca^{2+} channel function. *J Cell Biol* 169:435–445.
35. Chang, C.L. and Liou, J. 2016. Homeostatic regulation of the PI(4,5)P2-Ca^{2+} signaling system at ER-PM junctions. *Biochim Biophys Acta* 1861:862–873.
36. Mercer, J.C., Dehaven, W.I., Smyth, J.T., Wedel, B., Boyles, R.R., Bird, G.S., and Putney, J.W., Jr. 2006. Large store-operated calcium selective currents due to co-expression of Orai1 or Orai2 with the intracellular calcium sensor, Stim1. *J Biol Chem* 281:24979–24990.
37. Zhang, S.L., Kozak, J.A., Jiang, W., Yeromin, A.V., Chen, J., Yu, Y., Penna, A., Shen, W., Chi, V., and Cahalan, M.D. 2008. Store-dependent and -independent modes regulating Ca^{2+} release-activated Ca^{2+} channel activity of human Orai1 and Orai3. *J Biol Chem* 283:17662–17671.
38. Hoth, M. and Niemeyer, B.A. 2013. The neglected CRAC proteins: Orai2, Orai3, and STIM2. *Curr Top Membr* 71:237–271.
39. Fierro, L. and Parekh, A.B. 1999. On the characterisation of the mechanism underlying passive activation of the Ca^{2+} release-activated Ca^{2+} current I_{CRAC} in rat basophilic leukaemia cells. *J Physiol* 520(Pt 2):407–416.
40. Shilling, D., Mak, D.O., Kang, D.E., and Foskett, J.K. 2012. Lack of evidence for presenilins as endoplasmic reticulum Ca^{2+} leak channels. *J Biol Chem* 287:10933–10944.
41. Fasolato, C., Hoth, M., and Penner, R. 1993. A GTP-dependent step in the activation mechanism of capacitative calcium influx. *J Biol Chem* 268:20737–20740.
42. Hofer, A.M., Fasolato, C., and Pozzan, T. 1998. Capacitative Ca^{2+} entry is closely linked to the filling state of internal Ca^{2+} stores: A study using simultaneous measurements of I_{CRAC} and intraluminal $[Ca^{2+}]$. *J Cell Biol* 140:325–334.
43. Luik, R.M., Wang, B., Prakriya, M., Wu, M.M., and Lewis, R.S. 2008. Oligomerization of STIM1 couples ER calcium depletion to CRAC channel activation. *Nature* 454:538–542.
44. Imboden, J.B. and Stobo, J.D. 1985. Transmembrane signalling by the T cell antigen receptor. Perturbation of the T3-antigen receptor complex generates inositol phosphates and releases calcium ions from intracellular stores. *J Exp Med* 161:446–456.

45. Bird, G.S., Aziz, O., Lievremont, J.P., Wedel, B.J., Trebak, M., Vazquez, G., and Putney, J.W., Jr. 2004. Mechanisms of phospholipase C-regulated calcium entry. *Curr Mol Med* 4:291–301.
46. Gomperts, B.D., Kramer, I.M., and Tatham, P.E. 2003. *Signal Transduction*. Amsterdam, the Netherlands: Elsevier Academic Press.
47. Parekh, A.B. 2006. Electrophysiological recordings of Ca^{2+} currents. In *Calcium Signaling*, J.W. Putney, ed. Boca Raton, FL: Taylor & Francis, pp. 125–146.
48. Parekh, A.B. 1998. Slow feedback inhibition of calcium release-activated calcium current by calcium entry. *J Biol Chem* 273:14925–14932.
49. Mullins, F.M., Park, C.Y., Dolmetsch, R.E., and Lewis, R.S. 2009. STIM1 and calmodulin interact with Orai1 to induce Ca^{2+}-dependent inactivation of CRAC channels. *Proc Natl Acad Sci USA* 106:15495–15500.
50. Mullins, F.M. and Lewis, R.S. 2016. The inactivation domain of STIM1 is functionally coupled with the Orai1 pore to enable Ca^{2+}-dependent inactivation. *J Gen Physiol* 147:153–164.
51. Mullins, F.M., Yen, M., and Lewis, R.S. 2016. Orai1 pore residues control CRAC channel inactivation independently of calmodulin. *J Gen Physiol* 147:137–152.
52. Kew, J.N. and Davies, C.H. 2010. *Ion Channels: From Structure to Function*. Oxford, U.K.: Oxford University Press, 562pp.
53. Gallin, E.K. 1991. Ion channels in leukocytes. *Physiol Rev* 71:775–811.
54. Kozak, J.A., Kerschbaum, H.H., and Cahalan, M.D. 2002. Distinct properties of CRAC and MIC channels in RBL cells. *J Gen Physiol* 120:221–235.
55. Leinders, T. and Vijverberg, H.P. 1992. Ca^{2+} dependence of small Ca^{2+}-activated K^+ channels in cultured N1E-115 mouse neuroblastoma cells. *Pflugers Arch* 422:223–232.
56. Price, M., Lee, S.C., and Deutsch, C. 1989. Charybdotoxin inhibits proliferation and interleukin 2 production in human peripheral blood lymphocytes. *Proc Natl Acad Sci USA* 86:10171–10175.
57. Garcia, M.L., Hanner, M., and Kaczorowski, G.J. 1998. Scorpion toxins: Tools for studying K^+ channels. *Toxicon* 36:1641–1650.
58. Beeton, C., Wulff, H., Barbaria, J., Clot-Faybesse, O., Pennington, M., Bernard, D., Cahalan, M.D., Chandy, K.G., and Beraud, E. 2001. Selective blockade of T lymphocyte K^+ channels ameliorates experimental autoimmune encephalomyelitis, a model for multiple sclerosis. *Proc Natl Acad Sci USA* 98:13942–13947.
59. Akita, T., Fedorovich, S.V., and Okada, Y. 2011. Ca^{2+} nanodomain-mediated component of swelling-induced volume-sensitive outwardly rectifying anion current triggered by autocrine action of ATP in mouse astrocytes. *Cell Physiol Biochem* 28:1181–1190.
60. Rychkov, G.Y., Pusch, M., Roberts, M.L., Jentsch, T.J., and Bretag, A.H. 1998. Permeation and block of the skeletal muscle chloride channel, ClC-1, by foreign anions. *J Gen Physiol* 111:653–665.
61. Rychkov, G.Y., Litjens, T., Roberts, M.L., and Barritt, G.J. 2005. ATP and vasopressin activate a single type of store-operated Ca^{2+} channel, identified by patch-clamp recording, in rat hepatocytes. *Cell Calcium* 37:183–191.
62. Hermosura, M.C., Monteilh-Zoller, M.K., Scharenberg, A.M., Penner, R., and Fleig, A. 2002. Dissociation of the store-operated calcium current I_{CRAC} and the Mg-nucleotide-regulated metal ion current MagNuM. *J Physiol* 539:445–458.
63. Prakriya, M. and Lewis, R.S. 2002. Separation and characterization of currents through store-operated CRAC channels and Mg^{2+}-inhibited cation (MIC) channels. *J Gen Physiol* 119:487–507.
64. Chokshi, R., Matsushita, M., and Kozak, J.A. 2012. Sensitivity of TRPM7 channels to Mg^{2+} characterized in cell-free patches of Jurkat T lymphocytes. *Am J Physiol Cell Physiol* 302:C1642–C1651.

65. Chokshi, R., Matsushita, M., and Kozak, J.A. 2012. Detailed examination of Mg^{2+} and pH sensitivity of human TRPM7 channels. *Am J Physiol Cell Physiol* 302:C1004–C1011.
66. Bakowski, D. and Parekh, A.B. 2002. Monovalent cation permeability and Ca^{2+} block of the store-operated Ca^{2+} current I_{CRAC} in rat basophilic leukemia cells. *Pflugers Arch* 443:892–902.
67. Bakowski, D. and Parekh, A.B. 2002. Permeation through store-operated CRAC channels in divalent-free solution: Potential problems and implications for putative CRAC channel genes. *Cell Calcium* 32:379–391.
68. Prakriya, M. and Lewis, R.S. 2015. Store-operated calcium channels. *Physiol Rev* 95:1383–1436.
69. Bittner, K.C. and Hanck, D.A. 2008. The relationship between single-channel and whole-cell conductance in the T-type Ca^{2+} channel $Ca_V3.1$. *Biophys J* 95:931–941.
70. Zweifach, A. and Lewis, R.S. 1993. Mitogen-regulated Ca^{2+} current of T lymphocytes is activated by depletion of intracellular Ca^{2+} stores. *Proc Natl Acad Sci USA* 90:6295–6299.
71. McCloskey, M.A. and Zhang, L. 2000. Potentiation of Fcepsilon receptor I-activated Ca^{2+} current (I_{CRAC}) by cholera toxin: Possible mediation by ADP ribosylation factor. *J Cell Biol* 148:137–146.
72. Aromataris, E.C., Roberts, M.L., Barritt, G.J., and Rychkov, G.Y. 2006. Glucagon activates Ca^{2+} and Cl^- channels in rat hepatocytes. *J Physiol* 573:611–625.
73. Zhang, X., Zhang, W., Gonzalez-Cobos, J.C., Jardin, I., Romanin, C., Matrougui, K., and Trebak, M. 2014. Complex role of STIM1 in the activation of store-independent Orai1/3 channels. *J Gen Physiol* 143:345–359.
74. Kurachi, Y., Asano, Y., Takikawa, R., and Sugimoto, T. 1989. Cardiac Ca current does not run down and is very sensitive to isoprenaline in the nystatin-method of whole cell recording. *Naunyn Schmiedebergs Arch Pharmacol* 340:219–222.
75. Akaike, N. and Harata, N. 1994. Nystatin perforated patch recording and its applications to analyses of intracellular mechanisms. *Jpn J Physiol* 44:433–473.
76. Rae, J., Cooper, K., Gates, P., and Watsky, M. 1991. Low access resistance perforated patch recordings using amphotericin B. *J Neurosci Methods* 37:15–26.
77. D'Ambrosio, R. 2002. Perforated patch-clamp technique. In *Patch-Clamp Analysis*, W. Walz, A.A. Boulton, and G.B. Baker, eds. Totowa, NJ: Humana Press, pp. 195–216.
78. Zhang, L. and McCloskey, M.A. 1995. Immunoglobulin E receptor-activated calcium conductance in rat mast cells. *J Physiol* 483(Pt 1):59–66.
79. Kozak, J.A., Matsushita, M., Nairn, A.C., and Cahalan, M.D. 2005. Charge screening by internal pH and polyvalent cations as a mechanism for activation, inhibition, and rundown of TRPM7/MIC channels. *J Gen Physiol* 126:499–514.
80. Su, Z., Shoemaker, R.L., Marchase, R.B., and Blalock, J.E. 2004. Ca^{2+} modulation of Ca^{2+} release-activated Ca^{2+} channels is responsible for the inactivation of its monovalent cation current. *Biophys J* 86:805–814.
81. Zweifach, A. and Lewis, R.S. 1995. Slow calcium-dependent inactivation of depletion-activated calcium current. Store-dependent and -independent mechanisms. *J Biol Chem* 270:14445–14451.
82. Lioudyno, M.I., Kozak, J.A., Penna, A., Safrina, O., Zhang, S.L., Sen, D., Roos, J., Stauderman, K.A., and Cahalan, M.D. 2008. Orai1 and STIM1 move to the immunological synapse and are up-regulated during T cell activation. *Proc Natl Acad Sci USA* 105:2011–2016.
83. Hess, P. and Tsien, R.W. 1984. Mechanism of ion permeation through calcium channels. *Nature* 309:453–456.
84. McCleskey, E.W. and Almers, W. 1985. The Ca channel in skeletal muscle is a large pore. *Proc Natl Acad Sci USA* 82:7149–7153.
85. Carbone, E., Lux, H.D., Carabelli, V., Aicardi, G., and Zucker, H. 1997. Ca^{2+} and Na^+ permeability of high-threshold Ca^{2+} channels and their voltage-dependent block by Mg^{2+} ions in chick sensory neurones. *J Physiol* 504(Pt 1):1–15.

86. Lepple-Wienhues, A. and Cahalan, M.D. 1996. Conductance and permeation of monovalent cations through depletion-activated Ca^{2+} channels (I_{CRAC}) in Jurkat T cells. *Biophys J* 71:787–794.
87. Zweifach, A. and Lewis, R.S. 1996. Calcium-dependent potentiation of store-operated calcium channels in T lymphocytes. *J Gen Physiol* 107:597–610.
88. Christian, E.P., Spence, K.T., Togo, J.A., Dargis, P.G., and Patel, J. 1996. Calcium-dependent enhancement of depletion-activated calcium current in Jurkat T lymphocytes. *J Membr Biol* 150:63–71.
89. Prakriya, M. and Lewis, R.S. 2006. Regulation of CRAC channel activity by recruitment of silent channels to a high open-probability gating mode. *J Gen Physiol* 128:373–386.
90. Scrimgeour, N., Litjens, T., Ma, L., Barritt, G.J., and Rychkov, G.Y. 2009. Properties of Orai1 mediated store-operated current depend on the expression levels of STIM1 and Orai1 proteins. *J Physiol* 587:2903–2918.
91. Hoth, M. and Penner, R. 1992. Depletion of intracellular calcium stores activates a calcium current in mast cells. *Nature* 355:353–356.
92. Heinemann, S.H. and Conti, F. 1992. Nonstationary noise analysis and application to patch clamp recordings. *Methods Enzymol* 207:131–148.
93. Lingle, C.J. 2006. Empirical considerations regarding the use of ensemble-variance analysis of macroscopic currents. *J Neurosci Methods* 158:121–132.
94. Zweifach, A. and Lewis, R.S. 1995. Rapid inactivation of depletion-activated calcium current (I_{CRAC}) due to local calcium feedback. *J Gen Physiol* 105:209–226.
95. Yamashita, M. and Prakriya, M. 2014. Divergence of Ca^{2+} selectivity and equilibrium Ca^{2+} blockade in a Ca^{2+} release-activated Ca^{2+} channel. *J Gen Physiol* 143:325–343.
96. Kilch, T., Alansary, D., Peglow, M., Dorr, K., Rychkov, G., Rieger, H., Peinelt, C., and Niemeyer, B.A. 2013. Mutations of the Ca^{2+}-sensing stromal interaction molecule STIM1 regulate Ca^{2+} influx by altered oligomerization of STIM1 and by destabilization of the Ca^{2+} channel Orai1. *J Biol Chem* 288:1653–1664.
97. Saviane, C., Conti, F., and Pusch, M. 1999. The muscle chloride channel ClC-1 has a double-barreled appearance that is differentially affected in dominant and recessive myotonia. *J Gen Physiol* 113:457–468.
98. Hewavitharana, T., Deng, X., Soboloff, J., and Gill, D.L. 2007. Role of STIM and Orai proteins in the store-operated calcium signaling pathway. *Cell Calcium* 42:173–182.
99. Feske, S. 2009. ORAI1 and STIM1 deficiency in human and mice: Roles of store-operated Ca^{2+} entry in the immune system and beyond. *Immunol Rev* 231:189–209.
100. Baba, Y. and Kurosaki, T. 2009. Physiological function and molecular basis of STIM1-mediated calcium entry in immune cells. *Immunol Rev* 231:174–188.
101. Gwack, Y., Srikanth, S., Feske, S., Cruz-Guilloty, F., Oh-hora, M., Neems, D.S., Hogan, P.G., and Rao, A. 2007. Biochemical and functional characterization of Orai proteins. *J Biol Chem* 282:16232–16243.
102. Gwack, Y., Srikanth, S., Oh-Hora, M., Hogan, P.G., Lamperti, E.D., Yamashita, M., Gelinas, C. et al. 2008. Hair loss and defective T- and B-cell function in mice lacking ORAI1. *Mol Cell Biol* 28:5209–5222.
103. Lis, A., Peinelt, C., Beck, A., Parvez, S., Monteilh-Zoller, M., Fleig, A., and Penner, R. 2007. CRACM1, CRACM2, and CRACM3 are store-operated Ca^{2+} channels with distinct functional properties. *Curr Biol* 17:794–800.
104. DeHaven, W.I., Smyth, J.T., Boyles, R.R., and Putney, J.W., Jr. 2007. Calcium inhibition and calcium potentiation of Orai1, Orai2, and Orai3 calcium release-activated calcium channels. *J Biol Chem* 282:17548–17556.
105. Missiaen, L., Callewaert, G., De Smedt, H., and Parys, J.B. 2001. 2-Aminoethoxydiphenyl borate affects the inositol 1,4,5-trisphosphate receptor, the intracellular Ca^{2+} pump and the non-specific Ca^{2+} leak from the non-mitochondrial Ca^{2+} stores in permeabilized A7r5 cells. *Cell Calcium* 29:111–116.

106. Peinelt, C., Lis, A., Beck, A., Fleig, A., and Penner, R. 2008. 2-Aminoethoxydiphenyl borate directly facilitates and indirectly inhibits STIM1-dependent gating of CRAC channels. *J Physiol* 586:3061–3073.
107. Schindl, R., Bergsmann, J., Frischauf, I., Derler, I., Fahrner, M., Muik, M., Fritsch, R., Groschner, K., and Romanin, C. 2008. 2-Aminoethoxydiphenyl borate alters selectivity of Orai3 channels by increasing their pore size. *J Biol Chem* 283:20261–20267.
108. Gross, S.A., Wissenbach, U., Philipp, S.E., Freichel, M., Cavalie, A., and Flockerzi, V. 2007. Murine ORAI2 splice variants form functional Ca^{2+} release-activated Ca^{2+} (CRAC) channels. *J Biol Chem* 282:19375–19384.
109. Wissenbach, U., Philipp, S.E., Gross, S.A., Cavalie, A., and Flockerzi, V. 2007. Primary structure, chromosomal localization and expression in immune cells of the murine ORAI and STIM genes. *Cell Calcium* 42:439–446.
110. Fukushima, M., Tomita, T., Janoshazi, A., and Putney, J.W. 2012. Alternative translation initiation gives rise to two isoforms of Orai1 with distinct plasma membrane mobilities. *J Cell Sci* 125:4354–4361.
111. Inayama, M., Suzuki, Y., Yamada, S., Kurita, T., Yamamura, H., Ohya, S., Giles, W.R., and Imaizumi, Y. 2015. Orai1–Orai2 complex is involved in store-operated calcium entry in chondrocyte cell lines. *Cell Calcium* 57:337–347.
112. Philipp, S., Strauss, B., Hirnet, D., Wissenbach, U., Mery, L., Flockerzi, V., and Hoth, M. 2003. TRPC3 mediates T-cell receptor-dependent calcium entry in human T-lymphocytes. *J Biol Chem* 278:26629–26638.
113. Wissenbach, U. and Niemeyer, B.A. 2007. Trpv6. *Handb Exp Pharmacol* 179:221–234.
114. Wissenbach, U., Niemeyer, B.A., Fixemer, T., Schneidewind, A., Trost, C., Cavalie, A., Reus, K., Meese, E., Bonkhoff, H., and Flockerzi, V. 2001. Expression of CaT-like, a novel calcium-selective channel, correlates with the malignancy of prostate cancer. *J Biol Chem* 276:19461–19468.
115. Suzuki, Y., Kovacs, C.S., Takanaga, H., Peng, J.B., Landowski, C.P., and Hediger, M.A. 2008. Calcium channel TRPV6 is involved in murine maternal-fetal calcium transport. *J Bone Miner Res* 23:1249–1256.
116. Weissgerber, P., Kriebs, U., Tsvilovskyy, V., Olausson, J., Kretz, O., Stoerger, C., Mannebach, S. et al. 2012. Excision of Trpv6 gene leads to severe defects in epididymal Ca^{2+} absorption and male fertility much like single D541A pore mutation. *J Biol Chem* 287:17930–17941.
117. Jang, Y., Lee, Y., Kim, S.M., Yang, Y.D., Jung, J., and Oh, U. 2012. Quantitative analysis of TRP channel genes in mouse organs. *Arch Pharm Res* 35:1823–1830.
118. Zhou, Y. and Greka, A. 2016. Calcium-permeable ion channels in the kidney. *Am J Physiol Renal Physiol* 310:F1157–F1167.
119. Inada, H., Iida, T., and Tominaga, M. 2006. Different expression patterns of TRP genes in murine B and T lymphocytes. *Biochem Biophys Res Commun* 350:762–767.
120. Vennekens, R., Hoenderop, J.G., Prenen, J., Stuiver, M., Willems, P.H., Droogmans, G., Nilius, B., and Bindels, R.J. 2000. Permeation and gating properties of the novel epithelial Ca^{2+} channel. *J Biol Chem* 275:3963–3969.
121. Bödding, M. and Flockerzi, V. 2004. Ca^{2+} dependence of the Ca^{2+}-selective TRPV6 channel. *J Biol Chem* 279:36546–36552.
122. Bödding, M. 2005. Voltage-dependent changes of TRPV6-mediated Ca^{2+} currents. *J Biol Chem* 280:7022–7029.
123. Lambers, T.T., Weidema, A.F., Nilius, B., Hoenderop, J.G., and Bindels, R.J. 2004. Regulation of the mouse epithelial Ca^{2+} channel TRPV6 by the Ca^{2+}-sensor calmodulin. *J Biol Chem* 279:28855–28861.
124. Velisetty, P., Borbiro, I., Kasimova, M.A., Liu, L., Badheka, D., Carnevale, V., and Rohacs, T. 2016. A molecular determinant of phosphoinositide affinity in mammalian TRPV channels. *Sci Rep* 6:27652.

125. Cao, C., Zakharian, E., Borbiro, I., and Rohacs, T. 2013. Interplay between calmodulin and phosphatidylinositol 4,5-bisphosphate in Ca^{2+}-induced inactivation of transient receptor potential vanilloid 6 channels. *J Biol Chem* 288:5278–5290.
126. Fecher-Trost, C., Weissgerber, P., and Wissenbach, U. 2014. TRPV6 channels. *Handb Exp Pharmacol* 222:359–384.
127. Yue, L., Peng, J.B., Hediger, M.A., and Clapham, D.E. 2001. CAT1 manifests the pore properties of the calcium-release-activated calcium channel. *Nature* 410:705–709.
128. Hoenderop, J.G., Vennekens, R., Muller, D., Prenen, J., Droogmans, G., Bindels, R.J., and Nilius, B. 2001. Function and expression of the epithelial Ca^{2+} channel family: Comparison of mammalian ECaC1 and 2. *J Physiol* 537:747–761.
129. Fasolato, C., Hoth, M., Matthews, G., and Penner, R. 1993. Ca^{2+} and Mn^{2+} influx through receptor-mediated activation of nonspecific cation channels in mast cells. *Proc Natl Acad Sci USA* 90:3068–3072.
130. Fernando, K.C. and Barritt, G.J. 1995. Characterisation of the divalent cation channels of the hepatocyte plasma membrane receptor-activated Ca^{2+} inflow system using lanthanide ions. *Biochim Biophys Acta* 1268:97–106.
131. Nilius, B., Prenen, J., Vennekens, R., Hoenderop, J.G., Bindels, R.J., and Droogmans, G. 2001. Pharmacological modulation of monovalent cation currents through the epithelial Ca^{2+} channel ECaC1. *Br J Pharmacol* 134:453–462.
132. Al-Ansary, D., Bogeski, I., Disteldorf, B.M., Becherer, U., and Niemeyer, B.A. 2010. ATP modulates Ca^{2+} uptake by TRPV6 and is counteracted by isoform-specific phosphorylation. *FASEB J* 24:425–435.
133. Zakharian, E., Cao, C., and Rohacs, T. 2011. Intracellular ATP supports TRPV6 activity via lipid kinases and the generation of PtdIns(4,5) P(2). *FASEB J* 25:3915–3928.
134. Lee, J., Cha, S.K., Sun, T.J., and Huang, C.L. 2005. PIP2 activates TRPV5 and releases its inhibition by intracellular Mg^{2+}. *J Gen Physiol* 126:439–451.
135. Voets, T., Janssens, A., Prenen, J., Droogmans, G., and Nilius, B. 2003. Mg^{2+}-dependent gating and strong inward rectification of the cation channel TRPV6. *J Gen Physiol* 121:245–260.
136. Irnaten, M., Blanchard-Gutton, N., Praetorius, J., and Harvey, B.J. 2009. Rapid effects of 17beta-estradiol on TRPV5 epithelial Ca^{2+} channels in rat renal cells. *Steroids* 74:642–649.
137. Irnaten, M., Blanchard-Gutton, N., and Harvey, B.J. 2008. Rapid effects of 17beta-estradiol on epithelial TRPV6 Ca^{2+} channel in human T84 colonic cells. *Cell Calcium* 44:441–452.
138. Cha, S.K., Jabbar, W., Xie, J., and Huang, C.L. 2007. Regulation of TRPV5 single-channel activity by intracellular pH. *J Membr Biol* 220:79–85.
139. de Groot, T., Kovalevskaya, N.V., Verkaart, S., Schilderink, N., Felici, M., van der Hagen, E.A., Bindels, R.J., Vuister, G.W., and Hoenderop, J.G. 2011. Molecular mechanisms of calmodulin action on TRPV5 and modulation by parathyroid hormone. *Mol Cell Biol* 31:2845–2853.
140. Nilius, B., Vennekens, R., Prenen, J., Hoenderop, J.G., Bindels, R.J., and Droogmans, G. 2000. Whole-cell and single channel monovalent cation currents through the novel rabbit epithelial Ca^{2+} channel ECaC. *J Physiol* 527(Pt 2):239–248.
141. Alansary, D., Kilch, T., Holzmann, C., Peinelt, C., Hoth, M., and Lis, A. 2014. Measuring endogenous I_{CRAC} and ORAI currents with the patch-clamp technique. *Cold Spring Harb Protoc* 2014:630–637.
142. Virsolvy, A., Smith, P., Bertrand, G., Gros, L., Heron, L., Salazar, G., Puech, R., and Bataille, D. 2002. Block of Ca^{2+}-channels by alpha-endosulphine inhibits insulin release. *Br J Pharmacol* 135:1810–1818.

2 Studies of Structure–Function and Subunit Composition of Orai/STIM Channel

Marc Fahrner, Rainer Schindl, and Christoph Romanin

CONTENTS

2.1 INTRODUCTION

Among all known second messengers in eukaryotic cells, Ca^{2+} is one of the most versatile and is involved in a multitude of physiological and cellular processes including cell proliferation, growth, gene expression, muscle contraction, and exocytosis/secretion [1,2]. To act as an intracellular signal molecule, Ca^{2+} has to enter the cell at specific physiological/cellular situations and time points. One major pathway that allows Ca^{2+} entry into the cells involves the Ca^{2+} release–activated Ca^{2+} (CRAC) channels, which belong to the group of store-operated channels (SOC) [3–14]. In the beginning of the CRAC/SOC channel analysis, these channels were studied and characterized using mainly cells of the immune system, that is, T-lymphocytes and mast cells [9,10,14,15]. Finally, in 2005–2006, the major key players forming the functional CRAC channel complex were identified [16–27]: first, the stromal interaction molecule (STIM), which represents the Ca^{2+} sensor in the endoplasmic reticulum (ER), and second, Orai, which is located in the plasma membrane (PM) and builds the ion-conducting transmembrane (TM) protein complex. Feske and colleagues [16] had studied a defect in CRAC channel function linked to one form of hereditary severe combined immune deficiency (SCID) syndrome, which allowed the identification of the Orai1 (also initially termed CRACM1) channel protein and its mutated form (Orai1 R91W) in SCID patients. By successfully employing and combining a modified linkage analysis with single-nucleotide polymorphism arrays and a *Drosophila* RNA interference screen, light was shed on the gene and protein that forms the Ca^{2+} conducting CRAC channel [16]. Furthermore, the search for homologous proteins using a sequence database research revealed Orai1, Orai2, and Orai3 in higher vertebrates. The three members of the Orai protein family have been analyzed with bioinformatics methods showing that they represent TM proteins with 4 PM spanning domains connected by one intracellular and two extracellular loops and cytosolic N-and C-termini [16,20,28,29]. Several research groups have concentrated on the electrophysiological examination and characterization of Orai proteins revealing the typical high Ca^{2+} selectivity and low single-channel conductance, concluding that these proteins unequivocally represent the pore-forming entity of the CRAC channel.

The CRAC channel–activating protein—stromal interaction molecule (STIM)—has been presented and published by Liou et al. as well as Roos et al. in 2005 [18,19]. Screening about 2300 signaling proteins in *Drosophila* S2 cells and HeLa cells using an RNA interference–based gene knockdown approach, 2 homologous proteins highly involved in ER store depletion–mediated Ca^{2+} influx were elucidated—STIM1 and STIM2. These proteins serve as ER-resident Ca^{2+} sensors, which closely communicate with the CRAC channels upon Ca^{2+} depletion of the ER [18,19]. Both STIM1 and STIM2 are single-pass TM proteins with the N-terminus in the ER lumen and the larger C-terminal part facing the cytosol. The ER luminal part, which functions as a Ca^{2+} sensor of $[Ca^{2+}]_{ER}$, contains, among other parts, a Ca^{2+}-sensing EF hand followed by the α-helical TM domain. The larger cytosolic part of STIM is responsible for coupling to and activation of Orai channels [6,30–33]. Confocal microscopy images reveal an intracellular tubular distribution of STIM1 under resting conditions with full ER Ca^{2+} stores; however, a small percentage of STIM1 has also been detected in the PM [34]. Lowering the ER-intraluminal Ca^{2+} concentration represents the initial trigger for STIM1 activation. In the course of store depletion, Ca^{2+} is released from the STIM1 EF hand domain followed by STIM1 homomerization and translocation to the cell periphery into the so-called ER–PM junctions—regions where the ER membrane is in tight proximity to the plasma membrane. Low $[Ca^{2+}]_{ER}$ finally leads to the formation of oligomeric STIM1 clusters/punctae in these microdomains where Orai channels localize as well. This physical coupling of STIM1 to Orai channels therefore induces Ca^{2+} influx linked to specific downstream signaling and ER store refilling [30,35–39]. Besides the activation of CRAC channels, STIM1 has been shown to play a role in arachidonate as well as leukotriene C4–stimulated Ca^{2+} channels (see Chapter 11) as well as TRP channel regulation [40].

After the initial characterization of STIM and Orai with limited structural knowledge based on bioinformatics predictions, in 2012, the crystal structures of cytosolic fragments of STIM1 and full-length Orai were reported, allowing new and more focused studies of STIM1 and Orai related to their intra- and intermolecular interactions [41].

2.2 STIM1

In resting cells, ER membrane–resident STIM1 reveals a dynamic constitutive movement along microtubules, whereas store depletion results in the redistribution of STIM1 into clusters/punctae at ER–PM junctions [42,43]. Dissection of STIM1 for examination of specific ER luminal as well as cytosolic STIM1 domains has shown that cytosolic coiled coil (CC) domains and the S/P domain are involved in the constitutive movement, whereas the luminal portion, in combination with cytosolic domains, is essential for STIM1 oligomerization and cluster formation upon store depletion [43–45]. Furthermore, STIM1 directly binds to EB1, a microtubule plus end tracking molecule. Both STIM1 and EB1 are required for tip attachment complex–mediated ER tubule extension where an ER tubule elongates with a growing microtubule [46,47]. The trigger preceding cytosolic CC interactions resulting in STIM1 oligomerization has been localized to the ER luminal STIM1 part as its substitution using an artificial luminal cross-linker clearly replaced the store depletion–dependent activation mechanism of STIM1 [48]. In conclusion, the initial publications on the activation of STIM1 showed that the initial step is induced by ER store depletion resulting in STIM1 oligomerization on its luminal side consequently leading to oligomerization on its cytosolic side [18,19,27,30,38,43,49–51]. The STIM1 luminal part contains a canonical and a hidden EF hand followed by a sterile-α motif (SAM) [44,45,52] (Figure 2.1a). The EF hand structurally represents a helix-loop-helix motif with negatively charged residues (Asp and Glu) binding Ca^{2+} (K_d ~ 200–600 μM) in the high $[Ca^{2+}]_{ER}$ conditions. ER store depletion (lowering $[Ca^{2+}]_{ER}$) results in Ca^{2+} dissociation from the STIM1 EF hand, therefore destabilizing the entire EF-SAM entity [44]. CD-spectroscopy measurements of that crucial domain performed by Stathopulos and colleagues revealed that EF-SAM with bound Ca^{2+} results in high α-helicity in contrast to the apo-EF-SAM (without Ca^{2+}). Furthermore, the apo-EF-SAM domain proved to exist in at least dimeric form, whereas in the presence of Ca^{2+} the monomeric form is predominant [44,45]. An additional study by Covington and colleagues [53] using Förster resonance energy transfer (FRET) has shown that a STIM1 deletion mutant containing only the luminal part and the TM domain responds with FRET increase that depends on store depletion. Structurally, STIM2 is very similar to STIM1; however, comparisons of the EF-SAM domains in their ER luminal part reveal different degrees of structural stability explaining the observation that STIM2 reacts faster upon smaller decreases in $[Ca^{2+}]_{ER}$ [54–56]. Following the EF-SAM domain, a TM helix spans the ER membrane (Figure 2.1a). Ma et al. [57] were the first to identify a gain of function mutation in STIM1 TM. Screening the whole TM domain, the mutant STIM1 C227W exhibits the highest potency for eliciting constitutive Ca^{2+} influx. The function of C227W is to uncouple distinct activation steps, allowing the examination of STIM1 TM structural states without manipulating luminal or cytosolic functional components. Based on FRET and cross-linking experiments, it was concluded that C227W allows the close proximity of the N-terminal part of the TM domains, mimicking the store-depleted state with oligomerized EF-SAM domains. Furthermore, FRET analysis revealed a decreased CC1–SOAR (STIM-Orai activating region) interaction in the C227W mutant similar to store depletion–induced STIM1 activation [57] (Figure 2.2). By combining biochemical and bioinformatics

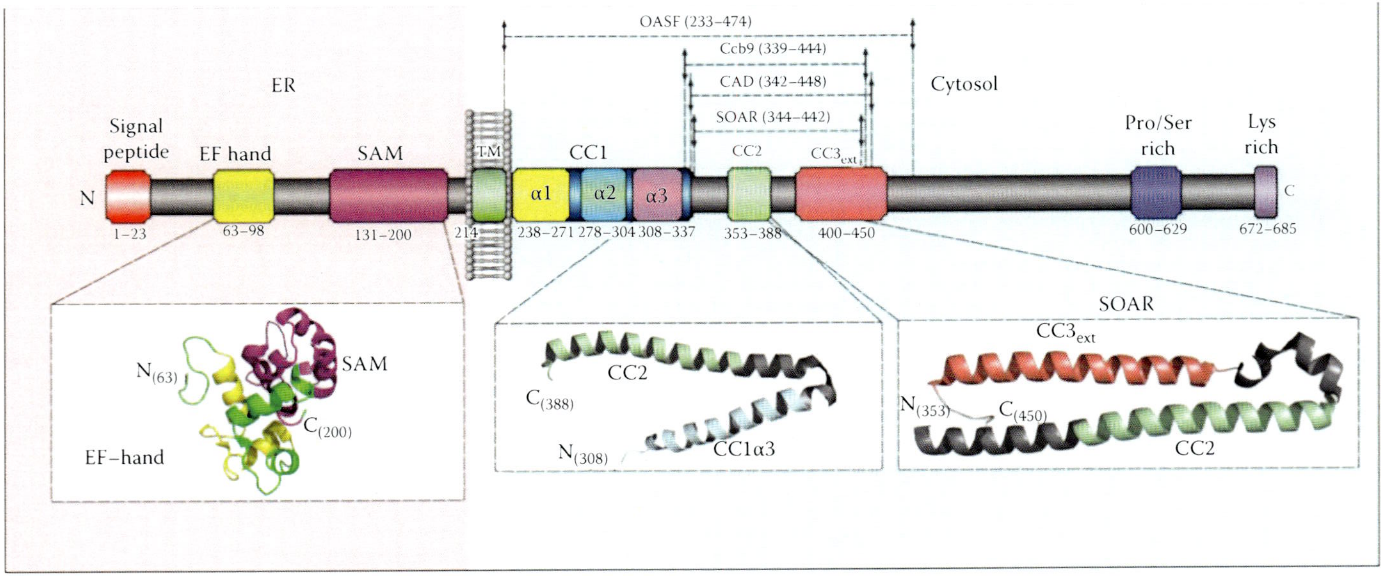

(a)

FIGURE 2.1 (a) Schematic depicting human STIM1 specifying essential structural/functional regions. Magnification of the NMR and crystal structures of the EF-SAM domain, the STIM1 CC1α3–CC2 fragment (312–387), and the STIM1 SOAR (344–444) fragment are shown in insets. *(Continued)*

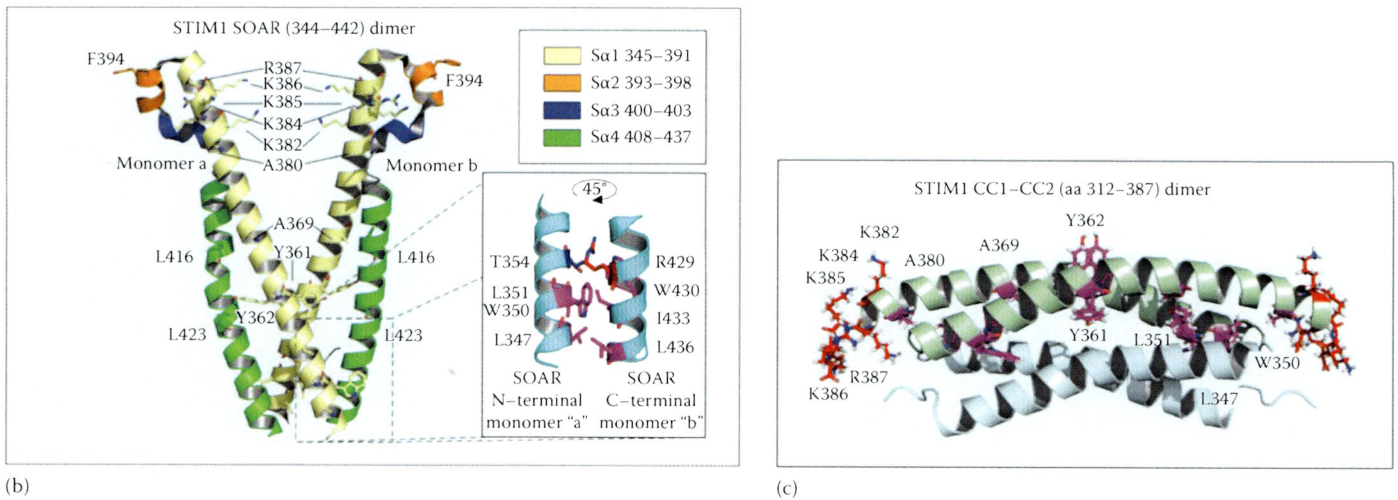

FIGURE 2.1 (*Continued*) (b) Crystallographic resolution of the dimeric STIM1 SOAR (344–444) structure, exhibiting a V-shaped form including the coiled-coil 2 (CC2) and 3 domains (CC3). The color code specifies Sα1, Sα2, Sα3, and Sα4. Inset shows the interface between SOAR N-terminal monomer "a" and C-terminal monomer "b." (c) NMR structure of a STIM1 CC1α3–CC2 dimer (312–387). The STIM1 CC1α3–CC2 monomers form an antiparallel dimer with Y361 as crossing point.

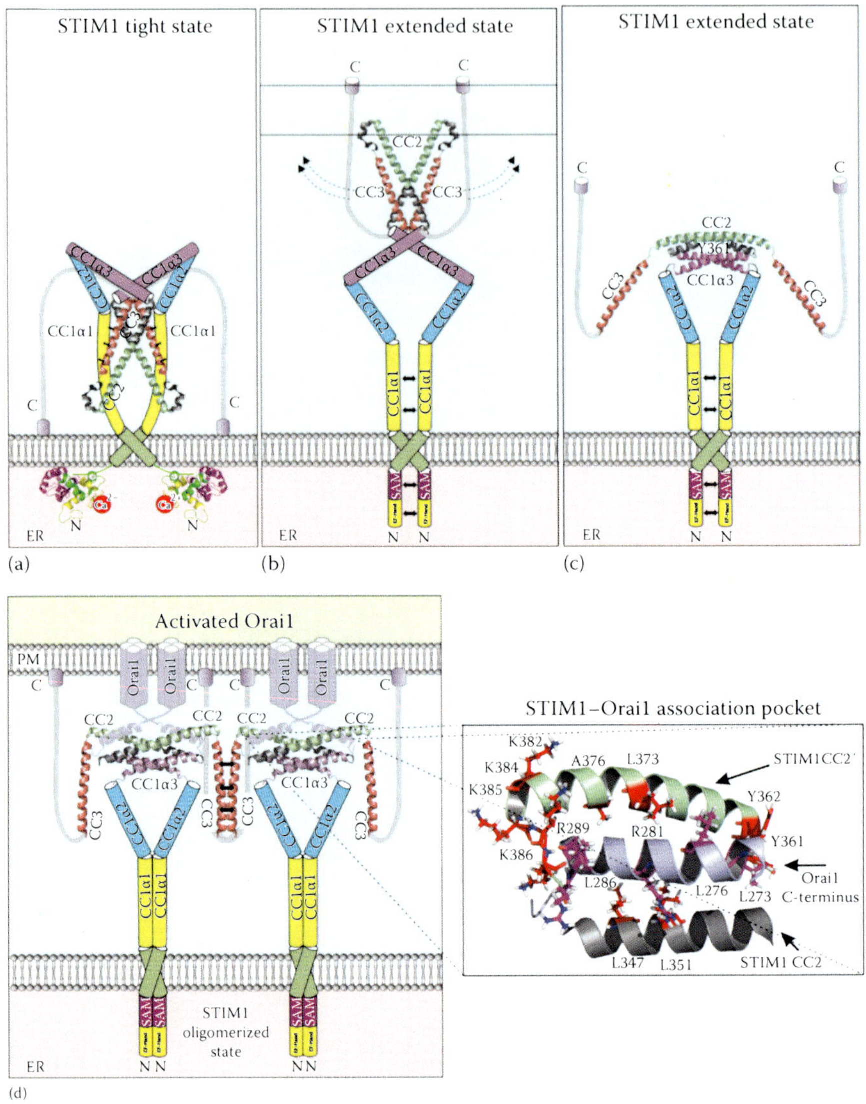

FIGURE 2.2 Cartoon of a hypothetical model depicting STIM1 conformational changes in the process of STIM1 activation and coupling to Orai upon store depletion. (a) STIM1 quiescent state is accomplished by STIM1 CC1/CC3 interaction in the presence of full ER Ca^{2+} stores. After store depletion, intraluminal STIM1 EF-SAM domains lose bound Ca^{2+} triggering a conformational rearrangement in the luminal part of STIM1, which changes the angle of the TM domains relaying the signal to the STIM1 cytosolic parts. (b) The following CC1α1 homomerization results in the release of the CC1/CC3 clamp leading to a more extended conformation with SOAR flipping to the top of the molecule. (c) Changes in angles of CC2 and CC1α3 in the dimeric STIM1 assembly and release of CC3 reveal SOAP (STIM–Orai association pocket) in combination with coupling to Orai C-terminus. (d) Additionally, STIM1 CC3 including SHD (STIM1 homomerization domain) connects STIM1 dimers forming higher-order oligomers. The inset depicts intermolecular interactions between STIM1 and Orai1 C-terminus.

approaches they further present a structural change of STIM1 TM domain dependent on the activation state. In their model, STIM1 TM dimers are not parallel but cross at a specific position, whereas the crossing angle is big enough to explain the separation of the luminal portions in a high ER Ca^{2+} condition. In contrast, store depletion results in a reduction of the crossing angle allowing the dimerization/oligomerization of the luminal STIM1 EF-SAM domains and consequent change of the relative positions of STIM1 cytosolic parts (Figure 2.2). Additionally, the TM domain contains three glycines (223, 225, 226) that yield high flexibility and are potentially involved in bend/kink formation representing a structural change in the course of signal transmission from the luminal to the cytosolic STIM1 parts [57,58].

The cytosolic portion following the TM domain includes three CC regions (CC1, CC2, and CC3), a CRAC modulatory domain (CMD), and a serine/proline- and a lysine-rich region (Figure 2.1a). Early studies by Huang et al. and Muik et al. demonstrated that the cytosolic expression of the STIM1 C-terminus is able to activate Orai1 [30,59]. As the STIM1 C-terminus (aa233–685) represents a large molecule, the goal has been laid on the identification of the STIM1 region representing the minimal fragment required to activate Orai channels. In a very narrow time frame, several Orai-activating STIM1 fragments were identified and published: (CRAC activation domain) CAD (aa342–448), SOAR (aa344–442), (Orai activating small fragment) OASF (aa233–450), and Ccb9 (aa339–444) [39,60–62] (Figure 2.1). All these Orai1-activating fragments contain CC2 (aa363–389), CC3 (aa399–423), and the STIM1 homomerization domain SHD (~aa421–450) for correct oligomerization and coupling to and activation of Orai channels [60]. At that time, no detailed knowledge about the structures of small Orai-activating fragments and only large segments within SOAR were suggested to form CC domains. It took 3 more years until Yang et al. presented the crystal structure of hSOAR ($345–444_{L374M,\ V419A,\ C437T}$) [63] (Figure 2.1b). They reported a dimeric assembly of the SOAR molecules with multiple intra- and intermolecular interactions. However, it is still not clear which activation state this structure represents. The monomeric SOAR consists of an antiparallel arrangement of CC2 and CC3 with two short α-helices linking these domains, resembling the capital letter “R” (Figure 2.1b). These regions are named Sα1 (345–391), Sα2 (393–398), Sα3 (400–403), and Sα4 (409–437). The dimeric SOAR arrangement represents an overall V-shape with N-terminal amino acids (T354, L351, W350, L347) from an SOAR monomer interacting with C-terminal residues (R429, W430, I433, L436) of the other monomer, respectively. Furthermore, the dimer contains a crossing point at position Y361 of both CC2 domains and a cluster of positively charged amino acids (K382, K384, K385, K386, R387) located on either tip of the V-shaped dimeric SOAR structure [63] (Figure 2.1c).

The STIM1 CC1 (aa238–343) domain, which is located between the TM domain and SOAR, includes three α-helical segments (CC1α1 aa238–271, CC1α2 aa278–304, and CC1α3 aa308–337) [32,64,65] and has been demonstrated to play a key role in keeping STIM1 in a locked tight inactive conformation when ER stores are full [57,63,65–67]. The decrease of $[Ca^{2+}]_{ER}$, which leads to luminal di-/oligomerization followed by closer proximity of the TM domains, triggers a conformational change in the cytosolic strand where CC1 adopts an extended conformation, thereby releasing SOAR [57,65,67] (Figure 2.2). Using FRET, Covington et al. examined the homomerization potential of different STIM1 C-terminal truncation mutants (i.e., STIM1-CC1 [aa1–344] and STIM1-CC1-CAD [aa1–448]), concluding that CC1 and CC3/SHD are both involved in homomerization [53]. However, further studies examining the role of CC1 have made clear that CC1 is a key domain in the STIM1 activation cascade integrating additional functions. One of the first hypothetical models of the control mechanism of CC1 has been proposed by Korzeniowski et al. [68]. They suggested a quiescent STIM1 state resulting from an autoinhibitory clamp based on attractive electrostatic interactions between a highly negatively charged segment within CC1α3 (aa308–337) and a positively charged region (aa382–387) within CC2 [68]. However, this hypothesis is not in line with the crystal structure of CC1-SOAR (from *Caenorhabditis elegans*) as the acidic region within CC1α3 (also called "inhibitory helix") is too distant from the basic region at the tip of the structure to form the proposed electrostatic clamp [63]. Based on their crystal structure, Yang et al. describe intramolecular interactions involving residues of CC1α3 and residues at the beginning of CC2 as well as the end of CC3 in the SOAR dimer. This dimerized conformation is suggested to represent the inactive, resting state where SOAR is occluded by the inhibitory helix [63]. Another study focusing on the CC1α3 describes its role involving intramolecular shielding leading to the STIM1 quiescent state, however, without the contribution of electrostatic or CC interactions, as deletions or substitutions within this segment revealed no clear effect [69]. Further examination of CC1α3 by multiple substitutions affecting its amphipathic character has led to the conclusion that the amphipathic nature of this segment has an impact on the regulation of STIM1 activation [69]. Another FRET-based approach developed by Romanin's group has allowed the monitoring of conformational changes within the cytosolic STIM1 region 233–474 (OASF) [66]. For this purpose, the OASF has been double labeled with yellow fluorescent protein (YFP) at the N-terminus and cyan fluorescent protein (CFP) at the C-terminus revealing different intramolecular FRET values depending on the OASF conformation. The wild-type OASF conformational sensor results in a robust FRET, suggesting a tightly packed structure representing the quiescent state of STIM1 when Ca^{2+} stores are full [66]. Another study performed by Hogan's group reported a tight conformation of the STIM1 C-terminus (233–685) based on Tb^{3+}-acceptor energy transfer measurements [67]. The obvious question arising is: what are the molecular steps that guide the cytosolic strand of STIM1 from the quiescent into the active form? To solve this task, several mutations as well as artificial cross-linking experiments have been performed. Point mutations within the OASF sensor with a high impact on the CC probability of CC1α1 (L251S) and CC3 (L416S L423S) result in decreased intramolecular FRET values consistent with an extended conformation representing the activated STIM1 state [66]. Introducing the same point mutations in full-length STIM1 and coexpressing these mutants with Orai1 in HEK cells has revealed constitutive inward Ca^{2+} currents independent of $[Ca^{2+}]_{ER}$. These results suggest that intramolecular CC1–CC3 interactions within STIM1 lock the quiescent state that is released upon store depletion or point mutations affecting the CC formation [65,66]. In an alternative approach, Zhou et al. succeeded in extending STIM1-CT by cross-linking of the CC1 domains. They also showed the "activating" mutation L251S resulting in a conformational extension of STIM1-CT in line with the OASF conformational sensor [67]. In summary, these results point to the high impact of CC1α1 as a key structure in the transition of STIM1 from the tight quiescent state to the extended active state [65–67]. The tight state has been suggested to involve an interaction between CC1 and CC3, whereas the active state via CC1 homomerization releases the CC1–CC3 clamp, leading to extended conformation.

Another FRET-based approach called "FIRE" (FRET Interaction in a Restricted Environment) was developed by Romanin's group to further dissect the STIM1 activation mechanism [65]. In that study, the role of CC1 in controlling the different cytosolic STIM1 activation states has been examined. The FIRE system has been engineered to mimic ER-targeted two-dimensional localization rather than using cytosolic expression. It consists of cytosolic STIM1 domains of interest linked via a flexible, 32 glycine linker to the ER STIM1 TM helix that is attached to a fluorescent protein (YFP or CFP) on the ER luminal side. The use of the FIRE system is accompanied by the advantages of a two-dimensional system, that is, low-affinity interactions of the protein domains of interest "survive" with higher probability in contrast to their cytosolic expression with three-dimensional degrees of freedom. Furthermore, FIRE allows the use of small fragments without the concern of steric hindrance as the fluorophores are attached on the ER luminal side. In case of an interaction of the peptides on the cytosolic side, the attached fluorophores on the ER luminal side achieve relatively close proximity, which is detected by FRET. In the study of Fahrner et al., using STIM1 fragments, the direct CC1α1–CC3 interaction has been established as a molecular determinant maintaining the quiescent STIM1 state. In line with the extended state, introduction of the activating L251S substitution in CC1α1 disrupted the interaction with CC3. CC1α3 is also involved in locking the inactive STIM1 state but not in a predominant manner as shown with deletion mutants [63,65]. Another result by this study has revealed a destabilizing role of CC1α2 that may be related to the Stormorken syndrome–associated mutant STIM1 R304W [65,70–72]. Possibly, this point mutation enhances the destabilizing force induced by CC1α2, leading to an extended and activated STIM1 that is able to couple to and activate Orai independent of $[Ca^{2+}]_{ER}$.

Based on all recent functional and structural STIM1 data, a hypothetical STIM1 activation model has been designed, including distinct STIM1 conformations reflecting the different steps of STIM1 activation upon store depletion [65] (Figure 2.2). In the STIM1 activation model, the inactive state is represented by a dimer where the intramolecular CC1α1 and CC1α3 interact with CC2/CC3, forming a clamp keeping STIM1 quiescent and tightly packed. Ca^{2+} depletion from the ER is the first step affecting the luminal STIM1 conformation [57,65]. Decreased $[Ca^{2+}]_{ER}$ results in luminal EF-SAM domain destabilization, which represents a signal that is transmitted via the TM domain to the cytosolic portion of STIM1 [45,52,57]. Ultimately, the CC1–CC3 interactions are released most likely accompanied by an increased CC1 homomerization and a conformational rearrangement of CC2 and CC3 (CAD/SOAR) resulting in the formation of STIM–Orai association pocket (SOAP) and CC3 oligomerization and clustering [57,65–67,73]. The current model does not directly involve CC1α2; however, the Stormorken syndrome–relevant STIM1 R304W point mutation [70–72], which is at the very end of CC1α2, induces the extended conformation consistent with constitutive Orai1 activation (M. Fahrner et al., unpublished data). How exactly this substitution is able to induce the extended STIM1 structure is still under examination.

2.3 ORAI

The Orai family consists of three homologous proteins named Orai1, Orai2, and Orai3, which reside in the PM and represent Ca^{2+}-selective ion channels allowing Ca^{2+} influx upon stimulation [16,20,74,75]. Each Orai molecule contains a cytosolic N- and C-terminus and four TM helices that are connected via two loops on the extracellular and one loop on the intracellular side [40] (Figure 2.3a). TM1 has the highest sequence similarity between the three Orai proteins, whereas the non-TM domains are less conserved. Orai1 and Orai3 share 34% sequence identity in the N-terminus and 46% in the C-terminus [76,77]. Another obvious difference applies to the third loop, which is much longer in Orai3. It has been shown by several groups that both the Orai C-terminus and the N-terminus are involved in functional coupling

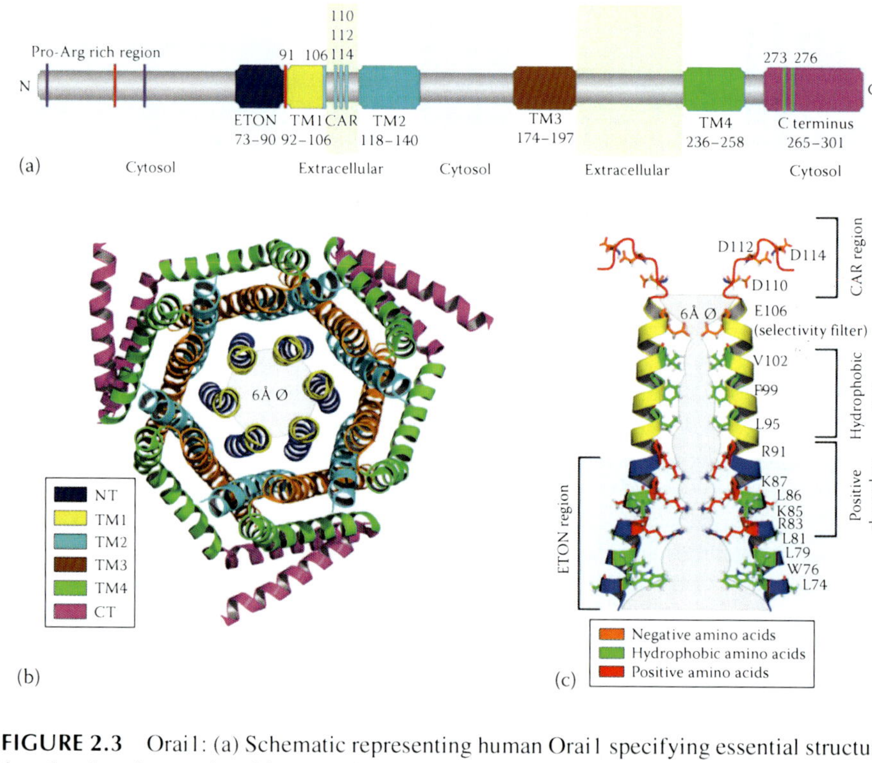

FIGURE 2.3 Orai1: (a) Schematic representing human Orai1 specifying essential structural/functional regions and residues required for Orai1 function. (b) Cartoon based on the x-ray crystallographic resolution of a *Drosophila* dOrai channel depicting the top view of Orai subunits revealing a hexameric assembly. The six transmembrane domains (TM) 1 form the inner ring surrounding the ion-conducting pore, while the concentric ring formed by domains TM2 and TM3 of the six subunits separate the TM1 ring from the lipid environment of the PM. The color code specifies the different helical regions of the Orai channel. (c) Cartoon showing two TM1 strands depicting the human Orai1 pore with the extracellular CAR region at the C-terminal end of TM1 and the cytosolic N-terminal helical extensions including the conserved ETON region. The selectivity filter, the hydrophobic core, and charged residues within the CAR and the ETON region are highlighted (amino acid numbering refers to human Orai1).

to STIM1. Upon activation by STIM1, Orai channels open, giving rise to inwardly rectifying Ca^{2+}currents with a low single-channel conductance [30,39,78–81].

The stoichiometry of Orai proteins forming a functional channel has been extensively investigated, suggesting that four Orai1 monomers constitute the Ca^{2+}-selective CRAC channel [4,76,82–86]. A heteromeric pentameric assembly, consisting of three Orai1 and two Orai3 subunits, has also been proposed for a Ca^{2+}-selective channel that is activated via arachidonic acid [87]. However, in 2012, dOrai (from *Drosophila melanogaster*) was crystallized, revealing an unexpected hexameric arrangement of the Orai subunits forming the functional channel [41]. In this architecture, three Orai dimers with crossing C-termini are combined to end in a quaternary structure with six Orai molecules per channel (Figure 2.3b). The ion-conducting pore is located right in the middle and is mainly created by TM1 of the six Orai subunits forming the first ring (Figure 2.3c). TM2 and TM3 form a ring surrounding TM1 and therefore separating TM1 from the PM lipid environment. TM4 contributes the third ring that is in tight contact with the components of the PM [41]. The cytosolic TM1 proximal N-terminal part of Orai represents the "Extended Transmembrane Orai1 N-terminal" region (ETON), which is an α-helical extension of TM1 reaching about 20 Å into the cytosol [78]. The helices of TM2 and TM3 expand into the cytosol too, to a smaller extent, and are connected via the intracellular loop 2, which is unfortunately not represented in the x-ray structure, as is the case for the two extracellularly located loops (loop 1 and loop 3), which are also missing in the structure [41]. Molecular modeling and molecular dynamics simulations have been performed for the missing parts of the crystal structure, suggesting a high flexibility of loop 3 [88]. The cytosolic extensions of TM4 representing the C-termini reveal two different kink angles in a highly conserved hinge region leading to an antiparallel dimeric interaction of the C-termini of two adjacent Orai molecules in the hexamer [41,73].

2.4 STIM1: ORAI INTERACTION

ER store depletion results in the activation of STIM1 that unfolds into an extended conformation, homo-oligomerizes forming clusters, translocates to the cell periphery, and eventually couples to Orai, thereby opening this channel [40]. The small Orai-activating fragment CAD has been shown to directly bind to the Orai1 C-terminus, which seems to be the major interaction site between these two proteins [39,61]. CAD–Orai1 N-terminus interaction was determined as well, whereas the interaction with Orai1 loop 2 has not been detected [39]. NMR structure of a STIM1 fragment (312–387) including CC1α3 and CC2 has been presented with and without an Orai1 C-terminal fragment (272–292), respectively, which for the first time has revealed a potential STIM1–Orai1 C-terminus interaction on a structural basis [73]. Here, two STIM1 fragments form an antiparallel dimeric structure that interacts with two antiparallel Orai1 C-terminal fragments in line with the Orai C-terminal conformation presented in the hexameric dOrai crystal structure [41,73]. The NMR structure of SOAP reveals the residues that are in close proximity to amino acids of the Orai1 C-terminus. With respect to STIM1, the side chains of L347, L351, Y362, L373, and A376 are involved in hydrophobic interactions with Orai1 C-terminal residues. Another part of STIM1 comprising the positively charged residues K382,

K384, K385, and K386 forms a highly charged region providing electrostatic complementarity to an acidic portion within the C-terminus. With respect to Orai1, on one hand, hydrophobic residues including L273 and L276 are involved in hydrophobic interactions, and on the other, negatively charged side chains of Orai1 (D284, D287, and D291) form electrostatic interactions with complementary STIM1 regions within SOAP, respectively. Point mutations of these hydrophobic as well as charged residues on either STIM1 or Orai1 have been reported to functionally disrupt the STIM1–Orai1 communication, however to a different extent, which is in agreement with the NMR structure [73]. Therefore, the STIM1–Orai1 C-terminal coupling is well described on a functional and structural level; however, the interaction between STIM1 and Orai1 N-terminus is only marginally defined. Possibly, F394 in STIM1 plays a critical role in binding to a hydrophobic counterpart of the ETON region as the point mutation F394H abolishes the activation of Orai1 [89].

Structurally, the Orai1 C-terminus adopts two different orientations in the crystal structure depending on the angles of the hinge region (aa261–265), which represents the link between TM4 and the cytosolic C-terminal part of Orai. With respect to an Orai dimer, the two C-termini cross each other at an angle of 152° [41]. The probability of forming CC for the Orai1 C-terminus is relatively weak, whereas for Orai2 and Orai3 it is about 17-fold higher [90]. Examination of strategically important positions of the CC domain (the hydrophobic "a" and "d" positions) of Orai1 using single point mutations (L273S and L276S) to decrease the CC probability, interferes with coupling to STIM1 [30,90]. Indeed, the NMR structure presenting SOAP as the STIM1–Orai1 binding domain is consistent with the previous functional experiments as positions L273 and L276 are engaged in hydrophobic interactions with STIM1 counterparts in SOAP [73]. Concerning Orai2 and Orai3, single point mutations do not reveal the same dominant effect, which may be explained by their higher C-terminal CC probabilities. Therefore, double mutations are required to robustly decrease the CC probability, resulting in full disruption of the communication with STIM1 [90]. Orai1 R281, L286, and R289 represent additional important residues in the Orai1 C-terminus contributing to SOAP. In accordance to the structural data, adjacent residues not involved in SOAP had no significant impact on channel activation [73]. The Orai1 C-terminal positions L273 and L276 are worth mentioning, as they are not solely involved in STIM1 binding within SOAP, but they also represent the crossing point of two antiparallel C-termini of each Orai dimer within the hexameric Orai. Substituting L273 and L276 in Orai1 with cysteines resulted in disulfide bond formation under oxidizing conditions in cross-linking experiments fully abolishing STIM1 binding. The addition of reducing agents breaking the disulfide bonds rescued STIM1 coupling [80,91]. The hinge region in Orai1 connecting TM4 and Orai C-terminus has a key role in defining the correct orientation and conformation of the C-terminus. As expected, mutations within the hinge region affect STIM1 coupling and Orai channel activation; however, to fully disrupt STIM1–Orai1 communication, additional Orai1 C-terminal mutations were necessary [80]. In summary, the hinge region is relevant for determining the optimal position of the C-termini for efficient coupling to STIM1, but the direct coupling of STIM1 to Orai C-terminal downstream regions is essential.

How do the crossing Orai C-termini reported in the crystal structure fit in with the NMR structure of SOAP? A closer look at the SOAP structure reveals antiparallel

Orai C-termini with an angle of 136° [73] in contrast to 152° measured in the crystal structure without STIM1 [41]. Possibly, the different angles arise from the different origins of Orai as dOrai was crystallized for the x-ray structure determination, whereas the NMR structure is based on the human Orai1. However, it has to be pointed out that the two structures also represent different activation states of Orai1. The crystal structure corresponds to the closed Orai state without STIM1, whereas the NMR SOAP structure is derived from a complex of STIM1/Orai1 C-termini. Therefore, the observed difference of angles could hypothetically reflect the activation state, that is, Orai1 C-termini undergo a mild angle modification upon coupling to STIM1, resulting in a conformational shift.

Biochemically an Orai1 N-terminus–STIM1 interaction is detectable too, but to a weaker extent [39,78]. As the ETON region (Orai1 aa73–90) represents the elongated TM1 and incorporates relevant binding residues, Orai1 N-terminal deletion mutants reveal different effects depending on the size of truncation. Full channel activation is observed in case of Orai1 Δ1–72, whereas Δ1–74 results in only 50% activation and Δ1–76 abolishes current activation completely [78]. Several studies analyzing the ETON region within full-length Orai1 have revealed an inhibition of coupling to STIM1 as well as channel activation by substituting various residues in ETON [78,89,92]. The positively charged side chains within ETON seemingly have various functions, that is, interaction with STIM1, stabilization of the elongated pore, and an electrostatic barrier to cations [78,93]. Although Orai1, Orai2, and Orai3 all contain the conserved ETON region, equivalent N-terminal truncations have revealed different effects in the three Orai homologs. Strikingly, Orai3 store-operated activation is still preserved even with extensive truncations that fully abolish Orai1 function [78]. Probably, additional structures present in Orai3 are able to compensate for the large N-terminal deletions; however, their position still remains elusive. In summary, residues within ETON most likely are engaged in communicating with STIM1; however, a structural view of this interaction is still missing.

Based on structural and functional data, the STIM1–Orai1 interaction is predominantly achieved by the binding of STIM1 C-terminus to Orai1 C-terminus as well as N-terminus. The presence of both cytosolic Orai strands is absolutely necessary for channel activation by STIM1 as deletions or mutations of either Orai N- or C-terminus result in fully nonfunctional channels [78–80,94]. However, the C-terminus seems to have a superior role in this coupling, as a deletion/mutation of the Orai C-terminal binding sites completely inhibits the interaction with STIM1 or STIM1 C-terminal fragments, whereas Orai N-terminal deletions or mutations lead to nonfunctional channels but still allow partial coupling to STIM1 [30,31]. Structural and biochemical evidence point to the fact that the Orai C-terminus is indeed the primary STIM1 coupling site and the Orai N-terminus most likely is involved in gating [30,93,95]. In addition to STIM1, calmodulin and cholesterol have been reported to regulate Orai1 via its N-terminus [96–99]. In summary, both N- and C-terminal cytosolic Orai strands are necessary for Orai channel activation; however, the detailed molecular choreography of STIM1–Orai coupling and activation still remains a mystery. A commonly accepted hypothetical picture includes as the first step STIM1 coupling to the Orai C-terminus consequently attaching STIM1 to Orai, thereby allowing the next interaction involving the Orai N-terminus for channel gating. This finally results in a bridging of Orai1 N- and C-termini by the CAD/SOAR domain of STIM1. In addition to the hypothetical interaction sequence, the Orai C-terminus could not only be responsible for STIM1 coupling but also for discrete conformational changes propagating through the Orai TM domains via rearrangements of TM4 upon interaction with STIM1; however, this is still speculative.

To shed light on the interaction of Orai1 N- and C-termini with STIM1 upon store depletion, various Orai1 proteins with CAD fragments linked to the cytosolic Orai1 N-terminal and C-terminal strands, respectively, have been explored [79,80,94]. The direct linkage results in a local enrichment of STIM1-CAD close to Orai1. Combining the Orai1–CAD fusion with N-terminal (Δ1–76) and C-terminal (Δ276–301) loss of function deletions of Orai1, respectively, yields slight compensation resulting in partial channel activation. However, a tethered CAD cannot compensate for more severe deletions pointing to the fact that the presence of Orai1 residues 77–90 (N-terminus) and 267–275 (C-terminus) is necessary for preserved activation via tethered CAD. On the other hand, single loss of function point mutations in either Orai1 N- or C-terminus lose their dominant effect in the presence of attached CAD [80,94]. Obviously, the local CAD enrichment results in a higher likelihood of coupling to and activation of Orai1 channels even in the presence of point mutations that otherwise lead to a loss of function. To fully destroy the CAD–Orai1 interaction using the Orai1–CAD fusion protein, mutations in both Orai1 N- and C-termini are necessary, emphasizing that both Orai1 cytosolic strands contribute to the interaction with the STIM1 C-terminus. Hence, an alternative to the sequential STIM1–Orai coupling model has been proposed by Palty et al. proposing the formation of a distinct STIM1 binding pocket by Orai1 N- and C-terminal sites [80,94].

The NMR 3D structure representing a dimerized STIM1 fragment (312–387) with two Orai1 C-termini (272–292) shows a clear picture of the interacting residues within the SOAP. Based on this holo-structure, point mutations have been introduced in STIM1 as well as Orai1 for functional electrophysiological analysis revealing strong agreement with the presented structural data [73]. It is important to note, however, that full-length Orai1 involves both C-terminus and N-terminus for functional coupling and activation [30,31,39,78], whereas the NMR holo-structure only contains a C-terminal fragment of Orai1. Furthermore, the STIM1 fragment lacks residues 388–442, which represent an important domain within SOAR [61]. To gain more evidence for the physiological STIM1–Orai1 interaction, a structure containing a larger STIM1 fragment together with Orai1 cytosolic strands or even the full-length Orai1 channel complex would be of high benefit. Moreover, the cytosolic loop2 of Orai could play a cooperative role in CRAC activation too, however this possibility needs to be specifically examined and analyzed.

A comparison of the SOAR crystal and the NMR STIM1 312–387 reveals different tertiary and quaternary structures probably reflecting different conformational activation states of the protein. Both share only a small overlapping part (residues 344–387) containing the crossing pivotal point around Y361; however, the angles between CC2 and CC2′ in the dimer are different in the two structures. The NMR structure of STIM1 fragments within SOAP fits well with the antiparallel crossing C-termini of Orai1 suggesting a conformation close to the STIM1-activated state, whereas the SOAR crystal possibly reflects a structure close to the STIM1-inactivated tight state [63,73].

The NMR resolution of the STIM1 312–387 + Orai1 272–292 structures allows the interpretation that one STIM1 dimer interacts with one Orai1 dimer during the coupling and activation process. However, several studies have been performed analyzing the stoichiometry of the oligomeric STIM1–Orai1 complex revealing divergent results. Electrophysiological experiments suggest that the extent of Orai1 activation depends on the number of STIM1 molecules present [100–102]. Furthermore, the CRAC current inactivation depends on the STIM1 protein number, as less STIM1 proteins interacting with Orai1 result in less CRAC channel inactivation [103]. Various approaches have been performed to elucidate the STIM1–Orai1 stoichiometry, including varying STIM1/Orai1 ratios or direct fusion proteins of Orai1 connected to CAD dimers. Results point to the fact that eight STIM1 molecules per four Orai1 subunits lead to maximal CRAC channel activation and inactivation [100,101]. Analyses of Orai1 and constitutively active Orai1 V102A mutants connected to single or tandem CAD molecules result in increased Ca^{2+} selectivity with an increased number of STIM1 fragments, showing that CRAC channel activation is not an "all or none" phenomenon but rather a gradual process depending on the number of Orai1 subunits in a tetramer with an undisturbed C-terminal STIM1 binding site [101]. In the past, most studies pointed to an Orai1 tetramer forming the CRAC channel activated by eight STIM1 molecules ending in a 2:1 STIM1:Orai1 ratio. However, the crystallized dOrai structure reveals a hexameric assembly and it is difficult to explain how eight STIM1 proteins interact with six Orai subunits. Based on the NMR structure of interacting dimeric STIM1 fragments with two Orai1 C-terminal fragments a STIM1:Orai1 ratio of 1:1 is most likely present in the active complex [73]. Connecting this view to the hexameric Orai channel suggests an interaction of six STIM1 proteins with six Orai1 subunits displaying a picture where three STIM1 dimers couple to three Orai1 dimers in the hexameric CRAC channel complex. An alternative hypothesis has been presented by Zhou et al. suggesting a new STIM1–Orai1 coupling model based on experiments using a STIM1 F394H mutant, which disrupts STIM1–Orai1 coupling and activation [89,104]. Fusing two SOAR molecules either containing two wild type or one wild type and one F394H SOAR mutant has revealed a similar interaction with and activation of Orai1. To disrupt the coupling to and activation of the CRAC channel, both SOAR domains have to carry the F394H substitution. Another experiment using a PM-anchored Orai1 C-terminus reveals an interaction with the SOAR tandem construct and ~50% decreased coupling in the case of the SOAR tandem with one F394H mutant monomer. These results suggest a unimolecular interaction involving one Orai monomer of the hexameric assembly and one STIM1 molecule within the STIM1 dimer [104]. This interpretation is based on the assumption that each SOAR monomer within a dimeric construct acts independently of the other in the Orai channel activation process, excluding a potential cooperative effect. This picture allows for the hypothetical interaction of the second STIM1 molecule of the dimer with an Orai subunit of an adjacent Orai hexameric channel. Connection of two Orai hexameric channels by one STIM1 dimer may result in a lattice structure of CRAC channels where STIM1 dimers grab neighboring Orai hexamers, consequently forming STIM1–Orai1 clusters at ER–PM junctions. However, one point difficult to explain is how the antiparallel crossing Orai C-termini reported in the dOrai crystal structure are released and structurally change position and orientation to fit the unimolecular interaction model proposed by Zhou et al. [104] (see Chapter 7). Potentially, one Orai dimer interacts with two STIM1 molecules but not from the same STIM1 dimer. The alternative to the unimolecular coupling and activation model is the bimolecular binding model presenting the interaction of one STIM1 dimer with one Orai1 dimer, as seen in the NMR structure [73]. This model would similarly enable cluster formation by the intrinsic ability of STIM1 to form higher-order oligomers via $CC3_{ext}$ upon store depletion that link STIM1 dimers between adjacent Orai channel complexes, meaning that Orai is not necessary for cluster formation, which follows the inherent STIM1 clustering propensity. However, further functional and structural evidence is required to shed light on the STIM1–Orai1 coupling process and oligomeric cluster formation.

2.5 ION CONDUCTION PATHWAY OF THE ORAI PORE

In the absence of STIM1 binding and store depletion, Orai1 channels remain in a closed conformation. However, specific mutations in the pore of Orai1 have been shown to yield constitutively active currents, suggesting that the pore itself is an essential gating domain [80,105–108]. To allow Ca^{2+} influx through the centrally located Orai1 pore by STIM1 requires energy to switch the Orai1 pore into a permeable state. The pore domain of Orai1 is formed by a ring of TM1 helices extending into the cytosol. The hexameric crystal structure of Orai1 reveals an external vestibule in the vicinity of the selectivity filter, a hydrophobic cavity, and a basic region [41] (Figure 2.3b and c). Such a long α-helical pore formation is further supported by cysteine cross-linking experiments [109,110]. Molecular dynamics simulations have revealed that Ca^{2+} ions can frequently bind to the external vestibule, named the Ca^{2+} accumulating region (CAR) [88] (Figure 2.3c). This external vestibule in Orai1 is formed by the TM1–TM2 loop, which includes three negatively charged residues (D110, D112, D114). D110 that is most centrally located to the pore contributes most frequently to Ca^{2+} ion binding. D112 instead not only binds Ca^{2+} but also basic residues of the longer extracellular loop 3, which affects Ca^{2+} binding and permeation [88]. Neither the negatively charged amino acids in loop 1 nor the positively charged residues in loop 3 are fully conserved among the three Orai isoforms [88]. Hence, these sequence differences may add to the plasticity of Ca^{2+} permeation of the Orai family members. The main role of the extracellular CAR in Orai channels is to raise local Ca^{2+} levels near the outer entrance of the pore and favor Ca^{2+} binding in the selectivity filter. CAR and the selectivity filter, which is exclusively formed by E106 residues, are only 1 nm apart [41,88] (Figure 2.3c). Even a conservative mutation E106 to D yields nonselective cation currents and widens the minimum pore diameter [111]. The relatively slow time of binding and unbinding of Ca^{2+} to the selectivity filter regulates the Ca^{2+} selectivity of the Orai1 channel [112]. This Ca^{2+} (un-) binding process likely also includes the closely located CAR segment. The more central part of the pore formed by hydrophobic residues hinders Ca^{2+} permeation in the closed conformation. Not surprisingly, a mutation of a rather bulky valine 102 to an alanine resulted in constitutively active Orai1 channels [105]. It is of note that these currents of Orai1-V102A are less Ca^{2+} selective, while the coexpression of STIM1 robustly increases their Ca^{2+} selectivity. These experiments suggest that the Ca^{2+} selectivity is not an intrinsic property of the Orai pore itself but rather is formed in a concerted manner with STIM1. The V102 residues may act as a hydrophobic gate upon the coupling of STIM1 with Orai1 to bend away from its pore blocking position and allow Ca^{2+} influx. Computer simulations of Orai1-V102 show that the hydrophobic barrier is lowered when this valine is changed to an alanine [113]. Consequently, water molecules can access the pore more easily and may help to shield Ca^{2+} charges and to allow Ca^{2+} influx. Additionally, STIM1 may induce additional pore conformations to the selectivity filter, as determined from terbium-binding experiments [114]. This observation also fits nicely with the altered Ca^{2+} selectivity of the pore mutant experiments in the absence and presence of STIM1 [105]. A second site that is essential for gating is the basic segment located at the cytosolic exit of the Orai1 TM1 pore. This is consistent with the disease-causing Orai1-R91W mutation that completely blocks the pore [16,95]. Remarkably, the crystal structure of dOrai was solved together with a small negatively charged plug that is surrounded by side chains of R91 and R87 [41]. It is yet unclear if such a negatively charged plug plays an essential physiological role, but Cl^- ions in the vicinity of the basic pore segment could play a similar role in preventing Ca^{2+} permeation. A second constitutive pore mutant, Orai1-G98P, has been shown to regulate the proposed R91 gate [106]. While the positive ring of R91 is, after the selectivity filter, the narrowest part of the wild-type Orai1 pore, the constitutively active Orai1-G98P extended the basic pore segment [106]. Again, the pore mutation not only affected gating but also the selectivity of the Orai1 channel [106]. Hence, an open Orai1 pore conformation might affect the whole pore, including the selectivity filter, the hydrophobic segment, and the basic segment. Within the conserved N-terminus of Orai1, a putative cholesterol binding site has recently been reported [99]. Upon cholesterol depletion, STIM1/Orai1-mediated as well as endogenous CRAC currents were enhanced without affecting the selectivity of the Ca^{2+} currents and retaining association of STIM1 to Orai1. Cholesterol binding to an Orai1 N-terminal peptide comprising the ETON region has been observed, and point mutations disrupting the cholesterol binding motif enhanced store-operated currents similar to cholesterol depletion. Since overall coupling of STIM1 to Orai1 remains unaffected by the amount of cholesterol in the plasma membrane, this lipid seems to interfere mainly with STIM1-dependent channel gating via the ETON region [99]. Clearly, further studies are required to resolve this unique channel modulation mechanism.

2.6 PERSPECTIVES

Within the past decade, major efforts have been made to identify the molecular components of the CRAC channel family together with the discovery of their 3D atomic structures. Nevertheless, several questions remain unresolved and need further examination. Regarding STIM1, different cytosolic C-terminal parts including a portion and the whole CAD/SOAR domain have been characterized by NMR and x-ray crystallography, revealing significant differences in their 3D atomic structures. In particular, larger STIM1 C-terminal fragments would be of benefit for the elucidation of intra- and intermolecular interactions that trigger the switch mechanism of

STIM1, resulting in its tight, inactive or its extended, active conformation. The use of point mutations like L251S or R426L would help locking the STIM1 C-terminal structures in their extended or tight conformation, respectively.

With respect to Orai1, the crystal structure shows the closed state, and hence, Orai1 point mutants like P245L that yield a constitutively active Orai1 channel would lock the open state and may help elucidate the open channel conformation. How exactly STIM1 triggers the gating of Orai channels into the open state is only partially understood. NMR resolution and characterization of the SOAP, including a STIM1 fragment and part of the Orai1 C-terminus, have suggested a direct STIM1–Orai1 coupling mechanism. However, the gating mechanism, which may involve other Orai1 cytosolic domains, is still not fully understood and needs further efforts to obtain structural data including these cytosolic Orai1 domains together with STIM1 C-terminal parts. Possibly, these studies may elucidate conformational rearrangements within Orai1 cytosolic strands as well as TM domains revealing aspects of the molecular Orai1 gating mechanism. The emerging field of optogenetics will further allow to obtain more detailed mechanistic insights of the STIM1–Orai communication via a sophisticated control by light (see Chapter 8). Finally, proteins potentially modulating the STIM1–Orai interaction and communication, like STIMATE [115], Septin [116], SARAF [117], or CRACR2A [118] (see Chapters 4 and 10), have to be taken into consideration, pointing to the complexities of the native CRAC channel system.

ACKNOWLEDGMENTS

We thank Isaac Jardin for editing the figures. This work was supported by the Austrian Science Fund (FWF project P28123 to M.F., FWF projects P26067 and P28701 to R.S., and FWF project P27263 to C.R.).

REFERENCES

1. Bootman, M.D., Collins, T.J., Peppiatt, C.M., Prothero, L.S., MacKenzie, L., De Smet, P., Travers, M. et al. 2001. Calcium signalling—An overview. *Semin Cell Dev Biol* 12:3–10.
2. Berridge, M.J., Bootman, M.D., and Roderick, H.L. 2003. Calcium signalling: Dynamics, homeostasis and remodelling. *Nat Rev Mol Cell Biol* 4:517–529.
3. Putney, J.W., Jr. 1986. A model for receptor-regulated calcium entry. *Cell Calcium* 7:1–12.
4. Derler, I., Madl, J., Schutz, G., and Romanin, C. 2012. Structure, regulation and biophysics of I_{CRAC}, STIM/Orai1. *Adv Exp Med Biol* 740:383–410.
5. Derler, I., Schindl, R., Fritsch, R., and Romanin, C. 2012. Gating and permeation of Orai channels. *Front Biosci* 17:1304–1322.
6. Fahrner, M., Derler, I., Jardin, I., and Romanin, C. 2013. The STIM1/Orai signaling machinery. *Channels (Austin)* 7:330–343.
7. Fahrner, M., Muik, M., Derler, I., Schindl, R., Fritsch, R., Frischauf, I., and Romanin, C. 2009. Mechanistic view on domains mediating STIM1-Orai coupling. *Immunol Rev* 231:99–112.
8. Frischauf, I., Schindl, R., Derler, I., Bergsmann, J., Fahrner, M., and Romanin, C. 2008. The STIM/Orai coupling machinery. *Channels (Austin)* 2:261–268.

9. Hoth, M. and Penner, R. 1992. Depletion of intracellular calcium stores activates a calcium current in mast cells. *Nature* 355:353–356.
10. Lewis, R.S. 1999. Store-operated calcium channels. *Adv Second Messenger Phosphoprotein Res* 33:279–307.
11. Parekh, A.B. 2008. Ca^{2+} microdomains near plasma membrane Ca^{2+} channels: Impact on cell function. *J Physiol* 586:3043–3054.
12. Parekh, A.B. 2010. Store-operated CRAC channels: Function in health and disease. *Nat Rev Drug Discov* 9:399–410.
13. Schindl, R., Muik, M., Fahrner, M., Derler, I., Fritsch, R., Bergsmann, J., and Romanin, C. 2009. Recent progress on STIM1 domains controlling Orai activation. *Cell Calcium* 46:227–232.
14. Zweifach, A. and Lewis, R.S. 1993. Mitogen-regulated Ca^{2+} current of T lymphocytes is activated by depletion of intracellular Ca^{2+} stores. *Proc Natl Acad Sci USA* 90:6295–6299.
15. Penner, R. and Fleig, A. 2004. Store-operated calcium entry: A tough nut to CRAC. *Sci STKE* 2004:pe38.
16. Feske, S., Gwack, Y., Prakriya, M., Srikanth, S., Puppel, S.H., Tanasa, B., Hogan, P.G., Lewis, R.S., Daly, M., and Rao, A. 2006. A mutation in Orai1 causes immune deficiency by abrogating CRAC channel function. *Nature* 441:179–185.
17. Prakriya, M., Feske, S., Gwack, Y., Srikanth, S., Rao, A., and Hogan, P.G. 2006. Orai1 is an essential pore subunit of the CRAC channel. *Nature* 443:230–233.
18. Roos, J., DiGregorio, P.J., Yeromin, A.V., Ohlsen, K., Lioudyno, M., Zhang, S., Safrina, O. et al. 2005. STIM1, an essential and conserved component of store-operated Ca^{2+} channel function. *J Cell Biol* 169:435–445.
19. Liou, J., Kim, M.L., Heo, W.D., Jones, J.T., Myers, J.W., Ferrell, J.E., Jr., and Meyer, T. 2005. STIM is a Ca^{2+} sensor essential for Ca^{2+}-store-depletion-triggered Ca^{2+} influx. *Curr Biol* 15:1235–1241.
20. Vig, M., Peinelt, C., Beck, A., Koomoa, D.L., Rabah, D., Koblan-Huberson, M., Kraft, S. et al. 2006. CRACM1 is a plasma membrane protein essential for store-operated Ca^{2+} entry. *Science* 312:1220–1223.
21. Vig, M., Beck, A., Billingsley, J.M., Lis, A., Parvez, S., Peinelt, C., Koomoa, D.L. et al. 2006. CRACM1 multimers form the ion-selective pore of the CRAC channel. *Curr Biol* 16:2073–2079.
22. Zhang, S.L., Yeromin, A.V., Zhang, X.H.F., Yu, Y., Safrina, O., Penna, A., Roos, J., Stauderman, K.A., and Cahalan, M.D. 2006. Genome-wide RNAi screen of Ca^{2+} influx identifies genes that regulate Ca^{2+} release-activated Ca^{2+} channel activity. *Proc Natl Acad Sci USA* 103:9357–9362.
23. Zhang, S.L., Yu, Y., Roos, J., Kozak, J.A., Deerinck, T.J., Ellisman, M.H., Stauderman, K.A., and Cahalan, M.D. 2005. STIM1 is a Ca^{2+} sensor that activates CRAC channels and migrates from the Ca^{2+} store to the plasma membrane. *Nature* 437:902–905.
24. Mercer, J.C., DeHaven, W.I., Smyth, J.T., Wedel, B., Boyles, R.R., Bird, G.S., and Putney, J.W., Jr. 2006. Large store-operated calcium selective currents due to co-expression of Orai1 or Orai2 with the intracellular calcium sensor, Stim1. *J Biol Chem* 281:24979–24990.
25. Peinelt, C., Vig, M., Koomoa, D.L., Beck, A., Nadler, M.J., Koblan-Huberson, M., Lis, A., Fleig, A., Penner, R., and Kinet, J.P. 2006. Amplification of CRAC current by STIM1 and CRACM1 (Orai1). *Nat Cell Biol* 8:771–773.
26. Soboloff, J., Spassova, M.A., Tang, X.D., Hewavitharana, T., Xu, W., and Gill, D.L. 2006. Orai1 and STIM reconstitute store-operated calcium channel function. *J Biol Chem* 281:20661–20665.
27. Luik, R.M., Wu, M.M., Buchanan, J., and Lewis, R.S. 2006. The elementary unit of store-operated Ca^{2+} entry: Local activation of CRAC channels by STIM1 at ER-plasma membrane junctions. *J Cell Biol* 174:815–825.

28. Feske, S., Skolnik, E.Y., and Prakriya, M. 2012. Ion channels and transporters in lymphocyte function and immunity. *Nat Rev Immunol* 12:532–547.
29. Hoth, M. and Niemeyer, B.A. 2013. The neglected CRAC proteins: Orai2, Orai3, and STIM2. *Curr Top Membr* 71:237–271.
30. Muik, M., Frischauf, I., Derler, I., Fahrner, M., Bergsmann, J., Eder, P., Schindl, R. et al. 2008. Dynamic coupling of the putative coiled-coil domain of ORAI1 with STIM1 mediates ORAI1 channel activation. *J Biol Chem* 283:8014–8022.
31. Li, Z., Lu, J., Xu, P., Xie, X., Chen, L., and Xu, T. 2007. Mapping the interacting domains of STIM1 and Orai1 in Ca^{2+} release-activated Ca^{2+} channel activation. *J Biol Chem* 282:29448–29456.
32. Soboloff, J., Rothberg, B.S., Madesh, M., and Gill, D.L. 2012. STIM proteins: Dynamic calcium signal transducers. *Nat Rev Mol Cell Biol* 13:549–565.
33. Stathopulos, P.B. and Ikura, M. 2013. Structural aspects of calcium-release activated calcium channel function. *Channels (Austin)* 7:344–353.
34. Spassova, M.A., Soboloff, J., He, L.P., Xu, W., Dziadek, M.A., and Gill, D.L. 2006. STIM1 has a plasma membrane role in the activation of store-operated Ca^{2+} channels. *Proc Natl Acad Sci USA* 103:4040–4045.
35. Barr, V.A., Bernot, K.M., Srikanth, S., Gwack, Y., Balagopalan, L., Regan, C.K., Helman, D.J. et al. 2008. Dynamic movement of the calcium sensor STIM1 and the calcium channel Orai1 in activated T-cells: Puncta and Distal Caps. *Mol Biol Cell* 19:2802–2817.
36. Cahalan, M.D. 2009. STIMulating store-operated Ca^{2+} entry. *Nat Cell Biol* 11:669–677.
37. Calloway, N., Vig, M., Kinet, J.P., Holowka, D., and Baird, B. 2009. Molecular clustering of STIM1 with Orai1/CRACM1 at the plasma membrane depends dynamically on depletion of Ca^{2+} stores and on electrostatic interactions. *Mol Biol Cell* 20:389–399.
38. Wu, M.M., Buchanan, J., Luik, R.M., and Lewis, R.S. 2006. Ca^{2+} store depletion causes STIM1 to accumulate in ER regions closely associated with the plasma membrane. *J Cell Biol* 174:803–813.
39. Park, C.Y., Hoover, P.J., Mullins, F.M., Bachhawat, P., Covington, E.D., Raunser, S., Walz, T., Garcia, K.C., Dolmetsch, R.E., and Lewis, R.S. 2009. STIM1 clusters and activates CRAC channels via direct binding of a cytosolic domain to Orai1. *Cell* 136:876–890.
40. Prakriya, M. and Lewis, R.S. 2015. Store-operated calcium channels. *Physiol Rev* 95:1383–1436.
41. Hou, X., Pedi, L., Diver, M.M., and Long, S.B. 2012. Crystal structure of the calcium release-activated calcium channel orai. *Science* 338:1308–1313.
42. Smyth, J.T., Dehaven, W.I., Bird, G.S., and Putney, J.W., Jr. 2008. Ca^{2+}-store-dependent and -independent reversal of Stim1 localization and function. *J Cell Sci* 121:762–772.
43. Baba, Y., Hayashit, K., Fujii, Y., Mizushima, A., Watarai, H., Wakamori, M., Numaga, T. et al. 2006. Coupling of STIM1 to store-operated Ca^{2+} entry through its constitutive and inducible movement in the endoplasmic reticulum. *Proc Natl Acad Sci USA* 103:16704–16709.
44. Stathopulos, P.B., Li, G.-Y., Plevin, M.J., Ames, J.B., and Ikura, M. 2006. Stored Ca^{2+} depletion-induced oligomerization of STIM1 via the EF-SAM region: An initiation mechanism for capacitive Ca^{2+} entry. *J Biol Chem* 284:728–732.
45. Stathopulos, P.B., Zheng, L., Li, G.Y., Plevin, M.J., and Ikura, M. 2008. Structural and mechanistic insights into STIM1-mediated initiation of store-operated calcium entry. *Cell* 135:110–122.
46. Grigoriev, I., Gouveia, S.M., van der Vaart, B., Demmers, J., Smyth, J.T., Honnappa, S., Splinter, D. et al. 2008. STIM1 is a MT-plus-end-tracking protein involved in remodeling of the ER. *Curr Biol* 18:177–182.

47. Honnappa, S., Gouveia, S.M., Weisbrich, A., Damberger, F.F., Bhavesh, N.S., Jawhari, H., Grigoriev, I. et al. 2009. An EB1-binding motif acts as a microtubule tip localization signal. *Cell* 138:366–376.
48. Luik, R.M., Wang, B., Prakriya, M., Wu, M.M., and Lewis, R.S. 2008. Oligomerization of STIM1 couples ER calcium depletion to CRAC channel activation. *Nature* 454:538–542.
49. Baba, Y., Nishida, K., Fujii, Y., Hirano, T., Hikida, M., and Kurosaki, T. 2008. Essential function for the calcium sensor STIM1 in mast cell activation and anaphylactic responses. *Nat Immunol* 9:81–88.
50. Malli, R., Naghdi, S., Romanin, C., and Graier, W.F. 2008. Cytosolic Ca^{2+} prevents the subplasmalemmal clustering of STIM1: An intrinsic mechanism to avoid Ca^{2+} overload. *J Cell Sci* 121:3133–3139.
51. Liou, J., Fivaz, M., Inoue, T., and Meyer, T. 2007. Live-cell imaging reveals sequential oligomerization and local plasma membrane targeting of stromal interaction molecule 1 after Ca^{2+} store depletion. *Proc Natl Acad Sci USA* 104:9301–9306.
52. Stathopulos, P.B. and Ikura, M. 2010. Partial unfolding and oligomerization of stromal interaction molecules as an initiation mechanism of store operated calcium entry. *Biochem Cell Biol* 88:175–183.
53. Covington, E.D., Wu, M.M., and Lewis, R.S. 2010. Essential role for the CRAC activation domain in store-dependent oligomerization of STIM1. *Mol Biol Cell* 21:1897–1907.
54. Brandman, O., Liou, J., Park, W.S., and Meyer, T. 2007. STIM2 is a feedback regulator that stabilizes basal cytosolic and endoplasmic reticulum Ca^{2+} levels. *Cell* 131:1327–1339.
55. Zheng, L., Stathopulos, P.B., Schindl, R., Li, G.Y., Romanin, C., and Ikura, M. 2011. Auto-inhibitory role of the EF-SAM domain of STIM proteins in store-operated calcium entry. *Proc Natl Acad Sci USA* 108:1337–1342.
56. Zheng, L., Stathopulos, P.B., Li, G.Y., and Ikura, M. 2008. Biophysical characterization of the EF-hand and SAM domain containing Ca^{2+} sensory region of STIM1 and STIM2. *Biochem Biophys Res Commun* 369:240–246.
57. Ma, G., Wei, M., He, L., Liu, C., Wu, B., Zhang, S.L., Jing, J. et al. 2015. Inside-out Ca signalling prompted by STIM1 conformational switch. *Nat Commun* 6: 7826.1–7826.14.
58. Dong, H., Sharma, M., Zhou, H.X., and Cross, T.A. 2012. Glycines: Role in alpha-helical membrane protein structures and a potential indicator of native conformation. *Biochemistry* 51:4779–4789.
59. Huang, G.N., Zeng, W., Kim, J.Y., Yuan, J.P., Han, L., Muallem, S., and Worley, P.F. 2006. STIM1 carboxyl-terminus activates native SOC, I(crac) and TRPC1 channels. *Nat Cell Biol* 8:1003–1010.
60. Muik, M., Fahrner, M., Derler, I., Schindl, R., Bergsmann, J., Frischauf, I., Groschner, K., and Romanin, C. 2009. A cytosolic homomerization and a modulatory domain within STIM1 C terminus determine coupling to ORAI1 channels. *J Biol Chem* 284:8421–8426.
61. Yuan, J.P., Zeng, W., Dorwart, M.R., Choi, Y.J., Worley, P.F., and Muallem, S. 2009. SOAR and the polybasic STIM1 domains gate and regulate Orai channels. *Nat Cell Biol* 11:337–343.
62. Kawasaki, T., Lange, I., and Feske, S. 2009. A minimal regulatory domain in the C terminus of STIM1 binds to and activates ORAI1 CRAC channels. *Biochem Biophys Res Commun* 385:49–54.
63. Yang, X., Jin, H., Cai, X., Li, S., and Shen, Y. 2012. Structural and mechanistic insights into the activation of Stromal interaction molecule 1 (STIM1). *Proc Natl Acad Sci USA* 109:5657–5662.

64. Cui, B., Yang, X., Li, S., Lin, Z., Wang, Z., Dong, C., and Shen, Y. 2013. The inhibitory helix controls the intramolecular conformational switching of the C-terminus of STIM1. *PLoS One* 8:e74735.1–e74735.11.
65. Fahrner, M., Muik, M., Schindl, R., Butorac, C., Stathopulos, P., Zheng, L., Jardin, I., Ikura, M., and Romanin, C. 2014. A coiled-coil clamp controls both conformation and clustering of stromal interaction molecule 1 (STIM1). *J Biol Chem* 289:33231–33244.
66. Muik, M., Fahrner, M., Schindl, R., Stathopulos, P., Frischauf, I., Derler, I., Plenk, P. et al. 2011. STIM1 couples to ORAI1 via an intramolecular transition into an extended conformation. *EMBO J* 30:1678–1689.
67. Zhou, Y., Srinivasan, P., Razavi, S., Seymour, S., Meraner, P., Gudlur, A., Stathopulos, P.B., Ikura, M., Rao, A., and Hogan, P.G. 2013. Initial activation of STIM1, the regulator of store-operated calcium entry. *Nat Struct Mol Biol* 20:973–981.
68. Korzeniowski, M.K., Manjarres, I.M., Varnai, P., and Balla, T. 2010. Activation of STIM1-Orai1 involves an intramolecular switching mechanism. *Sci Signal* 3:ra82.1–ra82.10.
69. Yu, F., Sun, L., Hubrack, S., Selvaraj, S., and Machaca, K. 2013. Intramolecular shielding maintains the ER Ca^{2+} sensor STIM1 in an inactive conformation. *J Cell Sci* 126:2401–2410.
70. Nesin, V., Wiley, G., Kousi, M., Ong, E.C., Lehmann, T., Nicholl, D.J., Suri, M. et al. 2014. Activating mutations in STIM1 and ORAI1 cause overlapping syndromes of tubular myopathy and congenital miosis. *Proc Natl Acad Sci USA* 111:4197–4202.
71. Misceo, D., Holmgren, A., Louch, W.E., Holme, P.A., Mizobuchi, M., Morales, R.J., De Paula, A.M. et al. 2014. A dominant STIM1 mutation causes Stormorken syndrome. *Hum Mutat* 35:556–564.
72. Morin, G., Bruechle, N.O., Singh, A.R., Knopp, C., Jedraszak, G., Elbracht, M., Bremond-Gignac, D. et al. 2014. Gain-of-function mutation in STIM1 (p.R304W) is associated with Stormorken Syndrome. *Hum Mutat* 35:1221–1232.
73. Stathopulos, P.B., Schindl, R., Fahrner, M., Zheng, L., Gasmi-Seabrook, G.M., Muik, M., Romanin, C., and Ikura, M. 2013. STIM1/Orai1 coiled-coil interplay in the regulation of store-operated calcium entry. *Nat Commun* 4:2963.1–2963.12.
74. Bogeski, I., Kummerow, C., Al-Ansary, D., Schwarz, E.C., Koehler, R., Kozai, D., Takahashi, N. et al. 2010. Differential redox regulation of ORAI ion channels: A mechanism to tune cellular calcium signaling. *Sci Signal* 3:ra24.1–ra24.10.
75. Frischauf, I., Schindl, R., Bergsmann, J., Derler, I., Fahrner, M., Muik, M., Fritsch, R., Lackner, B., Groschner, K., and Romanin, C. 2011. Cooperativeness of Orai cytosolic domains tunes subtype-specific gating. *J Biol Chem* 286:8577–8584.
76. Mignen, O., Thompson, J.L., and Shuttleworth, T.J. 2008. Both Orai1 and Orai3 are essential components of the arachidonate-regulated Ca^{2+}-selective (ARC) channels. *J Physiol* 586:185–195.
77. Shuttleworth, T.J. 2012. Orai3—The 'exceptional' Orai? *J Physiol* 590:241–257.
78. Derler, I., Plenk, P., Fahrner, M., Muik, M., Jardin, I., Schindl, R., Gruber, H.J., Groschner, K., and Romanin, C. 2013. The extended transmembrane Orai1 N-terminal (ETON) region combines binding interface and gate for Orai1 activation by STIM1. *J Biol Chem* 288:29025–29034.
79. McNally, B.A., Somasundaram, A., Jairaman, A., Yamashita, M., and Prakriya, M. 2013. The C- and N-terminal STIM1 binding sites on Orai1 are required for both trapping and gating CRAC channels. *J Physiol* 591: 2833–2850.
80. Palty, R., Stanley, C., and Isacoff, E.Y. 2015. Critical role for Orai1 C-terminal domain and TM4 in CRAC channel gating. *Cell Res* 25:963–980.
81. Zheng, H., Zhou, M.H., Hu, C., Kuo, E., Peng, X., Hu, J., Kuo, L., and Zhang, S.L. 2013. Differential roles of the C and N termini of Orai1 protein in interacting with stromal interaction molecule 1 (STIM1) for Ca^{2+} release-activated Ca^{2+} (CRAC) channel activation. *J Biol Chem* 288:11263–11272.

82. Demuro, A., Penna, A., Safrina, O., Yeromin, A.V., Amcheslavsky, A., Cahalan, M.D., and Parker, I. 2011. Subunit stoichiometry of human Orai1 and Orai3 channels in closed and open states. *Proc Natl Acad Sci USA* 108:17832–17837.
83. Penna, A., Demuro, A., Yeromin, A.V., Zhang, S.L., Safrina, O., Parker, I., and Cahalan, M.D. 2008. The CRAC channel consists of a tetramer formed by Stim-induced dimerization of Orai dimers. *Nature* 456:116–120.
84. Madl, J., Weghuber, J., Fritsch, R., Derler, I., Fahrner, M., Frischauf, I., Lackner, B., Romanin, C., and Schutz, G.J. 2010. Resting state Orai1 diffuses as homotetramer in the plasma membrane of live mammalian cells. *J Biol Chem* 285:41135–41142.
85. Maruyama, Y., Ogura, T., Mio, K., Kato, K., Kaneko, T., Kiyonaka, S., Mori, Y., and Sato, C. 2009. Tetrameric Orai1 is a teardrop-shaped molecule with a long, tapered cytoplasmic domain. *J Biol Chem* 284:13676–13685.
86. Ji, W., Xu, P., Li, Z., Lu, J., Liu, L., Zhan, Y., Chen, Y., Hille, B., Xu, T., and Chen, L. 2008. Functional stoichiometry of the unitary calcium-release-activated calcium channel. *Proc Natl Acad Sci USA* 105:13668–13673.
87. Mignen, O., Thompson, J.L., and Shuttleworth, T.J. 2009. The molecular architecture of the arachidonate-regulated Ca^{2+}-selective ARC channel is a pentameric assembly of Orai1 and Orai3 subunits. *J Physiol* 587:4181–4197.
88. Frischauf, I., Zayats, V., Deix, M., Hochreiter, A., Jardin, I., Muik, M., Lackner, B. et al. 2015. A calcium-accumulating region, CAR, in the channel Orai1 enhances Ca^{2+} permeation and SOCE-induced gene transcription. *Sci Signal* 8:ra131.1–ra131.13.
89. Wang, X., Wang, Y., Zhou, Y., Hendron, E., Mancarella, S., Andrake, M.D., Rothberg, B.S., Soboloff, J., and Gill, D.L. 2014. Distinct Orai-coupling domains in STIM1 and STIM2 define the Orai-activating site. *Nat Commun* 5:3183.1–3183.11.
90. Frischauf, I., Muik, M., Derler, I., Bergsmann, J., Fahrner, M., Schindl, R., Groschner, K., and Romanin, C. 2009. Molecular determinants of the coupling between STIM1 and Orai channels: Differential activation of Orai1-3 channels by a STIM1 coiled-coil mutant. *J Biol Chem* 284:21696–21706.
91. Tirado-Lee, L., Yamashita, M., and Prakriya, M. 2015. Conformational changes in the Orai1 C-terminus evoked by STIM1 binding. *PLoS One* 10:e0128622.1–e0128622.17.
92. Lis, A., Zierler, S., Peinelt, C., Fleig, A., and Penner, R. 2010. A single lysine in the N-terminal region of store-operated channels is critical for STIM1-mediated gating. *J Gen Physiol* 136:673–686.
93. Rothberg, B.S., Wang, Y., and Gill, D.L. 2013. Orai channel pore properties and gating by STIM: Implications from the Orai crystal structure. *Sci Signal* 6:pe9.1–pe9.9.
94. Palty, R. and Isacoff, E.Y. 2016. Cooperative binding of stromal interaction molecule 1 (STIM1) to the N and C termini of calcium release-activated calcium modulator 1 (Orai1). *J Biol Chem* 291:334–341.
95. Derler, I., Fahrner, M., Carugo, O., Muik, M., Bergsmann, J., Schindl, R., Frischauf, I., Eshaghi, S., and Romanin, C. 2009. Increased hydrophobicity at the N terminus/membrane interface impairs gating of the severe combined immunodeficiency-related ORAI1 mutant. *J Biol Chem* 284:15903–15915.
96. Liu, Y., Zheng, X., Mueller, G.A., Sobhany, M., Derose, E.F., Zhang, Y., London, R.E., and Birnbaumer, L. 2012. Crystal structure of calmodulin binding domain of Orai1 in complex with Ca^{2+}/Calmodulin displays a unique binding mode. *J Biol Chem* 287:43030–43041.
97. Mullins, F.M. and Lewis, R.S. 2016. The inactivation domain of STIM1 is functionally coupled with the Orai1 pore to enable Ca^{2+}-dependent inactivation. *J Gen Physiol* 147:153–164.
98. Mullins, F.M., Park, C.Y., Dolmetsch, R.E., and Lewis, R.S. 2009. STIM1 and calmodulin interact with Orai1 to induce Ca^{2+}-dependent inactivation of CRAC channels. *Proc Natl Acad Sci USA* 106:15495–15500.

99. Derler, I., Jardin, I., Stathopulos, P.B., Muik, M., Fahrner, M., Zayats, V., Pandey, S.K. et al. 2016. Cholesterol modulates Orai1 channel function. *Sci Signal* 9:ra10.1–ra10.10.
100. Hoover, P.J. and Lewis, R.S. 2011. Stoichiometric requirements for trapping and gating of Ca^{2+} release-activated Ca^{2+} (CRAC) channels by stromal interaction molecule 1 (STIM1). *Proc Natl Acad Sci USA* 108:13299–13304.
101. Li, Z., Liu, L., Deng, Y., Ji, W., Du, W., Xu, P., Chen, L., and Xu, T. 2011. Graded activation of CRAC channel by binding of different numbers of STIM1 to Orai1 subunits. *Cell Res* 21:305–315.
102. Scrimgeour, N., Litjens, T., Ma, L., Barritt, G.J., and Rychkov, G.Y. 2009. Properties of Orai1 mediated store-operated current depend on the expression levels of STIM1 and Orai1 proteins. *J Physiol* 587:2903–2918.
103. Scrimgeour, N.R., Wilson, D.P., Barritt, G.J., and Rychkov, G.Y. 2014. Structural and stoichiometric determinants of Ca release-activated Ca (CRAC) channel Ca-dependent inactivation. *Biochim Biophys Acta* 1838:1281–1287.
104. Zhou, Y., Wang, X., Loktionova, N.A., Cai, X., Nwokonko, R.M., Vrana, E., Wang, Y., Rothberg, B.S., and Gill, D.L. 2015. STIM1 dimers undergo unimolecular coupling to activate Orai1 channels. *Nat Commun* 6:8395.1–8395.10.
105. McNally, B.A., Somasundaram, A., Yamashita, M., and Prakriya, M. 2012. Gated regulation of CRAC channel ion selectivity by STIM1. *Nature* 482:241–245.
106. Zhang, S.L., Yeromin, A.V., Hu, J., Amcheslavsky, A., Zheng, H., and Cahalan, M.D. 2011. Mutations in Orai1 transmembrane segment 1 cause STIM1-independent activation of Orai1 channels at glycine 98 and channel closure at arginine 91. *Proc Natl Acad Sci USA* 108:17838–17843.
107. Srikanth, S., Jung, H.J., Ribalet, B., and Gwack, Y. 2010. The intracellular loop of Orai1 plays a central role in fast inactivation of Ca^{2+} release-activated Ca^{2+} channels. *J Biol Chem* 285:5066–5075.
108. Srikanth, S., Yee, M.K., Gwack, Y., and Ribalet, B. 2011. The third transmembrane segment of orai1 protein modulates Ca^{2+} release-activated Ca^{2+} (CRAC) channel gating and permeation properties. *J Biol Chem* 286:35318–35328.
109. Zhou, Y., Ramachandran, S., Oh-Hora, M., Rao, A., and Hogan, P.G. 2010. Pore architecture of the ORAI1 store-operated calcium channel. *Proc Natl Acad Sci USA* 107:4896–4901.
110. McNally, B.A., Yamashita, M., Engh, A., and Prakriya, M. 2009. Structural determinants of ion permeation in CRAC channels. *Proc Natl Acad Sci USA* 106:22516–22521.
111. Yamashita, M., Navarro-Borelly, L., McNally, B.A., and Prakriya, M. 2007. Orai1 mutations alter ion permeation and Ca^{2+}-dependent fast inactivation of CRAC channels: Evidence for coupling of permeation and gating. *J Gen Physiol* 130:525–540.
112. Yamashita, M. and Prakriya, M. 2014. Divergence of Ca^{2+} selectivity and equilibrium Ca^{2+} blockade in a Ca^{2+} release-activated Ca^{2+} channel. *J Gen Physiol* 143:325–343.
113. Dong, H., Fiorin, G., Carnevale, V., Treptow, W., and Klein, M.L. 2013. Pore waters regulate ion permeation in a calcium release-activated calcium channel. *Proc Natl Acad Sci USA* 110:17332–17337.
114. Gudlur, A., Quintana, A., Zhou, Y., Hirve, N., Mahapatra, S., and Hogan, P.G. 2014. STIM1 triggers a gating rearrangement at the extracellular mouth of the ORAI1 channel. *Nat Commun* 5:5164.
115. Jing, J., He, L., Sun, A., Quintana, A., Ding, Y., Ma, G., Tan, P. et al. 2015. Proteomic mapping of ER-PM junctions identifies STIMATE as a regulator of Ca^{2+} influx. *Nat Cell Biol* 17:1339–1347.

116. Sharma, S., Quintana, A., Findlay, G.M., Mettlen, M., Baust, B., Jain, M., Nilsson, R., Rao, A., and Hogan, P.G. 2013. An siRNA screen for NFAT activation identifies septins as coordinators of store-operated Ca^{2+} entry. *Nature* 499:238–242.
117. Palty, R., Raveh, A., Kaminsky, I., Meller, R., and Reuveny, E. 2012. SARAF inactivates the store operated calcium entry machinery to prevent excess calcium refilling. *Cell* 149:425–438.
118. Srikanth, S., Jung, H.J., Kim, K.D., Souda, P., Whitelegge, J., and Gwack, Y. 2010. A novel EF-hand protein, CRACR2A, is a cytosolic Ca^{2+} sensor that stabilizes CRAC channels in T cells. *Nat Cell Biol* 12:436–446.

3 Signaling ER Store Depletion to Plasma Membrane Orai Channels

Aparna Gudlur and Patrick Hogan

CONTENTS

3.1 INTRODUCTION

3.1.1 ER Calcium Repository

Cells maintain an inward Ca^{2+} gradient from the extracellular fluid to the cytosol, entrenched by the action of plasma membrane pumps/transporters that actively "pump out" Ca^{2+} and organellar pumps/transporters that mold the Ca^{2+} transients by storing it away in intracellular reservoirs [1,2]. External stimuli trigger Ca^{2+} influx driven by the electrochemical gradient between a plenteous extracellular reservoir of Ca^{2+} (millimolar ranges) and a Ca^{2+}-free cytosol. This Ca^{2+} surge is indispensible for its role as a second messenger, controlling a wide range of cellular functions like exocytosis, mast cell degranulation, and immune cell proliferation.

Sarcoplasmic/endoplasmic reticulum Ca^{2+} ATPase (SERCA) family proteins drive excess Ca^{2+} from the cytosol into the ER lumen. Low-affinity, high-capacity buffers like calsequestrin and calreticulin soak up the incoming Ca^{2+}, increasing the storage and Ca^{2+} buffering capacity of the ER [3,4]. Concurrently, Ca^{2+} overload and associated toxicity are prevented by the release channels inositol trisphosphate receptors (InsP3R) and ryanodine receptors. Maintaining ER Ca^{2+} concentration in the range of ~100 μM is essential for its protein synthesis and folding activities.

The concept of the ER as the "capacitator" of Ca^{2+} entry was pioneered by James Putney and found traction in experiments where blocking ER Ca^{2+} uptake using thapsigargin, a noncompetitive inhibitor of SERCA pump, led to Ca^{2+} influx [5,6]. Exhaustibility of ER Ca^{2+} reserve is an essential feature that orchestrates cellular signaling initiated by Ca^{2+} entry. Plasma membrane receptor stimulation generates a second messenger, inositol 1,4,5-trisphosphate (IP3), which signals ER Ca^{2+} release by organellar channels [7]. Diminution of free Ca^{2+} from ER stores is sensed and compensated by a highly selective influx of the extracellular Ca^{2+}, termed store-operated calcium entry (SOCE). Besides replenishment of the stores, initiation of specific local pathways that control downstream cellular responses is believed to be a driving force for SOCE [8].

3.1.2 CRAC Current and the Underlying Players

SOCE is measured as an inwardly rectifying Ca^{2+} current supporting a small but significant Ca^{2+} uptake (see Chapter 1). The current, I_{CRAC} (CRAC: acronym for Ca^{2+} release–activated Ca^{2+}), was first detected in Jurkat T cells and mast cells using a combination of whole-cell patch clamping and single-cell Ca^{2+} imaging [9,10]. Depletion of organellar stores, and not mere alteration in intracellular Ca^{2+}, trigger I_{CRAC}, substantiated by studies using IP3 or the Ca^{2+}-specific ionophore ionomycin in mast cells and using thapsigargin or intracellular Ca^{2+} chelators like BAPTA in Jurkat T cells and the revocable ER Ca^{2+} chelator TPEN ([11–13]; reviewed in [14]). The biophysical features of I_{CRAC} are indicative of a unique ion channel that conducts Ca^{2+} more precisely than any other characterized channel. In particular, it has a low conductance of <30 fS. This channel is ~1000-fold selective for Ca^{2+} over Na^+, blocked by low concentrations of Gd^{3+}, and detectably permeable to other divalent cations (Ca^{2+}, Ba^{2+}, Sr^{2+}). Both fast and slow inactivation of I_{CRAC}, attributed to either a local increase in Ca^{2+} levels or a gradual refilling of store, are Ca^{2+}-dependent feedback processes that regulate the channel activity [15,16]. CRAC channel can operate independent of the potential difference across the plasma membrane, necessitating the role of a ligand in gating [10,17].

CRAC channel identity remained elusive for years after I_{CRAC} was first recorded, and a pair of ER Ca^{2+} sensor and plasma membrane Ca^{2+} channel controlling the Ca^{2+} influx was predicted. Two independent genome-wide RNAi screens identified the ER component that controls SOCE as STIM1, an ER membrane–located Ca^{2+}-binding protein with the features of a signaling molecule [18,19]. *Drosophila* S2 cells that have mammal-like CRAC channel properties were instrumental in expounding the two molecular players responsible for I_{CRAC} [20]. Another set of coincident genome-wide RNAi studies identified *Drosophila* Orai (dOrai) as the plasma membrane component responsible for store-operated Ca^{2+} influx in these cells [21–23]. Accompanying genetic analysis of human severe combined immunodeficiency (SCID) patients devoid of SOCE activity and a restoration of Ca^{2+} uptake function by ectopic expression of human ORAI1 confirmed its role as the CRAC channel [21].

3.1.3 SOCE Current as an Outcome of STIM–ORAI Coupling

Coexpression of STIM1 and ORAI1 proteins is sufficient for generating SOCE current of large amplitude across various cellular systems [24–26]. Light microscopy studies with fluorescent proteins show a strong correlation between Ca^{2+} influx and a major redistribution of homogeneous ER-localized STIM1 into puncta in ER compartments adjoining the plasma membrane [27,28]. Implication from this finding is that STIM1 physically interacts with ORAI1 across a narrow gap between the ER and the plasma membrane.

ORAI1 protein forms a Ca^{2+}-permeable ion channel in the plasma membrane, and mutation of a conserved acidic residue in the pore can block Ca^{2+} flux through the channel [22,29] (Figure 3.1a). ER Ca^{2+} depletion is sensed by the Ca^{2+}-binding luminal EF-hand domain of STIM1, a disruption of which leads to a constitutive activation of STIM1 (Figure 3.1b) [19,28]. Following activation, STIM1 can recruit ORAI channels into clusters, and both puncta and clusters spatially correspond with the sites of Ca^{2+} entry [30]. Store depletion is essential for STIM1 redistribution [27,28,31–34], and interaction with STIM is sufficient for ORAI1 clustering and Ca^{2+} influx [30,31,35]. The obligatory role of STIM–ORAI interaction in generating SOCE current has been demonstrated through various biochemical studies, for the details of which, the reader is referred to other comprehensive literature reviews [36,37] (see Chapters 2 and 7).

The entire process of STIM1- and ORAI1-mediated Ca^{2+} influx calls for both proteins to switch from a high-Ca^{2+} inactive state to a low-Ca^{2+} active state. In the case of STIM, Ca^{2+}-binding status of the sensor itself can determine its conformation, while for ORAI1, STIM1 binding is essential for its gating. This conformational coupling can be regulated by several cellular factors, but STIM–ORAI interaction is adequate for SOCE. This chapter surveys the wealth of structural and biophysical studies to offer a mechanistic overview of STIM and ORAI functioning, focusing on the architectural changes these proteins undergo in response to store depletion.

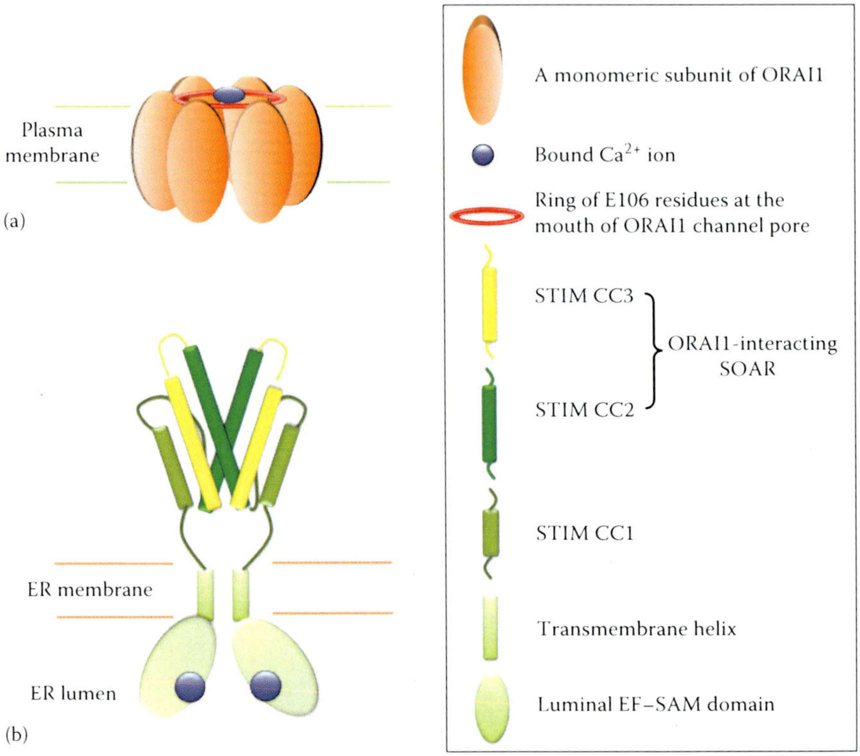

FIGURE 3.1 An illustration of ORAI1 and STIM1 proteins at rest under store replete conditions. (a) ORAI1 hexamer in the plasma membrane shown with Ca^{2+} bound to a ring of negatively charged E106 residues that mark the extracellular mouth of the channel. (b) Inactive STIM1 showing Ca^{2+} bound to the luminal EF–SAM domain and the cytoplasmic CC1 holding the ORAI1-interacting SOAR in a folded conformation. An index of feature description is given on the right. Only regions corresponding to STIM1 aa1–450 are depicted for clarity.

3.2 ORAI CHANNEL STRUCTURE

Mammalian ORAI1 (also known as CRACM1) is a ~33 kDa four-transmembrane cell surface protein [21,22]. Mammals express three ORAI-family proteins, namely, ORAI1, ORAI2, and ORAI3, with high sequence similarity in the transmembrane (TM) regions. ORAI homologs are named after the three mythological gatekeepers of heaven. ORAI1 forms an ion-conducting pore in the plasma membrane, and its subunit topology, confirmed experimentally, shows both N and C termini to be cytoplasmic [29,38,39] (see Chapter 2). ORAI1 is the principal ORAI-family protein contributing to store-operated Ca^{2+} entry in most mammalian cells studied.

3.2.1 Calcium Binding Site

ORAI proteins distinctively lack sequence similarity to not just other ion channels but also any other known protein. However, like other Ca^{2+} channels, ORAI proteins utilize the negative charge on a glutamate residue to bind and select for Ca^{2+} (Figure 3.2a). TM1 residue E106 (E178 in dOrai) is central to the Ca^{2+} binding and conductance of ORAI1, and alanine or glutamine substitutions result in a drastic loss of SOCE current [22,29,38]. E106D, a smaller negatively charged substitution, restores channel function only partially, with a concomitant increase in permeability to monovalent cations, emphasizing the need for a glutamate to

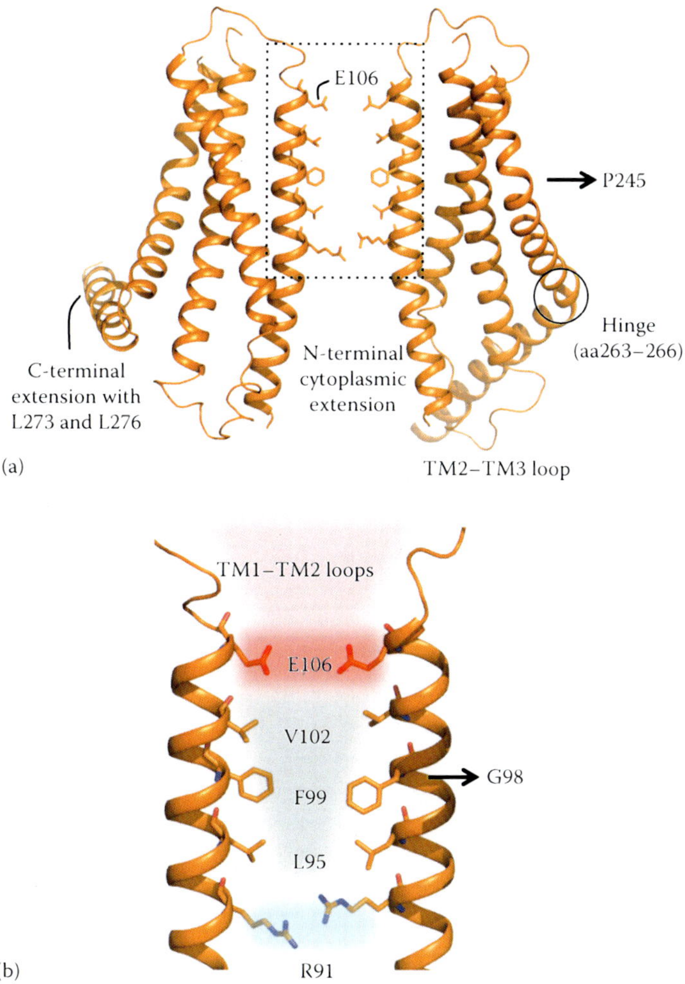

FIGURE 3.2 ORAI1 channel features highlighted in a homology model generated based on dOrai structure (PDB: 4HKR). (a) An overview of the channel showing only two subunits of the hexamer, the central line formed by TM1 helices, and the cytoplasmic extensions. Locations for channel features discussed in the text are marked. (b) Zoomed-in view of a section of the pore (region boxed in (a)), showing the Ca^{2+} binding site, a proposed vestibule comprised of aspartate residues in TM1–TM2 loop, nonpolar region with the gate at V102, and a part of the basic region with R91.

maintain ion selectivity [29]. Introduction of one copy of E106Q into a tandem tetramer of ORAI1 is sufficient to eliminate CRAC current [40]. E106C shows a high degree of oxidative cross-linking and coordinates Cd^{2+} (ionic size similar to Ca^{2+}) efficiently to result in a current blockade, demonstrating that the sidechains at position 106 are positioned to directly coordinate Ca^{2+} ions and hence form the Ca^{2+} binding site(s) [41,42].

3.2.2 Pore Features Mapped by Cd^{2+} Block and Disulfide Cross-Linking Experiments

Analysis of the packing of engineered cysteine substitutions against their counterparts from separate monomers of the ORAI1 channel has been instrumental in defining the ion permeation path. Cysteines replacing the TM1 residues along one helical face cross-link proficiently with oxidizing agents like copper phenanthroline and molecular iodine, demarcating the pore-facing amino acids [42]. The pore allows small probes like Cd^{2+} to penetrate and cause a current block due to coordination by the pore-lining cysteine substitutions while being impermeable to other cysteine-reactive probes with diameters >3 Å [41]. Both studies conclusively show a long narrow pore entirely composed of TM1 helices, with tightly packed E106 residues, a constricted nonpolar region spanning residues 99–104, and the TM1 apposition extending to at least residue 91 (Figure 3.2b). It is important to note that while oxidative cross-linking experiments report the architecture of resting channel pore, Cd^{2+} blocks the inward Ca^{2+} current through a conducting ORAI1 channel. Correspondence of the pore-facing residues in the two states implies that subtle changes in the pore can convert a closed channel into an open one.

3.2.3 *Drosophila* Orai Structure

The 3.35 Å crystal structure of the nonconducting dOrai channel (PDB: 4HKR) shows that TM1 helices indeed form a central pore independent of other TM helices, which are arranged concentrically around the pore [43]. The crystal structure, disuccinimidyl suberate cross-linking of the dOrai channel expressed in HEK293 cell membranes, and SEC-MALS measurements of the Orai channel mass indicate that the channel is hexameric. Although this conclusion differs from the tetramer model previously proposed on the basis of single-molecule photobleaching experiments and electrophysiology of concatenated ORAI1 constructs, a consensus has developed that this is the correct stoichiometry (discussed in Reference 44 and Chapters 2 and 14). The hexameric arrangement results in six negatively charged E178 residues (*Drosophila* equivalent of E106 discussed earlier) in close vicinity of each other, forming a Ca^{2+}-binding ring at the extracellular mouth of the pore [43]. Electron density of a putative Ca^{2+} ion bound to the glutamate ring reiterates the role of E106 as a Ca^{2+} binding site. Since the pore is devoid of any other acidic residue forming an ancillary cation binding site, one hypothesis to be tested is whether backbone carbonyls of residues adjacent to E106 are ligands that assist the inward flux of Ca^{2+} ions.

3.2.4 Pore from Outside to Inside

The extracellular accessibility to E106 and the ORAI1 channel pore is through a loop connecting TM1 and TM2, a region featuring clusters of conserved aspartates. Structural information for the TM1–TM2 loop is absent in the dOrai structure, possibly due to the typical flexibility associated with protein loop regions [43]. Although Cd^{2+} block analysis indicates high accessibility and mobility of the loop, experiments with larger cysteine-modifying reagents ~5–8 Å in diameter show an elevated block of D110C current, indicating a positioning of at least a part of the loop in line with the permeation pathway in the active state [41]. Alanine substitutions of all the loop aspartates result in altered ion selectivity and decrease substantially the luminescence intensity of Tb^{3+} bound in the vicinity of E106 in reconstituted ORAI1 channels [38,45,46a]. These findings and a molecular dynamics study simulating Ca^{2+} entry into ORAI1 pore indicate that TM1–TM2 loop might form a cation "focusing" vestibule that facilitates fast Ca^{2+} permeation [41,46a,47] (see Chapter 2).

The wide vestibule tapers at the mouth of the pore at the glutamate ring, juxtaposed to which is a constricted nonpolar section of the pore (Figure 3.2b). The hydrophobic residues of this region are well ordered in the engineered cysteine assays and show little variation in thermal motion in the *Drosophila* Orai structure [41–43]. As discussed in a later section on ORAI1 gating, this region evidently forms a barrier in the path of ion conductance and is presumably moved aside by STIM [46a,48].

A noteworthy feature of the pore captured in the crystallized state of dOrai is an anionic electron density trapped in the lysine–arginine-rich basic section that runs contiguous with the nonpolar region. The dOrai protein can cocrystallize with complex anions like iridium hexachloride, and mass spectrometric analysis of ions copurifying with dOrai detects considerable iron species, suggesting that binding an Fe-anion complex or other anions can overcome the strong repulsive forces and offer the constraint required for crystallization [43]. In a native closed channel, it is possible that common cytoplasmic anions remain bound in this region. However, cysteine cross-linking and Cd^{2+} block experiments find this basic region to be moderately accessible, suggesting flexibility in both states of the channel [41,42].

3.2.5 Cytoplasmic Extensions of ORAI1

A protracted cytoplasmic region was inferred from experiments where forced narrowing of ER–plasma membrane distance to ≤9 nm causes an exclusion of ORAI1 channels [49]. According to the dOrai structure, TM1 helix extends deeper into the cytoplasm than what is observed in chemical cross-linking studies and forms an N-terminal extension, contiguous with the pore (Figure 3.2a). The C-terminal extension is comprised of a helix that originates from the TM4 helix through a hinge formed by residues 263–266 of ORAI1 (aa306–309 in dOrai), and an intramembrane bend at TM4 residue P245 (P288 in dOrai) causes the C-terminal half of TM4 to run more or less parallel to the membrane with the extension angled away from the body of the channel (Figure 3.2a). Neighboring C-terminal extensions homodimerize through a hydrophobic patch containing L273 and L276 (I316 and L319 in dOrai) (Figure 3.2a). C-terminal dimerization confers upon ORAI the threefold down–up symmetry, as viewed from the cytoplasm, and forms the primary interaction site for STIM binding [32,43,50].

3.3 OVERVIEW OF STIM1 STRUCTURE

STIM1 is a multidomain protein anchored in the ER membrane through a single-pass transmembrane helix (Figure 3.1b). STIM1, an abbreviation for Stromal Interaction Molecule 1, was initially believed to be either a protein secreted by bone marrow stromal cells or a leukemia cell surface protein [51,52], but it has since been shown to act as an ER-localized regulator of SOCE [19]. STIM1 is largely localized in the ER, with a smaller fraction in the plasma membrane [52–54]. In addition to its well-studied role in SOCE, it has additional functions in both locations [55–57]. Mammals express two STIM proteins with complementary, and nonredundant, functions in Ca^{2+} signaling [58]. STIM1 and STIM2 are likely to share common mechanisms of sensing and communicating store depletion, given the high degree of sequence conservation between the proteins in their core functional domains.

The function of STIM proteins in Ca^{2+} sensing is manifest in their structure and orientation in the ER membrane, with a Ca^{2+}-sensing domain facing the ER lumen, and an extensible region and ORAI1-binding domain in the cytoplasm, enabling STIM to relay a Ca^{2+} depletion signal from the ER to the plasma membrane (reviewed in References 59 and 60 and Chapter 2).

3.3.1 Ca^{2+} SENSING BY STIM1 LUMINAL DOMAIN

The luminal "sensor" domain of STIM1 was predicted, from its sequence, to contain an EF-hand motif, a common helix–loop–helix structural element that binds Ca^{2+}. A solution nuclear magnetic resonance (NMR) structure of the luminal region of human STIM1 (residues 58–201; PDB: 2K60) in the presence of millimolar Ca^{2+}, which is likely to represent the conformation of STIM1 in the ER lumen in resting cells, revealed the presence of two EF-hands in tandem: a canonical EF-hand followed in the linear sequence by a noncanonical "hidden" one [61]. Although the hidden EF-hand retains a recognizable EF-hand fold, the expected Ca^{2+}-binding residues at positions 1 and 3 in the loop are replaced by Phe and Gly, respectively, consistent with the finding that only one Ca^{2+} is bound per luminal domain [34]. This leads to a model in which Ca^{2+} is bound in the canonical EF-hand site, and the second EF-hand provides structural stabilization, both through the β-strand interaction typical of paired EF-hands and through its participation in a hydrophobic pocket that partially envelops the STIM1 sterile alpha motif (SAM) domain that follows immediately in the linear sequence. Ca^{2+} binds to recombinant STIM(58–201) with a relatively low affinity, whether binding is monitored with $^{45}Ca^{2+}$ or by following changes in intrinsic protein fluorescence or far-UV circular dichroism, the K_d ranging between 200 and 600 μM [34].

3.3.2 CYTOPLASMIC DOMAIN OF STIM1

The cytoplasmic "effector" domain of STIM1, expressed as a soluble protein fragment, can activate native ORAI1 channels in Jurkat T cells [62] and recombinant

ORAI1 channels expressed in other cells [32,63,64]. The Ca^{2+} influx through ORAI1 channels under these conditions is constitutive and uncoupled from regulation by ER Ca^{2+} stores. "Effector" function of the STIM cytoplasmic domain is further verified by the ability of purified recombinant STIM1(233–685) to trigger Ca^{2+} efflux from lipid vesicles or artificial liposomes containing ORAI1 channels [46a,65]. The region sufficient to activate ORAI1 channels is $STIM^{SOAR}$ (residues 344–442; STIM1 Orai Activating Region) or a similar fragment containing SOAR [32,66,67].

Then why does STIM1 not activate ORAI1 channels in resting cells? The short answer is that SOAR is retained near the ER under these conditions. The first evidence that the cytoplasmic domain folds back on itself came from the determination of the distance in recombinant $STIM1^{CT}$ between a label placed adjacent to residue 233 and a label at the C terminus [68]. The measurement by luminescence resonance energy transfer (LRET) showed that the two sites are in close proximity, only 3–4 nm distant, implying that $STIM1^{CT}$ has a preferred conformation in which the polybasic tail would be near the ER rather than near the plasma membrane (see Chapter 7). Furthermore, the purified CC1 fragment (STIM1(233–343)) interacts with purified SOAR *in vitro* [68], and membrane-anchored $STIM^{EFSAM\text{-}TM\text{-}CC1}$ expressed in cells recruits cytosolic SOAR to the ER as evidenced by colocalization and by Förster resonance energy transfer (FRET) between the labeled proteins [69]. In the latter experiment, even a truncated EFSAM-TM-CC1(233–261) construct is effective, showing that SOAR is held very close to the ER membrane. These findings indicate that CC1 is not simply a tether linking SOAR to the ER but is intimately involved in maintaining the inactive conformation of STIM1, a point that will be taken up in the next section.

3.3.3 ORAI-Interacting Machinery

Critical features of the $STIM1^{CT}$ structure remain to be filled in, with definition of the structural interactions that retain SOAR near the ER being a prime goal. One piece of the puzzle is a crystal structure of human SOAR (PDB: 3TEQ) showing intramonomer antiparallel coiled coils that supercoil to form a V-shaped dimer [70]. There is a plausible argument that this structure might reflect the inactive conformation of the $STIM1^{SOAR}$ [50], but even if so, there is very limited information about how this subdomain is packed with the remainder of the $STIM1^{CT}$. An elegant analysis of peptide–peptide interactions using short fragments of $STIM1^{CT}$ [71] is consistent with available data from mutational studies and points to intramolecular contacts that might stabilize the inactive form of $STIM1^{CT}$ but has not so far provided sufficient constraints to predict a three-dimensional structure. A crystal structure of *Caenorhabditis elegans* SOAR (PDB: 3TER) captures in addition a short piece of the CC1 region, equivalent to residues 318–337 in human STIM1, making contact with the SOAR domain [70]. A caveat in extending this structural finding to the human protein is that the neighboring CC1 residues 310–317 and residues 275–300 of human STIM1 have no counterparts in the *C. elegans* STIM sequence, suggesting that the mode of CC1–SOAR packing in the human protein may be distinct. Another cautionary note is that in an NMR structure of human STIM1 residues 312–387 (PDB: 2MAJ), representing a part of CC1 and a part of SOAR, the positioning of the 313–340 helix differs appreciably from that of the 257–279 helix in the *C. elegans*

SOAR structure [50]. On the other hand, the STIM1(312–387) structure also displays an altered arrangement of helices when compared to the human SOAR structure, and the difference could be satisfactorily explained if the STIM1(312–387) structure represents an activated conformation of STIM1 [50], as discussed below.

3.4 STIM ACTIVE STATE

Experimental depletion of the resting ER Ca^{2+} stores triggers a development of CRAC current that correlates inversely to the level of Ca^{2+} remaining in the ER. Current is highly cooperative with the Ca^{2+} concentration and reaches its half-maximal value at ~170 μM Ca^{2+} [72]. Store depletion acts as a switch for STIM1 activation, a process that can be visualized by confocal, total internal reflection fluorescence (TIRF), and electron microscopy as a redistribution of appropriately tagged STIM1 [19,27,28]. The newly redistributed STIM1 is localized in "puncta" or ~100–300 nm wide regions where ER is closely apposed to the plasma membrane (reviewed in Reference 73). I_{CRAC} requires STIM1 puncta formation and shows a similar Ca^{2+} concentration dependence and cooperativity [72,74]. The studies discussed here have utilized STIM1 fragments or mutant proteins to gain valuable insights into the conformational changes in STIM1 that enable communication between the ER lumen and the plasma membrane. However, the precise sequence of events in STIM1 activation following Ca^{2+} store depletion and the detailed architecture of activated STIM1 are still active areas of investigation.

3.4.1 Activation of the Luminal Domain

The first response to a decline in store Ca^{2+} levels comes from the luminal EF-hand. Mutating one of the consensus Ca^{2+}-coordinating positions, D76, to alanine causes STIM1 to localize into puncta independent of store depletion [19]. Constitutive STIM1 activation is likewise observed with the E87Q mutant of the EF-hand, corroborating that an inability to bind Ca^{2+}, and by extrapolation a loss of bound Ca^{2+}, is responsible for STIM1 redistribution [28]. The high micromolar K_ds estimated for Ca^{2+} binding to isolated $STIM1^{EFSAM}$ or to the STIM1 EF-hand grafted into a scaffolding protein domain are arguably suited to sensing physiological changes in ER-luminal Ca^{2+} [34,75,76]. Direct evidence that Ca^{2+} depletion is coupled to a conformational change comes from far-UV CD spectra and NMR spectra of isolated recombinant $STIM^{EFSAM}$, which indicate a striking loss of α-helical structure in the absence of Ca^{2+}. This change in secondary structure is accompanied by a transition of $STIM^{EFSAM}$ from monomer to dimer and, in one study, the formation of higher aggregates [34,76]. In cells, an increase in intramolecular FRET efficiency between fluorophores attached to the luminal domain of STIM1 occurs upon store depletion and, notably, precedes the relocalization of STIM1 to puncta, reinforcing the idea that a STIM1 conformational change might drive subsequent relocalization [33,63]. The FRET change in cells could represent both the dimerization that has been observed with isolated STIM1 luminal domains and a further STIM1 oligomerization step as discussed in the following section.

3.4.2 Activation Involves Extension of STIM1 Cytoplasmic Domain

A key insight into the active conformation of the STIM cytoplasmic domain came from the constitutively activated L251S and L416S/L423S mutants of STIM1 [77]. When expressed in cells together with ORAI1, the mutant proteins are constitutively colocalized with ORAI1 channels in puncta and trigger a constitutive La^{3+}-blockable current. The mutations result in an extended conformation of the recombinant ORAI1-activating small fragment (OASF) (STIM1(233–474)) as assessed by intramolecular FRET. LRET analysis of the purified recombinant STIM1 cytoplasmic domain carrying the L251S mutation shows a physical extension of the molecule by at least several nanometers compared to wild-type $STIM1^{CT}$ [68]. LRET is not suitable for measuring distances greater than 10 nm, but it is arguable that STIM1 has been directly visualized in EM and cryo-electron tomography studies of cells overexpressing STIM1, as filaments spanning the 10–20 nm distance between cortical ER and plasma membrane [78,79]. These data support a model in which the ORAI-interacting SOAR/CAD domain of resting STIM1 is retained near the ER, whereas it is freed to interact with plasma membrane ORAI1 in the active form of the protein (Figure 3.3).

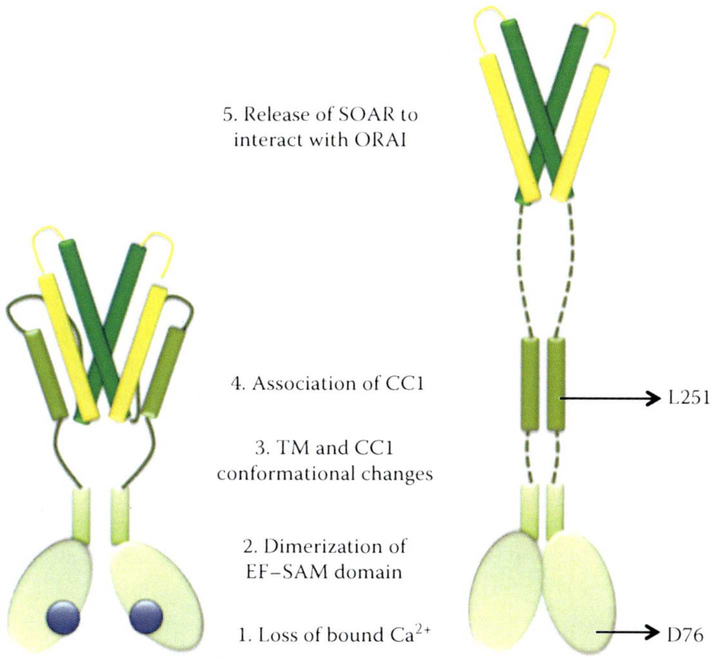

FIGURE 3.3 Proposed series of events in the process of STIM1 activation, based on the current research findings. Left panel shows a resting state STIM1 that undergoes at least the five steps listed in the figure, resulting in an active form of the protein that is proficient in ORAI1 binding (right panel). Domain organization is the same as in Figure 3.1. Positions of the D76 and L251 are marked. Only regions corresponding to STIM1 aa1–450 are depicted for clarity.

3.4.3 CC1 and the Release of ORAI1-Interacting Machinery

A series of experiments with engineered proteins dissected the mechanism underlying STIM1 activation. As noted earlier, recombinant STIM1^{CC1} interacts with immobilized MBP-tagged STIM1^{SOAR} *in vitro* [68]. The individual replacements L251S, L258G, and L261G in STIM1^{CC1} reduce this binding [68,69]. Parallel experiments in cells have shown that the mutations L258G and, to a lesser extent, L261G in full-length STIM1 resemble L251S in causing constitutive Ca^{2+} entry, suggesting that they also free SOAR to interact with plasma membrane ORAI [69]. It is unlikely that these mutations are disrupting a folded structure of the protein. Despite its designation as a predicted coiled coil, the isolated CC1 fragment is partially α-helical and partially unstructured, and monomeric [68]. Moreover the thermal melting curve for CC1 obtained by monitoring its CD spectrum offers no suggestion that intramolecular interactions stabilize parts of the α-helical secondary structure. Thus, it is most likely that L251, L258, and L261 engage SOAR directly.

The predicted propensity of CC1 to form a coiled coil comes into play in a further series of experiments. When STIM^{CC1} is forced to form dimers by oxidation of an N-terminally appended cysteine, there is an increase in α-helical content, along with a partial stabilization against thermal denaturation suggesting intradimer helix packing [68]. Dimerized STIM1^{CC1} binds less well to SOAR, and similar cross-linking of CC1 within the dimeric STIM1^{CT} fragment triggers an extended conformation of the protein as measured in LRET experiments [68]. It has not been rigorously proven that the conformational change detected upon dimerization of CC1 involves the formation of a partial coiled coil. However, this is the most straightforward explanation, in light of the facts that the residues L251, L258, and L261 are predicted to be buried in a coiled coil interface, that an L251S replacement in CC1 prevents the increase in α-helical content and a stabilization against thermal denaturation [68], and that each of the leucine replacements is effective in releasing SOAR [68,69].

The final link connecting depletion of ER Ca^{2+} stores and the STIM1^{CT} conformational change that releases SOAR comes from the FRET-based assay in cells in which STIM$^{EFSAM-TM-CC1}$ recruits cytosolic SOAR to the ER. The assay detects a considerable loss of FRET between ER-localized STIM$^{EFSAM-TM-CC1}$ and cytosolic SOAR after Ca^{2+} store depletion [69]. The current model for STIM1 activation [68] is that inactive STIM1 is a dimer because of its dimeric SOAR domains, that association of the ER-luminal domains within a STIM dimer brings together the CC1 segments, and that, by burying the critical CC1 residues that contact SOAR in the inactive conformation, this structural rearrangement of CC1 triggers the activating conformational change in STIM1^{CT}. Note that the model does not require that CC1 associate into a stable coiled coil. Temporary coiled coil formation in the segment containing residues 248–261 may trigger release of SOAR, and thereafter either an α-helical or partially α-helical extended CC1 chain would be of sufficient length to position SOAR in the vicinity of ORAI1 in the plasma membrane.

The demonstrated CC1–SOAR interaction implies that CC1 must have a partner surface in SOAR. A candidate region in SOAR was suggested by the early finding that L416S/L423S is an activating mutation [77]. Recent evidence that L416G, V419G, and L423G mutations in STIM1 all result in constitutive Ca^{2+} entry [69] gives additional support to this possibility. Indeed, studies using CC1 and SOAR fragments find an interaction between the CC1α1 helix (residues 233–276) and the SOAR helix $CC3_{430}$ (residues 388–430), and this interaction is abolished by the L251S substitution [71]. Special emphasis has been given to the role of CC1α1 in maintaining the inactive conformation, based on the finding that STIM1 with a deletion of residues 278–337 is still fully regulated by Ca^{2+}, and an L251S mutation in this protein results in constitutive current [71]. The emphasis is undoubtedly justified, but other parts of CC1 help to stabilize the inactive conformation of full-length STIM1. For example, the ^{318}EEELE^{322} > AAALA substitutions constitutively activate STIM1 and lead to physical extension of OASF and $STIM1^{CT}$ [68,77,80], and the single Y316A replacement causes appreciable though incomplete constitutive activation [81]. A full understanding of STIM1 activation will require the determination of how intramolecular packing changes throughout the CC1–SOAR region.

3.4.4 Higher-Order STIM1 Oligomerization

Because of the very steep Ca^{2+} concentration dependence of STIM relocalization to ER–plasma membrane junctions [72,74], it has been thought that STIM assembles into higher-order oligomers at some point during the activation process. Any oligomerization is likely to be through the SOAR/CAD domain [82,83], and a possible structural basis for oligomerization has been proposed in a documented CC3–CC3 interaction [71]. Persuasive as the data are, it should be recognized that all the studies thus far have been carried out with STIM1 fragments and with overexpressed STIM1. It remains to be tested whether the strength of STIM–STIM interactions is sufficient to drive the formation of higher-order oligomers of STIM1 in cells, at native levels of STIM1 and prior to the relocalization that concentrates STIM1 in ER–plasma membrane junctions.

3.5 ACTIVATION OF ORAI1 CHANNELS BY STIM1

SOCE is linked with the redistribution of ORAI1 to plasma membrane sites that correspond precisely to STIM puncta [30,31,35]. STIM1 activation and redistribution are independent of ORAI1 expression [27,28,31–33], whereas ORAI1 is recruited to ER–plasma membrane junctions only in the presence of activated STIM1 [30,31,35]. Close proximity between STIM1 and ORAI1 is evident in FRET measurements of appropriately labeled proteins, indicating that activated STIM1 at ER–plasma membrane junctions couples to ORAI1 itself or to a closely associated protein [63,84,85] (see Chapter 2).

SOAR/CAD decorates the plasma membrane of cells when ORAI1 is coexpressed, and SOAR/CAD coimmunoprecipitates with ORAI1, indicating a close interaction between the two proteins [32,66]. Purified $STIM1^{CT}$ protein triggers Ca^{2+} efflux from PC/PS liposomes into which purified ORAI1 has been reconstituted and from membrane vesicles isolated from yeast expressing human ORAI1, conclusively proving that direct STIM1–ORAI1 interactions are sufficient to gate the channel [46a,65]. Considerable work has been directed toward defining these interactions.

The current model favors multiple sites of contact between STIM and the ORAI1 channel [32,63,86–88]. This section considers the proposed sites of interaction and the role(s) attributed to each site in activation of ORAI1.

3.5.1 Interaction of Activated STIM1 with ORAI1

Truncated ORAI1 channels missing either aa267–301 at the C terminus or aa74–90 at the N terminus fail to generate CRAC currents, although only the C-terminal deletion shows a lack of ORAI1 redistribution [35,66]. The single replacement L273S in the C-terminal region of ORAI1 blocks both STIM–ORAI interaction and CRAC channel activation [63,66]. Purified $STIM1^{CT}$ can interact specifically with a GST-tagged C-terminal peptide of ORAI1 [65,66]. $ORAI1^{\Delta C}$ and other C-terminal mutants fail to show ORAI1 clustering and STIM–ORAI FRET after store depletion [63,84,89,90]. ORAI1 mutants L273D, L276D, and truncations of the C-terminal helix that remove aa272–283 indicate that the hydrophobic face of this helix is important for STIM interaction [32,84,91]. Although earlier studies focused on the role of STIM1 interaction with the ORAI1 C terminus in recruiting ORAI1 to ER–plasma membrane junctions, McNally et al. [86] have given persuasive evidence that mutations and deletions in the C-terminal region of ORAI1 affect channel gating as well as STIM–ORAI clustering.

The C-terminal ORAI1 segment implicated in STIM1 binding maps within a juxtamembrane cytoplasmic helix in the dOrai structure [43]. These juxtamembrane helices are connected to their TM4 helices by a short flexible segment that adopts two configurations in the crystals, permitting the cytoplasmic helices of adjacent ORAI1 monomers to pack in a short antiparallel coiled coil stabilized by L273–L276′ and L276–L273′ contacts [43]. STIM1 binding appears to displace the cytoplasmic C-terminal helices, since STIM1–ORAI1 interaction and CRAC current are disrupted when cysteine side chains of engineered L273C ORAI1 monomers are chemically cross-linked [87]. This cross-linking has been interpreted as locking the ORAI1 C-terminal helices in a STIM-inaccessible conformation resembling that in the *Drosophila* Orai crystal structure.

There are two distinct proposals on the mode of STIM–ORAI interaction. In one, based on the solution NMR structure of the complex of ORAI1(272–292) peptide with a STIM1(312–387) fragment, an active STIM1 dimer binds a pair of ORAI1 helices, supplanting the ORAI–ORAI interaction by providing two hydrophobic grooves on the surface of STIM1 to accommodate the ORAI1 helices [50]. The antiparallel orientation of the ORAI1 helices is maintained in this complex, but there is a shift in the relative positions of the ORAI1 helices compared to the dOrai crystal structure. An alternative proposal [43] is that the individual C-terminal ORAI1 helices extend away from the plasma membrane, unpaired, to interact with STIM. This possibility gains some support from the recent finding that SOAR dimers with an F394H substitution in only one subunit are fully able to bind and activate ORAI1 channels, while those with an F394H substitution in both subunits are severely impaired in interacting with ORAI [92] (see Chapter 7). No specific structural model for monomeric ORAI1 peptide binding to STIM has been put forward. Both proposals are consistent with the experimental finding that ORAI1 C-terminal helices locked in the crystal configuration cannot interact productively with STIM [87].

In the ORAI1 N-terminal region, the segment spanning residues 74–90, corresponding to a cytoplasmic extension of the TM1 helix in the dOrai structure [43], is implicated in channel gating. SOAR/CAD can elicit normal currents with ORAI1$^{\Delta 1-73}$ but no current with ORAI1$^{\Delta 1-88}$ [35,93,94], even though STIM1 still recruits the truncated proteins to ER–plasma membrane junctions [35,46a,63,84]. Further, specific mutations in the region of residues 81–85 reduce or disrupt CRAC current [46a,94,95]. What is unclear is whether this region contacts STIM1 directly during channel gating. Coimmunoprecipitation and pulldown experiments have demonstrated an interaction between STIM1 and the ORAI1 fragments ORAI1(48–91) and ORAI1(65–87) [32,65], and a fluorescence polarization assay established that STIM1^{CT} can bind the synthetic peptide ORAI1(66–91) [46a]. However, it has neither been proven nor disproven that such an interaction takes place in the context of the intact ORAI1 channel. Recent data indicate that the same mutations at residues 81–85 that compromise STIM1-activated current also block the constitutive current carried by an ORAI1 channel mutated in residues 261–265, the cytoplasmic hinge region following TM4 [46b]. Because the channel mutated at residues 261–265 adopts an open, Ca^{2+}-selective conformation in the absence of STIM1, these findings suggest that the N-terminal ORAI1 peptide interacts with another region of ORAI1, rather than with STIM1, during channel gating.

3.5.2 Gating of the ORAI1 Channel

STIM binding causes movements of ORAI1 transmembrane helices TM1 and TM4 that can be detected directly or inferred from biochemical experiments [46a,87,88]. The TM1 helix movements are part of the gating movement that opens the channel. The presence of a gate in the external region of the pore was first evident from the fact that the G98C sulfhydryl was inaccessible to a cysteine-modifying methanethiosulfonate reagent in resting ORAI1$^{G98C/E106D}$, but became accessible after store depletion [48]. This result indicated that a barrier to ion flux lies external to G98 and implied that STIM1 can trigger a movement in the vicinity of V102. Along those lines, replacement of either V102 or F99 with a small polar residue results in a constitutive and nonselective ion flux through the channel (Reference 48; Yamashita and Prakriya, personal communication). The simplest view is that these constitutively conducting channels are leaky channels that remain in the resting conformation. *In silico* simulations support this interpretation, indicating that water molecules in the narrowest part of the pore are less constrained in the V102A channel than in the wild-type channel and that the reduced local constraints on water reduce the energy barrier for Na^+ ions to pass through this region [96]. The simulations show that the change in Na^+ permeability occurs without appreciable displacement or rotation of the transmembrane helices. Experimentally, a truncated ORAI1 channel that deletes the entire N-terminal cytoplasmic region (residues 1–88) and replaces R91 with glycine is a closed channel, offering direct evidence for a gate within the nonpolar pore-lining segment of TM1, L95–V102 [46a].

The helix bundle consisting of the N-terminal cytoplasmic extensions of TM1 helices in the dOrai structure is a possible second location for the channel gate [43,90,98]. Blockade of the R91C channel by Cd^{2+} and by diamide cross-linking

[48,96] has been taken as evidence for such an internal gate, although the chemical cross-linking pattern of the ORAI1 region from S90C to L96C suggests that the helices have considerable flexibility [42] and hence that the helix bundle observed in the dOrai crystal structure might be just one representative of an ensemble of conformations. Hydrophobic residues substituted at R91, most notably the SCID mutation R91W [21], can block ORAI1 channel current and can even override the constitutive conductance of the V102C channel, presumably due to stabilization of the bundle of helices [48,90,97]. A key question, given the multiple positive charges on the individual wild-type helices in residues 77–91, is whether the helix bundle is stably "closed" in wild-type ORAI1. One conceivable way to stabilize the helix bundle would be by association of intracellular anions with the basic region of the pore, plausible evidence for which is seen in the dOrai structure [43]. Assuming that there is in fact a second, internal gate in the closed ORAI1 channel, the expectation is that both gates must open in a concerted way when STIM activates the channel.

Gating of ORAI1 involves conformational changes in other parts of the channel [86,87], including at residue P245 in the middle of the TM4 helix [88] and at a hinge region in the TM4 cytoplasmic extension [46b,87]. It seems very probable that all the ORAI1 transmembrane helices rearrange during gating. However, judging from the relatively small movements that have been detected in the pore-lining TM1 helix due to gating (References 41 and 42; Yamashita and Prakriya, personal communication), these rearrangements are likely to be subtle.

Detecting gating movements experimentally is important for further analyses of ORAI1 channel function, but the extremely small single-channel current has impeded the use of conventional electrophysiological approaches. Two new approaches have circumvented this roadblock. A Tb^{3+} luminescence assay that utilizes the intrinsic ability of Tb^{3+} to bind to the E106 Ca^{2+} site in ORAI1 channels clearly reports the gating movement at the mouth of the channel triggered by recombinant STIM1 [46a]. In cells, ORAI1 channels with genetically fused calcium indicators have been successful in measuring Ca^{2+} entry at the internal mouth of single ORAI1 channels [99]. These complementary tools can be used to study ORAI1 channel gating by STIM1 in parallel measurements made in cells and made with purified recombinant STIM1 and ORAI1 *in vitro*.

3.6 CONCLUSIONS

In summary, ER store depletion is a primary event in Ca^{2+} signaling that results in CRAC current. We have presented a glimpse of our current understanding of the structural changes that lead to activation of the ER Ca^{2+} sensor STIM1, its interaction with the plasma membrane Ca^{2+} channel ORAI1, and the gating of ORAI1. We have also pointed out where current knowledge of each of these processes is incomplete. Challenges such as resolving the structure of the entire STIM1 molecule in its resting state and its activated state, precisely defining the STIM–ORAI binding sites that control recruitment and gating of the channel, and solving the structure of the open ORAI1 channel physiologically gated by STIM1 will be met by further biophysical and biochemical studies of STIM1 and ORAI.

REFERENCES

1. McDonough PM, Button DC (1989) Measurement of cytoplasmic calcium concentration in cell suspensions: Correction for extracellular Fura-2 through use of Mn^{2+} and probenecid. *Cell Calcium*. 10:171–180.
2. Clapham DE (2007) Calcium signaling. *Cell*. 131:1047–1058.
3. Koch GL (1990) The endoplasmic reticulum and calcium storage. *Bioessays*. 12:527–531.
4. Verkhratsky A (2005) Physiology and pathophysiology of the calcium store in the endoplasmic reticulum of neurons. *Physiol Rev*. 85:201–279.
5. Putney JW Jr. (1986) A model for receptor-regulated calcium entry. *Cell Calcium*. 7:1–12.
6. Takemura H, Hughes AR, Thastrup O, Putney JW (1989) Activation of calcium entry by the tumour promoter thapsigargin in parotid acinar cells. *J Biol Chem*. 264:12266–12271.
7. Berridge MJ (2002) The endoplasmic reticulum: A multifunctional signaling organelle. *Cell Calcium*. 32:235–249.
8. Bird GS, Hwang SY, Smyth JT, Fukushima M, Boyles RR, Putney JW Jr. (2009) STIM1 is a calcium sensor specialized for digital signaling. *Curr Biol*. 19:1724–1729.
9. Lewis RS, Cahalan MD (1989) Mitogen-induced oscillations of cytosolic Ca^{2+} and transmembrane Ca^{2+} current in human leukemic T cells. *Cell Regul*. 1:99–112.
10. Hoth M, Penner R (1993) Calcium release-activated calcium current in rat mast cells. *J Physiol*. 465:359–386.
11. Hoth M, Penner R (1992) Depletion of intracellular calcium stores activates a calcium current in mast cells. *Nature*. 355:353–356.
12. Zweifach A, Lewis RS (1993) Mitogen-regulated Ca^{2+} current of T lymphocytes is activated by depletion of intracellular Ca^{2+} stores. *Proc Natl Acad Sci USA*. 90:6295–6299.
13. Hofer AM, Fasolato C, Pozzan T (1998) Capacitative Ca^{2+} entry is closely linked to the filling state of internal Ca^{2+} stores: A study using simultaneous measurements of I_{CRAC} and intraluminal $[Ca^{2+}]$. *J Cell Biol*. 140:325–334.
14. Putney JW, Bird GS (1993) The inositol phosphate-calcium signaling system in nonexcitable cells. *Endocr Rev*. 14:610–631.
15. Zweifach A, Lewis RS (1995) Slow calcium-dependent inactivation of depletion-activated calcium current. Store-dependent and -independent mechanisms. *J Biol Chem*. 270:14445–14451.
16. Fierro L, Parekh AB (1999) Fast calcium-dependent inactivation of calcium release-activated calcium current (CRAC) in RBL-1 cells. *J Membr Biol*. 168:9–17.
17. Prakriya M, Lewis RS (2003) CRAC channels: Activation, permeation, and the search for a molecular identity. *Cell Calcium*. 33:311–321.
18. Roos J, DiGregorio PJ, Yeromin AV, Ohlsen K, Lioudyno M, Zhang S, Safrina O et al. (2005) STIM1, an essential and conserved component of store-operated Ca^{2+} channel function. *J Cell Biol*. 169:435–445.
19. Liou J, Kim ML, Heo WD, Jones JT, Myers JW, Ferrell JE Jr., Meyer T (2005) STIM is a Ca^{2+} sensor essential for Ca^{2+}-store-depletion-triggered Ca^{2+} influx. *Curr Biol*. 15:1235–1241.
20. Yeromin AV, Roos J, Stauderman KA, Cahalan MD (2004) A store-operated calcium channel in *Drosophila* S2 cells. *J Gen Physiol*. 123:167–182.
21. Feske S, Gwack Y, Prakriya M, Srikanth S, Puppel SH, Tanasa B, Hogan PG, Lewis RS, Daly M, Rao A (2006) A mutation in Orai1 causes immune deficiency by abrogating CRAC channel function. *Nature*. 441:179–185.
22. Vig M, Peinelt C, Beck A, Koomoa DL, Rabah D, Koblan-Huberson M, Kraft S et al. (2006) CRACM1 is a plasma membrane protein essential for store-operated Ca^{2+} entry. *Science*. 312:1220–1223.

23. Zhang SL, Yeromin AV, Zhang XH, Yu Y, Safrina O, Penna A, Roos J, Stauderman KA, Cahalan MD (2006) Genome-wide RNAi screen of Ca^{2+} influx identifies genes that regulate Ca^{2+} release-activated Ca^{2+} channel activity. *Proc Natl Acad Sci USA*. 103:9357–9362.
24. Peinelt C, Vig M, Koomoa DL, Beck A, Nadler MJ, Koblan-Huberson M, Lis A, Fleig A, Penner R, Kinet JP (2006) Amplification of CRAC current by STIM1 and CRACM1 (Orai1). *Nat Cell Biol*. 8:771–773.
25. Soboloff J, Spassova MA, Tang XD, Hewavitharana T, Xu W, Gill DL (2006) Orai1 and STIM reconstitute store-operated calcium channel function. *J Biol Chem*. 281:20661–20665.
26. Mercer JC, Dehaven WI, Smyth JT, Wedel B, Boyles RR, Bird GS, Putney JW Jr. (2006) Large store-operated calcium selective currents due to coexpression of Orai1 or Orai2 with the intracellular calcium sensor, Stim1. *J Biol Chem*. 281:24979–24990.
27. Wu MM, Buchanan J, Luik RM, Lewis RS (2006) Ca^{2+} store depletion causes STIM1 to accumulate in ER regions closely associated with the plasma membrane. *J Cell Biol*. 174(6):803–813.
28. Zhang SL, Yu Y, Roos J, Kozak JA, Deerinck TJ, Ellisman MH, Stauderman KA, Cahalan MD (2005) STIM1 is a Ca^{2+} sensor that activates CRAC channels and migrates from the Ca^{2+} store to the plasma membrane. *Nature*. 437:902–905.
29. Prakriya M, Feske S, Gwack Y, Srikanth S, Rao A, Hogan PG (2006) Orai1 is an essential pore subunit of the CRAC channel. *Nature*. 443:230–233.
30. Luik RM, Wu MM, Buchanan J, Lewis RS (2006) The elementary unit of store-operated Ca^{2+} entry: Local activation of CRAC channels by STIM1 at ER-plasma membrane junctions. *J Cell Biol*. 174:815–825.
31. Xu P, Lu J, Li Z, Yu X, Chen L, Xu T (2006) Aggregation of STIM1 underneath the plasma membrane induces clustering of Orai1. *Biochem Biophys Res Commun*. 350:969–976.
32. Park CY, Hoover PJ, Mullins FM, Bachhawat P, Covington ED, Raunser S, Walz T, Garcia KC, Dolmetsch RE, Lewis RS (2009) STIM1 clusters and activates CRAC channels via direct binding of a cytosolic domain to Orai1. *Cell*. 136:876–890.
33. Liou J, Fivaz M, Inoue T, Meyer T (2007) Live-cell imaging reveals sequential oligomerization and local plasma membrane targeting of stromal interaction molecule 1 after Ca^{2+} store depletion. *Proc Natl Acad Sci USA*. 104:9301–9306.
34. Stathopulos PB, Li GY, Plevin MJ, Ames JB, Ikura M (2006) Stored Ca^{2+} depletion-induced oligomerization of stromal interaction molecule 1 (STIM1) via the EF-SAM region: An initiation mechanism for capacitive Ca^{2+} entry. *J Biol Chem*. 281:35855–35862.
35. Li Z, Lu J, Xu P, Xie X, Chen L, Xu T (2007) Mapping the interacting domains of STIM1 and Orai1 in Ca^{2+} release-activated Ca^{2+} channel activation. *J Biol Chem*. 282:29448–29456.
36. Hogan PG, Lewis RS, Rao A (2010) Molecular basis of calcium signaling in lymphocytes: STIM and ORAI. *Annu Rev Immunol*. 28:491–533.
37. Gudlur A, Zhou Y, Hogan PG (2013) STIM-ORAI interactions that control the CRAC channel. *Curr Top Membr*. 71:33–58.
38. Yeromin AV, Zhang SL, Jiang W, Yu Y, Safrina O, Cahalan MD (2006) Molecular identification of the CRAC channel by altered ion selectivity in a mutant of Orai. *Nature*. 443:226–229.
39. Gwack Y, Srikanth S, Feske S, Cruz-Guilloty F, Oh-hora M, Neems DS, Hogan PG, Rao A (2007) Biochemical and functional characterization of Orai proteins. *J Biol Chem*. 282:16232–16243.
40. Mignen O, Thompson JL, Shuttleworth TJ (2008) Orai1 subunit stoichiometry of the mammalian CRAC channel pore. *J Physiol*. 586:419–425.

41. McNally BA, Yamashita M, Engh A, Prakriya M (2009) Structural determinants of ion permeation in CRAC channels. *Proc Natl Acad Sci*. 106:22516–22521.
42. Zhou Y, Ramachandran S, Oh-Hora M, Rao A, Hogan PG (2010) Pore architecture of the ORAI1 store-operated calcium channel. *Proc Natl Acad Sci USA*. 107:4896–4901.
43. Hou X, Pedi L, Diver MM, Long SB (2012) Crystal structure of the calcium release-activated calcium channel Orai. *Science*. 338:1308–1313.
44. Amcheslavsky A, Wood ML, Yeromin AV, Parker I, Freites JA, Tobias DJ, Cahalan MD (2015) Molecular biophysics of Orai store-operated Ca^{2+} channels. *Biophys J*. 108:237–246.
45. Yamashita M, Navarro-Borelly L, McNally BA, Prakriya M (2007) Orai1 mutations alter ion permeation and Ca^{2+}-dependent fast inactivation of CRAC channels: Evidence for coupling of permeation and gating. *J Gen Physiol*. 130:525–540.
46. (a) Gudlur A, Quintana A, Zhou Y, Hirve N, Mahapatra S, Hogan PG (2014) STIM1 triggers a gating rearrangement at the extracellular mouth of the ORAI1 channel. *Nat Commun*. 5:5164; (b) Zhou Y, Cai X, Loktionova NA, Wang X, Nwokonko RM, Wang X, Wang Y, Rothberg BS, Trebak M, Gill DL (2016) The STIM1 binding site nexus remotely controls Orai1 channel gating. *Nat Commun*. 7:13725.
47. Frischauf I, Zayats V, Deix M, Hochreiter A, Jardin I, Muik M, Lackner B et al. (2015) A calcium-accumulating region, CAR, in the channel Orai1 enhances Ca^{2+} permeation and SOCE-induced gene transcription. *Sci Signal*. 8:ra131.
48. McNally BA, Somasundaram A, Yamashita M, Prakriya M (2012) Gated regulation of CRAC channel ion selectivity by STIM1. *Nature*. 482:241–245.
49. Varnai P, Toth B, Toth DJ, Hunyady L, Balla T (2007) Visualization and manipulation of plasma membrane-endoplasmic reticulum contact sites indicates the presence of additional molecular components within the STIM1–Orai1 complex. *J Biol Chem*. 282:29678–29690.
50. Stathopulos PB, Schindl R, Fahrner M, Zheng L, Gasmi-Seabrook GM, Muik M, Romanin C, Ikura M. (2013) STIM1/Orai1 coiled-coil interplay in the regulation of store-operated calcium entry. *Nat Commun*. 4:2963.
51. Oritani K, Kincade PW (1996) Identification of stromal cell products that interact with pre-B cells. *J Cell Biol*. 134:771–782.
52. Manji SS, Parker NJ, Williams RT, van Stekelenburg L, Pearson RB, Dziadek M, Smith PJ (2000) STIM1: A novel phosphoprotein located at the cell surface. *Biochim Biophys Acta*. 1481:147–155.
53. Williams RT, Senior PV, Van Stekelenburg L, Layton JE, Smith PJ, Dziadek MA (2002) Stromal interaction molecule 1 (STIM1), a transmembrane protein with growth suppressor activity, contains an extracellular SAM domain modified by N-linked glycosylation. *Biochim Biophys Acta*. 1596:131–137.
54. Spassova MA, Soboloff J, He LP, Xu W, Dziadek MA, Gill DL (2006) STIM1 has a plasma membrane role in the activation of store-operated Ca^{2+} channels. *Proc Natl Acad Sci USA*. 103:4040–4045.
55. Mignen O, Thompson JL, Shuttleworth TJ (2007) STIM1 regulates Ca^{2+} entry via arachidonate-regulated Ca^{2+}-selective (ARC) channels without store depletion or translocation to the plasma membrane. *J Physiol*. 579:703–715.
56. Lefkimmiatis K, Srikanthan M, Maiellaro I, Moyer MP, Curci S, Hofer AM (2009) Store-operated cyclic AMP signaling mediated by STIM1. *Nat Cell Biol*. 11:433–442.
57. Park CY, Shcheglovitov A, Dolmetsch R (2010) The CRAC channel activator STIM1 binds and inhibits L-type voltage-gated calcium channels. *Science*. 330:101–105.
58. Collins SR, Meyer T (2011) Evolutionary origins of STIM1 and STIM2 within ancient Ca^{2+} signaling systems. *Trends Cell Biol*. 21:202–211.
59. Cahalan MD (2009) STIMulating store-operated Ca^{2+} entry. *Nat Cell Biol*. 11:669–677.

60. Soboloff J, Rothberg BS, Madesh M, Gill DL (2012) STIM proteins: Dynamic calcium signal transducers. *Nat Rev Mol Cell Biol.* 13:549–565.
61. Stathopulos PB, Zheng L, Li GY, Plevin MJ, Ikura M (2008) Structural and mechanistic insights into STIM1-mediated initiation of store-operated calcium entry. *Cell.* 135:110–122.
62. Huang GN, Zeng W, Kim JY, Yuan JP, Han L, Muallem S, Worley PF (2006) STIM1 carboxyl-terminus activates native SOC, I(crac) and TRPC1 channels. *Nat Cell Biol.* 8:1003–1010.
63. Muik M, Frischauf I, Derler I, Fahrner M, Bergsmann J, Eder P, Schindl R et al. (2008) Dynamic coupling of the putative coiled-coil domain of ORAI1 with STIM1 mediates ORAI1 channel activation. *J Biol Chem.* 283:8014–8022.
64. Zhang SL, Kozak JA, Jiang W, Yeromin AV, Chen J, Yu Y, Penna A, Shen W, Chi V, Cahalan MD (2008) Store-dependent and -independent modes regulating Ca^{2+} release-activated Ca^{2+} channel activity of human Orai1 and Orai3. *J Biol Chem.* 283:17662–17671.
65. Zhou Y, Meraner P, Kwon HT, Machnes D, Oh-hora M, Zimmer J, Huang Y, Stura A, Rao A, Hogan PG (2010) STIM1 gates the store-operated calcium channel ORAI1 in vitro. *Nat Struct Mol Biol.* 17:112–116.
66. Yuan JP, Zeng W, Dorwart MR, Choi YJ, Worley PF, Muallem S (2009) SOAR and the polybasic STIM1 domains gate and regulate Orai channels. *Nat Cell Biol.* 11:337–343.
67. Kawasaki T, Lange I, Feske S (2009) A minimal regulatory domain in the C terminus of STIM1 binds to and activates ORAI1 CRAC channels. *Biochem Biophys Res Commun.* 385:49–54.
68. Zhou Y, Srinivasan P, Razavi S, Seymour S, Meraner P, Gudlur A, Stathopulos PB, Ikura M, Rao A, Hogan PG (2013) Initial activation of STIM1, the regulator of store-operated calcium entry. *Nat Struct Mol Biol.* 20:973–981.
69. Ma G, Wei M, He L, Liu C, Wu B, Zhang SL, Jing J et al. (2015) Inside-out Ca^{2+} signalling prompted by STIM1 conformational switch. *Nat Commun.* 6:7826.
70. Yang X, Jin H, Cai X, Li S, Shen Y (2012) Structural and mechanistic insights into the activation of Stromal interaction molecule 1 (STIM1). *Proc Natl Acad Sci USA.* 109:5657–5662.
71. Fahrner M, Muik M, Schindl R, Butorac C, Stathopulos P, Zheng L, Jardin I, Ikura M, Romanin C (2014) A coiled-coil clamp controls both conformation and clustering of stromal interaction molecule 1 (STIM1). *J Biol Chem.* 289:33231–33244.
72. Luik RM, Wang B, Prakriya M, Wu MM, Lewis RS (2008) Oligomerization of STIM1 couples ER calcium depletion to CRAC channel activation. *Nature.* 454:538–542.
73. Hogan PG (2015) The STIM1-ORAI1 microdomain. *Cell Calcium.* 58:357–367.
74. Brandman O, Liou J, Park WS, Meyer T (2007) STIM2 is a feedback regulator that stabilizes basal cytosolic and endoplasmic reticulum Ca^{2+} levels. *Cell.* 131:1327–1339.
75. Huang Y, Zhou Y, Wong HC, Chen Y, Chen Y, Wang S, Castiblanco A, Liu A, Yang JJ (2009) A single EF-hand isolated from STIM1 forms dimer in the absence and presence of Ca^{2+}. *FEBS J.* 276:5589–5597.
76. Furukawa Y, Teraguchi S, Ikegami T, Dagliyan O, Jin L, Hall D, Dokholyan NV et al. (2014) Intrinsic disorder mediates cooperative signal transduction in STIM1. *J Mol Biol.* 426:2082–2097.
77. Muik M, Fahrner M, Schindl R, Stathopulos P, Frischauf I, Derler I, Plenk P et al. (2011) STIM1 couples to ORAI1 via an intramolecular transition into an extended conformation. *EMBO J.* 30:1678–1689.
78. Fernandez-Busnadiego R, Saheki Y, De Camilli P (2015) Three-dimensional architecture of extended synaptotagmin-mediated endoplasmic reticulum-plasma membrane contact sites. *Proc Natl Acad Sci USA.* 112:E2004–E2013.

79. Perni S, Dynes JL, Yeromin AV, Cahalan MD, Franzini-Armstrong C (2015) Nanoscale patterning of STIM1 and Orai1 during store-operated Ca^{2+} entry. *Proc Natl Acad Sci USA*. 112:E5533–E5554.
80. Korzeniowski MK, Manjarres IM, Varnai P, Balla T (2010) Activation of STIM1–Orai1 involves an intramolecular switching mechanism. *Sci Signal*. 3:ra82.
81. Yu J, Zhang H, Zhang M, Deng Y, Wang H, Lu J, Xu T, Xu P (2013) An aromatic amino acid in the coiled-coil 1 domain plays a crucial role in the auto-inhibitory mechanism of STIM1. *Biochem J*. 454:401–409.
82. Muik M, Fahrner M, Derler I, Schindl R, Bergsmann J, Frischauf I, Groschner K, Romanin C (2009) A cytosolic homomerization and a modulatory domain within STIM1 C terminus determine coupling to ORAI1 channels. *J Biol Chem*. 284:8421–8426.
83. Covington ED, Wu MM, Lewis RS (2010) Essential role for the CRAC activation domain in store-dependent oligomerization of STIM1. *Mol Biol Cell*. 21:1897–1907.
84. Navarro-Borelly L, Somasundaram A, Yamashita M, Ren D, Miller RJ, Prakriya M (2008) STIM1–Orai1 interactions and Orai1 conformational changes revealed by live-cell FRET microscopy. *J Physiol*. 586:5383–5401.
85. Barr VA, Bernot KM, Srikanth S, Gwack Y, Balagopalan L, Regan CK, Helman DJ et al. (2008) Dynamic movement of the calcium sensor STIM1 and the calcium channel Orai1 in activated T-cells: Puncta and distal caps. *Mol Biol Cell*. 19:2802–2817.
86. McNally BA, Somasundaram A, Jairaman A, Yamashita M, Prakriya M (2013) The C- and N-terminal STIM1 binding sites on Orai1 are required for both trapping and gating CRAC channels. *J Physiol*. 591:2833–2850.
87. Tirado-Lee L, Yamashita M, Prakriya M (2015) Conformational changes in the Orai1 C-terminus evoked by STIM1 Binding. *PLoS One*. 10:e0128622.
88. Palty R, Stanley C, Isacoff EY (2015) Critical role for Orai1 C-terminal domain and TM4 in CRAC channel gating. *Cell Res*. 25:963–980.
89. Frischauf I, Muik M, Derler I, Bergsmann J, Fahrner M, Schindl R, Groschner K, Romanin C (2009) Molecular determinants of the coupling between STIM1 and Orai channels: Differential activation of Orai1–3 channels by a STIM1 coiled-coil mutant. *J Biol Chem*. 284:21696–21706.
90. Derler I, Fahrner M, Carugo O, Muik M, Bergsmann J, Schindl R, Frischauf I, Eshaghi S, Romanin C (2009) Increased hydrophobicity at the N terminus/membrane interface impairs gating of the severe combined immunodeficiency-related ORAI1 mutant. *J Biol Chem*. 284:15903–15915.
91. Lee KP, Yuan JP, Zeng W, So I, Worley PF, Muallem S (2009) Molecular determinants of fast Ca^{2+}-dependent inactivation and gating of the Orai channels. *Proc Natl Acad Sci USA*. 106:14687–14692.
92. Zhou Y, Wang X, Wang X, Loktionova NA, Cai X, Nwokonko RM, Vrana E, Wang Y, Rothberg BS, Gill DL (2015) STIM1 dimers undergo unimolecular coupling to activate Orai1 channels. *Nat Commun*. 6:8395.
93. Yuan JP, Zeng W, Huang GN, Worley PF, Muallem S (2007) STIM1 heteromultimerizes TRPC channels to determine their function as store-operated channels. *Nat Cell Biol*. 9:636–645.
94. Lis A, Zierler S, Peinelt C, Fleig A, Penner R (2010) A single lysine in the N-terminal region of store-operated channels is critical for STIM1-mediated gating. *J Gen Physiol*. 136:673–686.
95. Bergsmann J, Derler I, Muik M, Frischauf I, Fahrner M, Pollheimer P, Schwarzinger C, Gruber HJ, Groschner K, Romanin C (2011) Molecular determinants within N terminus of Orai3 protein that control channel activation and gating. *J Biol Chem*. 286:31565–31575.

96. Dong H, Fiorin G, Carnevale V, Treptow W, Klein ML (2013) Pore waters regulate ion permeation in a calcium release-activated calcium channel. *Proc Natl Acad Sci USA*. 110:17332–17337.
97. Zhang SL, Yeromin AV, Hu J, Amcheslavsky A, Zheng H, Cahalan MD (2011) Mutations in Orai1 transmembrane segment 1 cause STIM1-independent activation of Orai1 channels at glycine 98 and channel closure at arginine 91. *Proc Natl Acad Sci USA*. 108:17838–17843.
98. Derler I, Plenk P, Fahrner M, Muik M, Jardin I, Schindl R, Gruber HJ, Groschner K, Romanin C (2013) The extended transmembrane Orai1 N-terminal (ETON) region combines binding interface and gate for Orai1 activation by STIM1. *J Biol Chem*. 288:29025–29034.
99. Dynes JL, Amcheslavsky A, Cahalan MD (2016) Genetically targeted single-channel optical recording reveals multiple Orai1 gating states and oscillations in calcium influx. *Proc Natl Acad Sci USA*. 113:440–445.

4 Modulation of Orai1 and STIM1 by Cellular Factors

Jin Seok Woo, Sonal Srikanth, and Yousang Gwack

CONTENTS

4.1 INTRODUCTION

Store-operated Ca^{2+} entry (SOCE) is a ubiquitous mechanism in eukaryotic cells to elevate the intracellular Ca^{2+} concentration and stimulate downstream signaling pathways. SOCE is especially important for Ca^{2+} entry in cells with immune receptors, including T cells, B cells, and mast cells. Under resting conditions, cytoplasmic Ca^{2+} concentration ($[Ca^{2+}]$) is very low (~100 nM) in T cells, while that in the endoplasmic reticulum (ER), which serves as an intracellular Ca^{2+} store, is much higher (~4,000–10,000-fold higher—0.4–1.0 mM) [1,2]. Extracellular $[Ca^{2+}]$ reaches almost 2 mM concentration, establishing a huge $[Ca^{2+}]$ gradient between the extracellular space and the cytoplasm (~20,000-fold). Therefore, dynamic regulation of Ca^{2+} flow occurs constantly to maintain these gradients even in resting T cells. When T cells are activated, there is a sustained increase in intracellular Ca^{2+} concentration ($[Ca^{2+}]_i$), which is initiated by emptying of the ER Ca^{2+} stores. The increase in cytoplasmic $[Ca^{2+}]$ by depletion of ER Ca^{2+} stores can be minor especially in T cells, due to small volume of the ER. Instead, ER Ca^{2+} depletion induces Ca^{2+} entry via store-operated Ca^{2+} (SOC) channels, which raises $[Ca^{2+}]_i$ up to micromolar concentrations [3–6]. Therefore, SOCE via the Ca^{2+} release–activated Ca^{2+} (CRAC) channels is a primary mechanism for activation of Ca^{2+} signaling in T cells.

Upon pathogen infection or self-peptide presentation in autoimmunity, antigen-presenting cells (APCs, e.g., dendritic cells or B cells) present antigens on their surface together with major histocompatibility complex class II to activate CD4+ helper T cells. Antigen engagement of T cell receptors (TCRs) triggers a conformational change of TCRαβ chain, which induces a cascade of tyrosine phosphorylation events mediated by CD3ζ chain–ZAP70 (zeta chain–associated protein kinase 70) complexes [7,8]. This results in phosphorylation of a signaling adaptor Lat, which dissociates from CD3ζ chain–ZAP70 complex and activates phospholipase C-γ (PLCγ). In turn, PLCγ hydrolyzes phosphatidylinositol 4,5-bisphosphate (PIP_2) into inositol trisphosphate (IP_3) and diacylglycerol. IP_3 binds to the IP_3 receptor on the ER membrane and releases Ca^{2+} from the ER into the cytoplasm, and this depletion leads to activation of CRAC channels. The CRAC channel is a prototype SOC channel, well characterized in immune cells. Because ER Ca^{2+} store especially in T cells is limited, SOCE via CRAC channels is important to maintain elevated levels of $[Ca^{2+}]_i$, which are required for activation of downstream signaling pathways including the protein kinase C, extracellular signal–regulated kinases, or nuclear factor of activated T cell (NFAT) pathways to affect the transcriptional programs for generating a productive immune response (see Chapter 5). Defective function or lack of expression of the CRAC channel components causes severe combined immune deficiency in humans [9]. Hence, an in-depth understanding of CRAC channel–mediated Ca^{2+} signaling in T cells is crucial for developing drug therapies for immune deficiency or inflammatory disorders.

Identification of essential components of CRAC channels revealed a unique mechanism of its activation, which is mediated by protein interactions. Genome-wide RNAi screens identified Orai1 as a pore subunit of the CRAC channels [10–13]. Prior to identification of Orai1, limited RNAi screens in *Drosophila* and HeLa cells had identified STIM1, a Ca^{2+}-binding protein localized predominantly in the ER as an important regulator of SOCE [14–16]. STIM1 senses $[Ca^{2+}]_{ER}$ via its N-terminal EF-hands and gates Orai1 by direct interaction. The EF-hand of STIM1 has a low affinity for Ca^{2+}, between 0.2 and 0.6 mM [17], and remains Ca^{2+} bound at rest. Under resting conditions, Orai1 and STIM1 are homogeneously distributed at the plasma membrane (PM) and the ER membrane, respectively. Upon ER Ca^{2+} depletion triggered by TCR stimulation, STIM1 loses Ca^{2+} binding, multimerizes, translocates to the ER–PM junctions, mediates clustering of Orai1 on the PM, and stimulates Ca^{2+} entry (Figure 4.1a) [14–16]. Detailed studies have identified a minimal domain of STIM1 necessary for activation of Orai1 as the CRAC activation domain (CAD)/STIM1–Orai1 activating region (SOAR) that directly binds to the cytosolic N and C termini of Orai1 [18,19] (see Chapter 2). This region, containing coiled-coil (CC) domains 2 and 3 of STIM1 is located in its cytoplasmic C terminus (Figure 4.1a). Further studies showed that Ca^{2+}-bound STIM1 under resting conditions exhibits a folded structure mediated by intramolecular protein interaction between the positively charged residues within its CAD/SOAR domain and the negatively charged, autoinhibitory region preceding the CAD/SOAR domain, located in the CC1 region [20]. STIM1 activation requires unfolding of this intramolecular interaction to allow the basic residues within the CAD/SOAR domain to interact with the acidic residues within the C terminus of Orai1 [20]. While Orai1 and STIM1 are the major

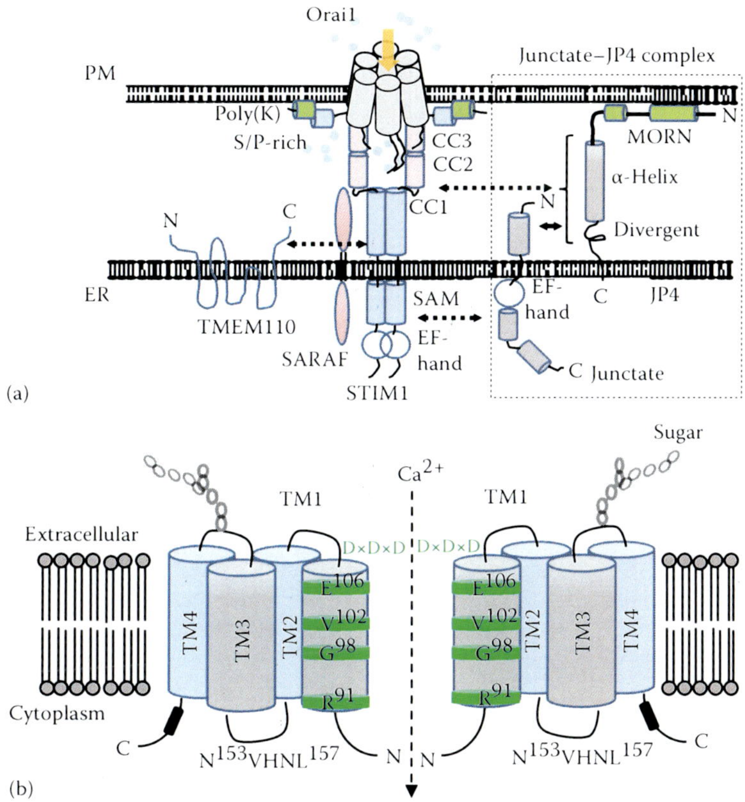

FIGURE 4.1 Molecular mechanisms of CRAC channel regulation. (a) Schematic showing protein interactions among Orai1, STIM1, and auxiliary proteins at the ER–PM junctions. Under resting conditions, Orai1 and STIM1 are distributed at the PM and the ER. Upon store depletion, STIM1 oligomerizes by sensing ER Ca^{2+} depletion with its ER-luminal EF-hand domain and translocates to form clusters at the ER–PM junctions predetermined by the junctate–JP4 complex in T cells. By physical interaction with Orai1 through the CAD/SOAR domain (CC2 and CC3, indicated in red), STIM1 recruits and activates Orai1 in the ER–PM junctions. Junctate also contains Ca^{2+}-sensing motif in the ER lumen, and ER Ca^{2+} depletion increases its interaction with STIM1. SARAF and TMEM110 have been shown to translocate together with STIM1 to the ER–PM junctions. SARAF interacts with STIM1 CAD/SOAR domain to modulate slow CDI of CRAC channels, while TMEM110 directly interacts with the CC1 region of STIM1 via its cytoplasmic C terminus. While both the N and C termini of TMEM110 face the cytoplasm, the transmembrane topology (4 or 5 TMs) is based on prediction by bioinformatics programs (TMHMM2.0, TMPred, and TOPCONS). (b) Schematic of Orai1. Orai1 has four transmembrane segments (TM1–TM4). It has two extracellular domains, and the second extracellular domain between TM3 and TM4 contains the asparagine (N^{223}) residue involved in glycosylation. The TM1 lines the pore and the residues in TM1 involve Ca^{2+} selectivity and gating. Orai1 contains three intracellular domains including the N terminus, intracellular loop, and C-terminal CC domain that are important for protein interactions and channel activation/inactivation. Arginine at position 91 was mutated in patients (R91W) with nonfunctional CRAC channels suffering from severe combined immunodeficiency. The schematic indicates a dimeric form for convenience of representation.

TABLE 4.1
Summary of Associating Partners of Orai1 and STIM1

	Interacting Partners	Binding Region (Orai1/STIM1)	Function	References
Orai1	*Modulators*			
	CRACR2A-c	N terminus	Stabilization of Orai1–STIM1 protein complex	[33]
	Calmodulin	N terminus	Ca^{2+}-dependent inactivation	[30] but see also [32]
	α-SNAP	C terminus	Formation of functional CRAC channel	[36]
	Adenylyl cyclase 8 (AC8)	N terminus	cAMP signaling	[37]
	Vesicle components			
	Caveolin	N terminus	Orai1 recycling	[38,39]
	CRACR2A-a	N terminus	Stabilization of Orai1–STIM1 protein complex	[34]
	Channels			
	Orai2/Orai3	TM	Channel formation	[13]
	TRPC channel	N.D.	Store-/agonist-operated Ca^{2+} entry	[65–68] but see also [69]
	Store independent			
	SPCA2	N and C termini	Store-independent gating	[41]
	SARAF (TMEM66)	N.D.	Inactivation of ARC channels	[57]
STIM1	*Modulators*			
	CRACR2A-c	Cytoplasmic	Stabilization of Orai1–STIM1 protein complex	[33]
	SARAF (TMEM66)	CAD/SOAR	Ca^{2+}-dependent inactivation	[57]
	α-SNAP	CAD/SOAR	Formation of functional CRAC channel	[36]
	ERp57	ER luminal	Negative regulation of STIM1	[77]
	Golli MBP	Cytoplasmic	Negative regulation of STIM1	[84]
	Polycystin1-P100	N.D.	Inhibition of STIM1 translocation	[85]
	Stanniocalcin 2 (STC2)	N.D.	Negative regulation of STIM1	[86]
	POST (TMEM20)	N.D.	Inhibition of PMCA	[87]
	EB1	Cytoplasmic	ER tubulation	[74,75]
	Channel or pump			
	TRPC channel	Cytoplasmic	Activated by STIM1	[65–68] but see also [69]
	PMCA	Cytoplasmic	Inhibited by STIM1	[71,72]
	$Ca_v1.2$	CAD/SOAR	Inhibited by STIM1	[59,60]
	ARC channels	N.D.	Store-independent activation by STIM1	[43]
	Junctional proteins			
	Junctate	ER luminal	Recruiting STIM1 to junctions	[50]
	Junctophilin-4 (JP4)	CC1+2	Recruiting STIM1 to junctions	[51]
	TMEM110	CC1	Recruiting STIM1 to junctions/ formation of junctions	[54,55]
	Phospholamban (PLN)	N.D.	Regulation of SR Ca^{2+}	[88]

Notes: Orai1 contains three intracellular domains including the N terminus, intracellular loop, and C-terminal CC domain. The interacting partners for each domain are summarized. The associating partners of STIM1 are also indicated. N.D. (not determined) indicates that their binding domains are not known.

components of CRAC channels, multiple auxiliary proteins have been shown to regulate CRAC channel function. Some of these have been reviewed in detail before [21,22] and are only briefly summarized in Table 4.1. In this chapter, we focus on the recently identified molecules regulating the function of Orai1 and STIM1.

4.2 MODULATORS OF ORAI1 VIA PROTEIN INTERACTION

As demonstrated by STIM1, protein interactions broadly regulate gating, Ca^{2+} selectivity, intracellular localization, and clustering of Orai1. The pore subunit of the CRAC channels comprises of Orai1 multimers with each monomer containing four transmembrane segments (TM1–TM4). The monomers contain the N and C termini facing the cytoplasm as well as an intracellular loop between TM2 and TM3, which is important for channel inactivation (Figure 4.1b) [23–27]. The residues R91, G98, V102, and E106 within TM1 of Orai1 and $D^{110}xD^{112}xD^{114}$ facing the extracellular milieu line the pore and are important for Ca^{2+} selectivity and gating [3–6,28,29]. Therefore, interacting partners can potentially induce conformational changes to regulate gating and Ca^{2+} selectivity in a STM1-dependent or STM1-independent manner. In addition to direct regulation of channel gating, it is also possible that interacting partners of Orai1 regulate its localization, clustering, posttranslational modification, or degradation. Various aspects of these regulatory mechanisms are currently under investigation.

It was previously proposed that calmodulin (CaM) binds to the N terminus of Orai1 at elevated $[Ca^{2+}]_i$ and potentiates a negative feedback to induce Ca^{2+}-dependent inactivation (CDI) to inhibit SOCE [30]. However, recent crystal structure of the *Drosophila melanogaster* Orai protein showed that side chains of the residues corresponding to W76 and Y80 of Orai1, previously shown to interact with CaM, are facing the lumen of the pore, and this can potentially create steric hindrance between CaM and other Orai TMs [31]. Careful analyses of CDI of Orai1 in the presence of various CaM mutants defective in Ca^{2+} binding suggested that CaM binding may not be important for CDI [32]. Mullins et al. found that the residues W76 and Y80 are actually involved in conformational changes within the pore and accordingly proposed a conformational gating model for induction of CDI in Orai channels [32]. Hence, the role of CaM in regulating CRAC channels, particularly CDI, needs further investigation. In addition to STIM1 and CaM, other proteins interacting with the N terminus of Orai1 include a novel cytoplasmic EF-hand-containing protein, CRAC channel regulator 2A (CRACR2A, EFCAB4B or FLJ33805), which was identified from large-scale affinity protein purification using Orai1 as bait [33]. CRACR2A has two splice isoforms, CRACR2A-a (~80 kDa) and CRACR2A-c (45 kDa). The short isoform CRACR2A-c is cytoplasmic and forms a ternary complex with Orai1 and STIM1 to stabilize their interaction after store depletion. Accordingly, its depletion decreases STIM1 clustering at the ER–PM junctions and, hence, SOCE. This interaction with Orai1 and STIM1 is $[Ca^{2+}]_i$ dependent, with low $[Ca^{2+}]_i$ favoring association and high $[Ca^{2+}]_i$ favoring its dissociation, and thus fine-tunes Ca^{2+} entry. The long isoform CRACR2A-a encodes a large Rab GTPase (Figure 4.2) [34,35]. Recently, α-SNAP was identified as a cytosolic factor that interacts with both Orai1 and STIM1 [36]. The original function of α-SNAP is disassembly of the SNARE (NSF attachment protein receptor) complex, a cellular machinery used for vesicle

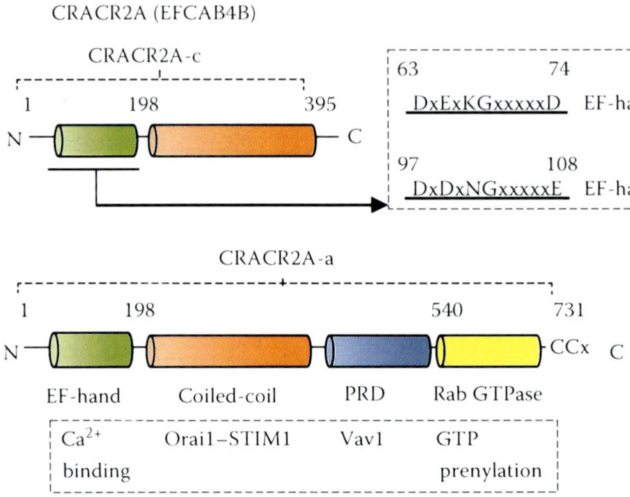

FIGURE 4.2 Schematic depicting the different isoforms of CRACR2A proteins. Schematic showing the predicted domain structure of human CRACR2A proteins. *CRACR2A* gene encodes two splice isoforms, CRACR2A-a and CRACR2A-c. Both proteins share EF-hand motifs (dotted box), coiled-coil domains (CC1 and CC2), and leucine-rich region (LR), which interact with the Orai1–STIM1 complex to regulate Ca^{2+} entry. CRACR2A-a contains additional PRD that interacts with Vav1 and a Rab GTPase domain with an unconventional prenylation site at the C terminus (CCx; x, any amino acid).

fusion. Different from its original function, α-SNAP, a predominantly cytoplasmic protein, physically interacts with the cytosolic CAD/SOAR domain of STIM1 and the C-terminal tail of Orai1. Through this interaction, α-SNAP regulates an active molecular rearrangement within Orai1–STIM1 clusters to obtain the STIM1/Orai1 ratio required for optimal activation of CRAC channels, without affecting the rate of STIM1 translocation into the ER–PM junctions. Accordingly, after store depletion, α-SNAP-deficient cells stably expressing Orai1 and STIM1 exhibited an increase in density of Orai1 in clusters without altering STIM1 density, leading to a reduced ratio of STIM1/Orai1 in individual clusters, thereby reducing SOCE. A direct binding to the C terminus of Orai1 and the CAD/SOAR domain of STIM1 suggests that α-SNAP may regulate CRAC channel gating, which has not been investigated so far.

Protein interactions are not only important for CRAC channel regulation but also essential for activation of signaling pathways downstream of Orai1. Cyclic adenosine monophosphate (cAMP), an important second messenger, is generated by adenylyl cyclase (AC)-mediated cleavage of adenosine triphosphate and in turn activates downstream protein kinase A pathway. A recent study demonstrated a direct interaction between AC8 and Orai1 N terminus using Forster resonance energy transfer (FRET) and GST pulldown and immunoprecipitation analyses [37]. This interaction places AC8 close to Ca^{2+} microdomains to mediate its Ca^{2+}-dependent activation. These studies reveal that the interaction with Orai1 is important not only for positive or negative feedback to regulate CRAC channels but also for crosstalk between SOCE and other signaling pathways.

4.3 VESICULAR COMPONENTS IN REGULATION OF ORAI1

Orai1-interacting molecules also play a role in its trafficking. Machaca and colleagues used *Xenopus* oocyte as a model and showed that during meiosis, SOCE is inactivated due to internalization of Orai1 into an intracellular vesicular compartment and inhibition of STIM1 clustering [38]. In a follow-up study, the authors showed that Orai1 internalization occurred via a caveolin- and dynamin-dependent endocytic pathway. The authors mapped a caveolin-binding site in the N terminus of Orai1 and showed that a significant fraction of total Orai1 is intracellular, and at rest, Orai1 actively recycles between an endosomal compartment and the PM [39]. Vesicles are also involved in translocation of Orai1 from the intracellular pool into the PM after store depletion. Recently, the same group showed that a large portion of Orai1 (~60%) exists in intracellular vesicles rather than the PM during the steady state in CHO and HEK293 cells [40]. A subset of these vesicles localize in close proximity to the PM and fuse to it after store depletion, increasing surface Orai1 levels. Furthermore, this study also showed that Orai1 trafficking to the PM after store depletion is dependent on interaction with STIM1 (in a "trafficking trap" mechanism). Accordingly, depletion of STIM1 blocked enrichment of surface Orai1 after store depletion. Conversely, overexpressed STIM1 caused intracellular trapping of Orai1-containing vesicles, resulting in reduced SOCE. Therefore, an optimal ratio between Orai1 and STIM1 is not only important for gating but also for enrichment of Orai1 to the PM.

Recently, we identified a long isoform of CRACR2A, CRACR2A-a, which is localized to the proximal Golgi area and vesicles, and uncovered its role in TCR signaling pathways including the SOCE pathway [34]. While the regions for Ca^{2+} binding and interaction with Orai1 and STIM1 are conserved between the two isoforms (Figure 4.2), CRACR2A-a has an additional proline-rich domain (PRD) and a Rab GTPase domain in its C terminus and is enriched in lymphoid organs. CRACR2A-a is unique because it clearly distinguishes itself from small Rab GTPases (~20 kDa) due to its large size (~85 kDa) and presence of multiple functional domains [34,35]. Interestingly, the Rab GTPase domain establishes the localization of CRACR2A-a in a GTP-/GDP-binding and prenylation-dependent manner. GTP-bound and prenylated CRACR2A-a localizes within vesicles close to the *trans*-Golgi network, whereas GDP-bound or unprenylated CRACR2A-a is cytosolic and rapidly degraded. Prenylation of CRACR2A-a involves geranylgeranylation at an unconventional site (CCx, x; any amino acid) in the C terminus. Upon TCR stimulation, CRACR2A-a translocates into the immunological synapse via interaction of its PRD with Vav1, a proximal TCR signaling molecule, to activate SOCE and the Ca^{2+}–NFAT and the JNK signaling pathways. CRACR2A-a also translocates into the ER–PM junctions via vesicle trafficking after passive ER Ca^{2+} store depletion and recovers SOCE in Jurkat T cells depleted of both the isoforms, similar to CRACR2A-c. Because CRACR2A-a retains the Orai–STIM interaction domain, one can assume that it supports SOCE by interacting with both Orai1 and STIM1, similar to CRACR2A-c. The molecular mechanism of translocation of CRACR2A-a and activation of SOCE upon passive store depletion remains to be investigated.

4.4 STORE-INDEPENDENT REGULATION OF ORAI1 VIA PROTEIN INTERACTION

Only a few cases of store-independent regulation of Orai1 have been identified so far. An isoform of the secretory pathway Ca^{2+} ATPase, SPCA2, was shown to enhance mammary tumor cell growth by raising $[Ca^{2+}]_i$ via a direct interaction with the N and C termini of Orai1 in a STIM1 and store-independent manner [41]. Another example of store-independent interaction is the one between STIM1 and arachidonate-regulated Ca^{2+} (ARC) channels, which are activated by low concentrations of arachidonic acid. While CRAC channels are formed of homomultimers of Orai1, ARC channels are heteromers of Orai1 and Orai3 monomers and are opened by a pool of STIM1 that constitutively resides in the PM [42,43]. Albarran et al. showed that in addition to the ER, SOCE-associated regulatory factor (SARAF) also localizes to the PM (see Section 4.6). Addition of the ARC channel agonist, arachidonic acid, increases the association of PM-resident SARAF with Orai1 to negatively regulate Ca^{2+} entry via the ARC channels [44]. Accordingly, knockdown of SARAF increases arachidonic acid–induced Ca^{2+} entry, while its overexpression decreases arachidonic acid–induced Ca^{2+} entry (see Chapter 11). These studies describe a novel and interesting aspect of Orai channel activation via store depletion–independent mechanisms, and future studies are needed to uncover other mechanisms regulating Orai and STIM function in a store-independent manner.

4.5 STIM1-INTERACTING MOLECULES AT THE ER–PM JUNCTIONS

Orai1 and STIM1 cluster at preexisting junctions of the ER and the PM, a space of 10–25 nm [45,46]. In excitable cells (e.g., muscle cells), proteins localized to the junctions between the PM and ER/sarcoplasmic reticulum (SR) membrane form a structural foundation for Ca^{2+} dynamics essential for excitation–contraction coupling [2,47]. Various biochemical screening approaches have identified junctophilins, mitsugumins, sarcalumenin, junctin, and junctate as components of these junctions [47–49]. Recent studies have shown that homologues and isoforms of these junctional proteins are also expressed in T cells. Srikanth et al. identified the EF-hand-containing protein, junctate, as an interactor of STIM1 [50]. Junctate localization defined the sites of accumulation of CRAC channel components since after store depletion, Orai1 and STIM1 accumulated at junctions that were already marked by junctate (Figure 4.1a). In a recent study, Woo et al. identified an important role of another junctional protein, JP4, in regulation of SOCE in T cells [51]. Junctophilin family consists of four genes, JP1, JP2, JP3, and JP4, that are expressed in a tissue-specific manner and are known to form ER–PM junctions in excitable cells including skeletal muscle, cardiac, and neuronal cells [48,52]. Junctophilins contain eight repeats of the membrane occupation and recognition nexus motifs that bind to phospholipids in the N terminus and a C-terminal ER membrane–spanning TM [48,53]. Depletion of JP4 inhibited STIM1 recruitment into the ER–PM junctions and significantly decreased SOCE. Biochemical analyses showed a direct interaction of JP4 cytoplasmic domain with CC1 and CC2 regions of STIM1. JP4 was also shown to interact with the N-terminal cytoplasmic region of junctate. Therefore, this study demonstrates that junctate–JP4 complex is an important component of the ER–PM junctions in T cells that synergistically recruits STIM1 into these junctions by protein interaction (Figure 4.1a). When overexpressed, STIM1 alone is sufficient to establish the ER–PM junctions using its C-terminal polylysine regions. However, in a physiological condition when the concentration of STIM1 is low or STIM1 has a defect in its phospholipid-binding capacity (e.g., low [PIP_2] in the PM), interaction of STIM1 with the junctate–JP4 complex can be important for efficient assembly of a functional CRAC channel complex at the junctions. This junctional protein complex also acts as a determinant of the site of Ca^{2+} entry because it is preassembled at the regions where Orai1 and STIM1 accumulate after store depletion.

Two studies identified transmembrane protein 110 (TMEM110 or STIM-activating enhancer) as a positive regulator of Ca^{2+} influx by the Orai1 and STIM1 complex using biotin-labeled protein purification and a genome-wide RNAi screen, respectively [54,55]. TMEM110 is a multipass ER-resident protein with its N and C termini facing the cytoplasm. Jiang et al. showed that TMEM110 physically interacted with the CC1 region of STIM1 and induced its active conformation to interact with Orai1 [54] (Figure 4.1a). The CC1 region of STIM1 contains an acidic amino acid motif that binds to the Orai1-interacting CAD/SOAR fragment, blocking its interaction with Orai1 in an autoinhibitory manner. Interaction of TMEM110 with the CC1 region of STIM1 facilitated the release of this autoinhibition in STIM1. Furthermore, this study also showed that depletion of TMEM110 had a modest influence on the frequency of the ER–PM junctions with 8%–12% decrease in cortical ER (ER that is in close proximity to the PM). Thus, the observed phenotype of more than 60% reduction in SOCE in TMEM110 KO cells was predominantly ascribed to its direct interaction with STIM1 to relieve its autoinhibition. The molecular mechanism proposed in this study is different from the one reported by Hogan and colleagues, who showed that siRNA-mediated depletion of TMEM110 significantly reduced the density of ER–PM junctions by >60% in HeLa/HEK293 cells, both under resting conditions and after store depletion. Importantly, artificial expansion of the junctions by overexpression of a yeast junctional protein lst2, which is unlikely to affect STIM1 autoinhibition, significantly rescued STIM1 translocation and SOCE. Therefore, this study concluded that TMEM110 is important for maintenance of the ER–PM junctions involved in SOCE in resting conditions and for dynamic remodeling of these junctions after store depletion. Although the molecular mechanism proposed in each study is different, the common conclusion from both studies is that TMEM110 is localized at the ER–PM junctions and is essential for translocation of STIM1, and thereby for SOCE.

In another study, extended synaptotagmin proteins (E-Syts) were shown to play a critical role in the formation of ER–PM junctions in HeLa cells [56]. Using siRNA-mediated depletion of all three E-Syt proteins, the authors showed a >50% reduction in the density of ER–PM junctions; however, E-Syts-dependent junctions were completely dispensable for SOCE. The authors did observe a significant decrease in accumulation of Orai1 and STIM1 at the ER–PM junctions; however, this decrease in accumulation did not significantly affect SOCE. Collectively, these studies suggest that specialized proteins like TMEM110, junctate, or JP4 may play a significant role in the formation and/or dynamic regulation of junctions specifically involved in SOCE, which are likely to be distinct from those involved in other functions of ER–PM junctions including PIP_2 replenishment and lipid transfer.

4.6 MODULATORS OF STIM1 FUNCTION

An ER-resident protein SARAF was identified as an interacting partner of STIM1 that facilitates the Ca^{2+}-dependent slow inactivation of CRAC channels [57]. By screening a cDNA overexpression library, Palty et al. identified SARAF as a candidate that reduces mitochondrial Ca^{2+} accumulation. It turned out that this overexpression strategy was useful specially to identify inhibitors of Orai1 and STIM1 function because siRNA-based screens are excellent to pick up positive regulators, but not inhibitory proteins. SARAF plays multiple roles in regulation of basal and ER [Ca^{2+}] as well as SOCE. Human SARAF encodes a 339-amino acid protein containing a single transmembrane segment (aa 173–195) with its N terminus facing the ER lumen (aa 1–172) and its C terminus facing the cytoplasm (aa 196–339) (Figure 4.1a). It shares similarities with STIM1 in global domain structure together with the presence of positively charged residues, which may interact with the PM phospholipids and a serine/proline-rich domain in its C-terminal end. Depletion of SARAF increased intracellular Ca^{2+} concentration and enhanced SOCE after store depletion, whereas its overexpression showed an opposite effect. Therefore, SARAF plays a negative role in SOCE with different modes of action: first, it interacts with the inactive form of STIM1 in the resting condition to stabilize its inactive state in the ER; second, it translocates to the ER–PM junctions together with STIM1 to induce CDI of Orai1 channels; and finally it also facilitates dissociation of clustered STIM1 proteins. More detailed structure–function studies identified a C-terminal inhibitory domain (CTID, aa 448–530) within STIM1 that regulates SARAF–STIM1 interaction [58]. STIM1 CTID is located at the C-terminal region of the Orai1-interacting CAD/SOAR domain, and interestingly, deletion of CTID from full-length STIM1 resulted in constitutively active Orai1 channels. CTID does not bind to SARAF directly but mediates the interaction of SARAF with the CAD/SOAR region. Therefore, this study highlights the important role of STIM1 and SARAF in CDI of Orai1, which is necessary to avoid excessive Ca^{2+} entry and its ensuing outcomes including cell death.

4.7 STIM1 AS A REGULATOR FOR NON-CRAC CHANNEL-RELATED FUNCTIONS

Two independent studies have shown an interaction between STIM proteins and $Ca_v1.2$ [59,60]. STIM1 interacted with $Ca_v1.2$ and inhibited the channel activity in a short term and blocked its surface expression in a long term [59,60]. Both studies showed a direct interaction between the CAD/SOAR domain of STIM1 and the C terminus of the $Ca_v1.2$ channel. These results provide an interesting scenario where the same region of STIM1 plays exactly opposite roles in regulation of CRAC channels and $Ca_v1.2$ channels. This must be a useful strategy in excitable cells where timing and activities of $Ca_v1.2$ and SOC channels can be simultaneously regulated.

However, in T cells, the activation of CRAC channels and Ca_v1 channels occurs simultaneously; hence, it would be interesting to understand how STIM1 interacts with both these channels at the same time. Recently, several reports have shown a positive role of Ca_v1 family of Ca^{2+} channels in activation, homeostatic proliferation, and cytokine production by CD4+ and CD8+ T cells [61–64]. It would be interesting to examine the activity of Ca_v1 channels in T cells isolated from mice lacking the expression of STIM proteins. In addition to CRAC and Ca_v channels, STIM1 also interacts with and activates transient receptor potential type C (TRPC) channels via an electrostatic interaction between dilysine motif in its polybasic tail and a conserved diaspartate motif in TRPC1 (and other TRPC channels) [65–69]. One report shows expression of TRPC3 among all the TRP channels in human T cells; however, the authors did not observe any significant defect in T cell activation, SOCE, or proliferation in the presence of physiological amounts of extracellular [Ca^{2+}] [70]. Only under Ca^{2+}-limiting conditions, with an extracellular [Ca^{2+}] of ~30–50 μM, the authors observed very mild reduction in TCR stimulation–induced SOCE and proliferation of human T cells. Hence, the role of TRPC channels in T cell activation and proliferation awaits further detailed investigation.

STIM1 negatively regulates plasma membrane Ca^{2+} ATPase (PMCA) pump directly or indirectly via a novel 10-transmembrane segment–containing protein, POST (partner of STIM1, TMEM20). Ritchie et al. showed that STIM1 interacts with PMCA and inhibits its Ca^{2+} clearance function, while Quintana et al. suggested that Ca^{2+} buffering by mitochondria in the immunological synapse inhibits PMCA activation [71,72]. In addition, while POST expression did not affect CRAC currents, POST negatively regulated PMCA activity. These studies suggest that both stimulation of CRAC channels and inhibition of PMCA activity at the immunological synapse may be important for generation of sustained, local Ca^{2+} entry required for NFAT activation. As discussed earlier, STIM1 also functions in a completely store-independent manner. STIM1 in the PM interacts with and activates ARC channels independent of Ca^{2+} stores as previously described [73]. Another function of STIM1, which is independent from store depletion, is growth of microtubules. When a growing microtubule tip comes across the ER membrane, STIM1 bound to a microtubule tip–binding protein, EB1, and pulled out a new ER tubule through the "tip attachment complex" mechanism [74,75]. This interaction of STIM1 with EB1 does not affect SOCE and does not require store depletion [74]. Together, STIM1 plays an important role not only in gating of CRAC channels but also in the function of other Ca^{2+} transport proteins including Ca_v, TRPC, PMCA, or ARC channels as well as SOCE-independent functions such as growth of microtubule tips [74,75].

4.8 METHODS USED TO IDENTIFY INTERACTING PARTNERS OF ORAI1 AND STIM1

The history of studying Orai1 and STIM1 began with the identification of the SOCE mechanism and measurement of CRAC currents by electrophysiological methods of whole-cell patch-clamp recording by depleting Ca^{2+} stores with Ca^{2+}-chelating reagents or a blocker for the sarcoplasmic/endoplasmic reticulum Ca^{2+} ATPase

(SERCA) pump, thapsigargin [76] (see Chapter 1). The CRAC channels were later shown to be present in multiple immune cell types including T cells and mast cells, identified by their unique biophysical properties; however, the molecular components remained unknown for more than two decades. Orai1 and STIM1 were identified by siRNA-mediated functional screens when the discovery of siRNA allowed large-scale loss-of-function studies. Limited RNAi screens in *Drosophila* and HeLa cells identified STIM1 [14–16], and genome-wide RNAi screens in *Drosophila* cells identified *Drosophila* Orai first, which led to revealing its human homologues; among them, Orai1 as a pore subunit of the CRAC channels [10–13]. Two types of readout were used for these screens: direct measurement of SOCE by Ca^{2+} indicators (e.g., Fluo-4) and the use of translocation of GFP-fused NFAT into the nucleus as a readout for functional SOCE pathway. The benefits of using SOCE measurement as readout include a high chance of picking up direct components and identification of potential regulators of SOCE. Shortcomings can be potentially high false positives due to difficulties in plate handling for live-cell Ca^{2+} measurement, and a less robust signal-to-noise ratio between resting and stimulated conditions. On the contrary, NFAT translocation assay as readout yielded robust results due to the ease of plate handling after fixing the cells and very high signal-to-noise ratio that resulted in <30 candidates from a whole genome-wide screen. The shortcomings include the absolute necessity for a secondary screen using Ca^{2+} measurement because it is possible that the depletion of non-SOCE-involved components of the NFAT pathway (e.g., calcineurin) can influence its nuclear translocation.

After identification of Orai1 and STIM1, numerous interacting partners have been identified from protein purifications using affinity tags (e.g., FLAG) [33], biotin-labeling techniques [54], plasmon resonance assays [77], and targeted or genome-scale high-throughput screens [55]. Functions of interacting partners of Orai1 and STIM1 in SOCE were validated by single-cell Ca^{2+} measurement as well as electrophysiological tools of CRAC current measurements when they act on channel gating or inactivation. In addition, SOCE levels can be measured by flow cytometry using ratiometric Ca^{2+} indicators such as Indo-1 to measure responses from a large population of cells. For example, one can label a mixture of cells with various cell type–specific markers (e.g., CD4+ T cells, CD8+ T cells, and B cells) and measure SOCE from all of them simultaneously. Later, during data analysis, one can specifically examine the cell type of their choice based on the expression of various surface markers. However, it is technically difficult to accurately determine the basal and ER Ca^{2+} levels using this method. Single-cell ratiometric Ca^{2+} imaging using microscopy is the method of choice to accurately measure basal, ER [Ca^{2+}], as well as SOCE. The physiological outcomes of Ca^{2+} signaling can be measured by checking nuclear translocation or dephosphorylation of NFATc2, or induction of NFATc1 expression [78,79]. To determine the roles of these associating molecules in translocation, clustering, and dissociation of Orai1 and STIM1, high-resolution confocal and total internal reflection fluorescence (TIRF) microscopy imaging are being commonly used. TIRF imaging is a very useful tool to determine the rate of association and dissociation of Orai1 and STIM1 at the ER–PM junctions and monitor if interacting partners alter this rate or translocate together with Orai/STIM. The shortcomings of these imaging techniques are an extensive use of overexpression system because these tools require fluorescently labeled Orai1/STIM1 and interacting molecules. It is difficult to uncover roles of these regulators in SOCE, when Orai1 and STIM1 are overexpressed, because overexpression of Orai1 and STIM1 are necessary and sufficient to restore CRAC currents in various cell types. The regulators are likely to play an important role in modulating the function of Orai/STIM when their expression is limiting, as observed in physiological conditions. These problems can be partly resolved by selecting cells with mild expression of Orai1 and STIM1 for imaging or using stable cell lines, which express proteins at much lower levels than transient transfection, by generation of mutants defective in dominant functions (e.g., STIM1ΔK) [50], and also by comparing phenotypes with loss-of-function studies using more physiological experimental settings without overexpression, for example, checking the effect on endogenous Orai1 or STIM1.

To determine protein interactions of regulators with Orai1 and STIM1, GST pull-down and immunoprecipitation analyses have been widely used to establish a foundation for other techniques. When combined with mutational analyses within specific domains, it can provide useful information about the site of interaction between various molecules and unravel the working mechanism of regulators. These are common and conventional methods that have been used to determine protein interactions for many decades and are still widely used due to their straightforwardness, sensitivity, and reproducibility. The shortcoming of these techniques is that they are very laborious, and it is difficult to quantify the results from multiple independent experiments (e.g., biological replicates). In addition, standardization of these biochemical techniques often requires trial and error with different kinds and concentrations of detergents, concentrations for monovalent or divalent cations, incubation duration, and the amount of proteins. Negative control proteins play a very important role here, because both Orai1 and STIM1 contain transmembrane segments. Appropriate controls/mutants will be important to validate authentic protein interactions. These techniques can be combined with cell-based assays such as FRET or biomolecular fluorescence complementation [80,81]. These techniques allow validating protein interactions in intact cells, but this is only possible when detailed topology of candidate molecules is clearly known for appropriate design of experiments. Therefore, these methods cannot be the best choice for any primary or secondary screening to identify novel proteins with unknown topology. In summary, the usage of these biochemical and imaging techniques to determine protein interactions has tremendously advanced our understanding of the mechanism of CRAC channel regulation, and they have proven to be very powerful tools, especially when combined with relevant functional assays.

4.9 CONCLUSIONS AND PERSPECTIVES

The activation mechanism of CRAC channels provides a new paradigm of channel activation that solely depends on protein interactions. The studies on conformational changes of STIM1 from the folded inactive state to the open active state provide a fundamental insight into the mechanism of how signal is transmitted from sensing ER Ca^{2+} depletion to activation of Orai1. Identification of Orai1 and STIM1 as components of the CRAC channels has allowed the uncovering the ubiquitous role of SOCE in multiple cell types including immune cells, platelets, keratinocytes, osteoblasts, cardiac myocytes, skeletal muscle cells, and even neuronal cells. Some regulators of Orai1 and STIM1 also show tissue-selective expression patterns, which may determine the unique properties of CRAC channels in T cells. For example, CRACR2A and junctophilin-4 are highly abundant in T cells to selectively support CRAC channel activities [34,51]. While we mostly focus on protein interactions in regulation of Orai1 and STIM1 functions here, second messengers such as cAMP, the phospholipid PIP_2, or reactive oxygen species are also able to regulate Orai1 and STIM1 functions. Furthermore, posttranslational mechanisms such as phosphorylation and glycosylation covered elsewhere [7] also play an important role in regulation of Orai1 and STIM1 functions and should be considered as topics for future investigations.

Ion channels are considered excellent drug targets because many of them are localized at the cell surface providing a relatively easily accessible target for the drug. There is a notion that Ca^{2+} signaling is broadly involved in most cellular activities because Ca^{2+} is a ubiquitous second messenger for many signaling pathways. However, numerous studies argue that cells have an amazing capability to distinguish minor differences in the agonists, amplitude, timing, duration, oscillation frequency, and location (e.g., microdomain) of Ca^{2+} signaling and to generate unique downstream responses [82,83] (see Chapter 5). Cell type–specific Ca^{2+} response can also be mediated by unique expression of regulatory proteins as discussed earlier. The next challenging question is to develop better therapy for immune and inflammatory disorders by modulation of Ca^{2+} signaling based on these studies, and the core of such trials is to understand the accurate composition of Orai1–STIM1 complex with its auxiliary proteins, specifically formed in immune cells.

ACKNOWLEDGMENTS

This work was supported by the National Institutes of Health grants AI-083432 and AI-109059 (Y.G.) and a scientist development grant from the American Heart Association, 12SDG12040188 (S.S.).

REFERENCES

1. Lioudyno MI, Kozak JA, Penna A, Safrina O, Zhang SL, Sen D, Roos J, Stauderman KA, Cahalan MD. 2008. Orai1 and STIM1 move to the immunological synapse and are up-regulated during T cell activation. *Proc Natl Acad Sci USA* 105: 2011–2016.
2. Berridge MJ, Bootman MD, Roderick HL. 2003. Calcium signalling: Dynamics, homeostasis and remodelling. *Nat Rev Mol Cell Biol* 4: 517–529.
3. Lewis RS. 2011. Store-operated calcium channels: New perspectives on mechanism and function. *Cold Spring Harb Perspect Biol* 3: a003970.
4. Putney JW. 2009. Capacitative calcium entry: From concept to molecules. *Immunol Rev* 231: 10–22.
5. Cahalan MD. 2009. STIMulating store-operated Ca^{2+} entry. *Nat Cell Biol* 11: 669–677.
6. Hogan PG, Lewis RS, Rao A. 2010. Molecular basis of calcium signaling in lymphocytes: STIM and ORAI. *Annu Rev Immunol* 28: 491–533.
7. Srikanth S, Gwack Y. 2013. Orai1-NFAT signalling pathway triggered by T cell receptor stimulation. *Mol Cell* 35: 182–194.

8. Wang H, Kadlecek TA, Au-Yeung BB, Goodfellow HE, Hsu LY, Freedman TS, Weiss A. 2010. ZAP-70: An essential kinase in T-cell signaling. *Cold Spring Harb Perspect Biol* 2: a002279.
9. Feske S. 2010. CRAC channelopathies. *Pflugers Arch* 460: 417–435.
10. Feske S, Gwack Y, Prakriya M, Srikanth S, Puppel SH, Tanasa B, Hogan PG, Lewis RS, Daly M, Rao A. 2006. A mutation in Orai1 causes immune deficiency by abrogating CRAC channel function. *Nature* 441: 179–185.
11. Vig M, Peinelt C, Beck A, Koomoa DL, Rabah D, Koblan-Huberson M, Kraft S et al. 2006. CRACM1 is a plasma membrane protein essential for store-operated Ca^{2+} entry. *Science* 312: 1220–1223.
12. Zhang SL, Yeromin AV, Zhang XH, Yu Y, Safrina O, Penna A, Roos J, Stauderman KA, Cahalan MD. 2006. Genome-wide RNAi screen of Ca^{2+} influx identifies genes that regulate Ca^{2+} release-activated Ca^{2+} channel activity. *Proc Natl Acad Sci USA* 103: 9357–9362.
13. Gwack Y, Srikanth S, Feske S, Cruz-Guilloty F, Oh-hora M, Neems DS, Hogan PG, Rao A. 2007. Biochemical and functional characterization of Orai proteins. *J Biol Chem* 282: 16232–16243.
14. Liou J, Kim ML, Heo WD, Jones JT, Myers JW, Ferrell JE, Jr., Meyer T. 2005. STIM is a Ca^{2+} sensor essential for Ca^{2+}-store-depletion-triggered Ca^{2+} influx. *Curr Biol* 15: 1235–1241.
15. Roos J, DiGregorio PJ, Yeromin AV, Ohlsen K, Lioudyno M, Zhang S, Safrina O et al. 2005. STIM1, an essential and conserved component of store-operated Ca^{2+} channel function. *J Cell Biol* 169: 435–445.
16. Zhang SL, Yu Y, Roos J, Kozak JA, Deerinck TJ, Ellisman MH, Stauderman KA, Cahalan MD. 2005. STIM1 is a Ca^{2+} sensor that activates CRAC channels and migrates from the Ca^{2+} store to the plasma membrane. *Nature* 437: 902–905.
17. Stathopulos PB, Li GY, Plevin MJ, Ames JB, Ikura M. 2006. Stored Ca^{2+} depletion-induced oligomerization of stromal interaction molecule 1 (STIM1) via the EF-SAM region: An initiation mechanism for capacitive Ca^{2+} entry. *J Biol Chem* 281: 35855–35862.
18. Yuan JP, Zeng W, Dorwart MR, Choi YJ, Worley PF, Muallem S. 2009. SOAR and the polybasic STIM1 domains gate and regulate Orai channels. *Nat Cell Biol* 11: 337–343.
19. Park CY, Hoover PJ, Mullins FM, Bachhawat P, Covington ED, Raunser S, Walz T, Garcia KC, Dolmetsch RE, Lewis RS. 2009. STIM1 clusters and activates CRAC channels via direct binding of a cytosolic domain to Orai1. *Cell* 136: 876–890.
20. Korzeniowski MK, Manjarres IM, Varnai P, Balla T. 2010. Activation of STIM1-Orai1 involves an intramolecular switching mechanism. *Sci Signal* 3: ra82.
21. Srikanth S, Gwack Y. 2012. Orai1, STIM1, and their associating partners. *J Physiol* 590: 4169–4177.
22. Srikanth S, Gwack Y. 2013. Molecular regulation of the pore component of CRAC channels, Orai1. *Curr Top Membr* 71: 181–207.
23. Mignen O, Thompson JL, Shuttleworth TJ. 2008. Orai1 subunit stoichiometry of the mammalian CRAC channel pore. *J Physiol* 586: 419–425.
24. Penna A, Demuro A, Yeromin AV, Zhang SL, Safrina O, Parker I, Cahalan MD. 2008. The CRAC channel consists of a tetramer formed by Stim-induced dimerization of Orai dimers. *Nature* 456: 116–120.
25. Demuro A, Penna A, Safrina O, Yeromin AV, Amcheslavsky A, Cahalan MD, Parker I. 2011. Subunit stoichiometry of human Orai1 and Orai3 channels in closed and open states. *Proc Natl Acad Sci USA* 108: 17832–17837.
26. Madl J, Weghuber J, Fritsch R, Derler I, Fahrner M, Frischauf I, Lackner B, Romanin C, Schutz GJ. 2010. Resting state Orai1 diffuses as homotetramer in the plasma membrane of live mammalian cells. *J Biol Chem* 285: 41135–41142.

27. Ji W, Xu P, Li Z, Lu J, Liu L, Zhan Y, Chen Y, Hille B, Xu T, Chen L. 2008. Functional stoichiometry of the unitary calcium-release-activated calcium channel. *Proc Natl Acad Sci USA* 105: 13668–13673.
28. Zhang SL, Yeromin AV, Hu J, Amcheslavsky A, Zheng H, Cahalan MD. 2011. Mutations in Orai1 transmembrane segment 1 cause STIM1-independent activation of Orai1 channels at glycine 98 and channel closure at arginine 91. *Proc Natl Acad Sci USA* 108: 17838–17843.
29. McNally BA, Somasundaram A, Yamashita M, Prakriya M. 2012. Gated regulation of CRAC channel ion selectivity by STIM1. *Nature* 482: 241–245.
30. Mullins FM, Park CY, Dolmetsch RE, Lewis RS. 2009. STIM1 and calmodulin interact with Orai1 to induce Ca^{2+}-dependent inactivation of CRAC channels. *Proc Natl Acad Sci USA* 106: 15495–15500.
31. Hou X, Pedi L, Diver MM, Long SB. 2012. Crystal structure of the calcium release-activated calcium channel Orai. *Science* 338: 1308–1313.
32. Mullins FM, Yen M, Lewis RS. 2016. Orai1 pore residues control CRAC channel inactivation independently of calmodulin. *J Gen Physiol* 147: 137–152.
33. Srikanth S, Jung HJ, Kim KD, Souda P, Whitelegge J, Gwack Y. 2010. A novel EF-hand protein, CRACR2A, is a cytosolic Ca^{2+} sensor that stabilizes CRAC channels in T cells. *Nat Cell Biol* 12: 436–446.
34. Srikanth S, Kim KD, Gao Y, Woo JS, Ghosh S, Calmettes G, Paz A, Abramson J, Jiang M, Gwack Y. 2016. A large Rab GTPase encoded by CRACR2A is a component of subsynaptic vesicles that transmit T cell activation signals. *Sci Signal* 9: ra31.
35. Wilson LA, McKeown L, Tumova S, Li J, Beech DJ. 2015. Expression of a long variant of CRACR2A that belongs to the Rab GTPase protein family in endothelial cells. *Biochem Biophys Res Commun* 456: 398–402.
36. Miao Y, Miner C, Zhang L, Hanson PI, Dani A, Vig M. 2013. An essential and NSF independent role for alpha-SNAP in store-operated calcium entry. *Elife* 2: e00802.
37. Willoughby D, Everett KL, Halls ML, Pacheco J, Skroblin P, Vaca L, Klussmann E, Cooper DM. 2012. Direct binding between Orai1 and AC8 mediates dynamic interplay between Ca^{2+} and cAMP signaling. *Sci Signal* 5: ra29.
38. Yu F, Sun L, Machaca K. 2009. Orai1 internalization and STIM1 clustering inhibition modulate SOCE inactivation during meiosis. *Proc Natl Acad Sci USA* 106: 17401–17406.
39. Yu F, Sun L, Machaca K. 2010. Constitutive recycling of the store-operated Ca^{2+} channel Orai1 and its internalization during meiosis. *J Cell Biol* 191: 523–535.
40. Hodeify R, Selvaraj S, Wen J, Arredouani A, Hubrack S, Dib M, Al-Thani SN, McGraw T, Machaca K. 2015. A STIM1-dependent 'trafficking trap' mechanism regulates Orai1 plasma membrane residence and Ca^{2+} influx levels. *J Cell Sci* 128: 3143–3154.
41. Feng M, Grice DM, Faddy HM, Nguyen N, Leitch S, Wang Y, Muend S et al. 2010. Store-independent activation of Orai1 by SPCA2 in mammary tumors. *Cell* 143: 84–98.
42. Mignen O, Thompson JL, Shuttleworth TJ. 2009. The molecular architecture of the arachidonate-regulated Ca^{2+}-selective ARC channel is a pentameric assembly of Orai1 and Orai3 subunits. *J Physiol* 587: 4181–4197.
43. Mignen O, Thompson JL, Shuttleworth TJ. 2007. STIM1 regulates Ca^{2+} entry via arachidonate-regulated Ca^{2+}-selective (ARC) channels without store depletion or translocation to the plasma membrane. *J Physiol* 579: 703–715.
44. Albarran L, Lopez JJ, Woodard GE, Salido GM, Rosado JA. 2016. Store-operated Ca^{2+} entry-associated regulatory factor (SARAF) plays an important role in the regulation of arachidonate-regulated Ca^{2+} (ARC) channels. *J Biol Chem* 291: 6982–6988.
45. Wu MM, Buchanan J, Luik RM, Lewis RS. 2006. Ca^{2+} store depletion causes STIM1 to accumulate in ER regions closely associated with the plasma membrane. *J Cell Biol* 174: 803–813.

46. Varnai P, Toth B, Toth DJ, Hunyady L, Balla T. 2007. Visualization and manipulation of plasma membrane-endoplasmic reticulum contact sites indicates the presence of additional molecular components within the STIM1–Orai1 complex. *J Biol Chem* 282: 29678–29690.
47. Carrasco S, Meyer T. 2011. STIM proteins and the endoplasmic reticulum-plasma membrane junctions. *Annu Rev Biochem* 80: 973–1000.
48. Takeshima H, Komazaki S, Nishi M, Iino M, Kangawa K. 2000. Junctophilins: A novel family of junctional membrane complex proteins. *Mol Cell* 6: 11–22.
49. Weisleder N, Takeshima H, Ma J. 2008. Immuno-proteomic approach to excitation–contraction coupling in skeletal and cardiac muscle: Molecular insights revealed by the mitsugumins. *Cell Calcium* 43: 1–8.
50. Srikanth S, Jew M, Kim KD, Yee MK, Abramson J, Gwack Y. 2012. Junctate is a Ca^{2+}-sensing structural component of Orai1 and stromal interaction molecule 1 (STIM1). *Proc Natl Acad Sci USA* 109: 8682–8687.
51. Woo JS, Srikanth S, Nishi M, Ping P, Takeshima H, Gwack Y. 2016. Junctophilin-4, a component of the endoplasmic reticulum-plasma membrane junctions, regulates Ca^{2+} dynamics in T cells. *Proc Natl Acad Sci USA* 113: 2762–2767.
52. Nishi M, Sakagami H, Komazaki S, Kondo H, Takeshima H. 2003. Coexpression of junctophilin type 3 and type 4 in brain. *Brain Res Mol Brain Res* 118: 102–110.
53. Garbino A, van Oort RJ, Dixit SS, Landstrom AP, Ackerman MJ, Wehrens XH. 2009. Molecular evolution of the junctophilin gene family. *Physiol Genomics* 37: 175–186.
54. Jing J, He L, Sun A, Quintana A, Ding Y, Ma G, Tan P et al. 2015. Proteomic mapping of ER-PM junctions identifies STIMATE as a regulator of Ca^{2+} influx. *Nat Cell Biol* 17: 1339–1347.
55. Quintana A, Rajanikanth V, Farber-Katz S, Gudlur A, Zhang C, Jing J, Zhou Y, Rao A, Hogan PG. 2015. TMEM110 regulates the maintenance and remodeling of mammalian ER-plasma membrane junctions competent for STIM-ORAI signaling. *Proc Natl Acad Sci USA* 112: E7083–E7092.
56. Giordano F, Saheki Y, Idevall-Hagren O, Colombo SF, Pirruccello M, Milosevic I, Gracheva EO, Bagriantsev SN, Borgese N, De Camilli P. 2013. PI(4,5)P(2)-dependent and Ca^{2+}-regulated ER-PM interactions mediated by the extended synaptotagmins. *Cell* 153: 1494–1509.
57. Palty R, Raveh A, Kaminsky I, Meller R, Reuveny E. 2012. SARAF inactivates the store operated calcium entry machinery to prevent excess calcium refilling. *Cell* 149: 425–438.
58. Jha A, Ahuja M, Maleth J, Moreno CM, Yuan JP, Kim MS, Muallem S. 2013. The STIM1 CTID domain determines access of SARAF to SOAR to regulate Orai1 channel function. *J Cell Biol* 202: 71–79.
59. Wang Y, Deng X, Mancarella S, Hendron E, Eguchi S, Soboloff J, Tang XD, Gill DL. 2010. The calcium store sensor, STIM1, reciprocally controls Orai and $Ca_V1.2$ channels. *Science* 330: 105–109.
60. Park CY, Shcheglovitov A, Dolmetsch R. 2010. The CRAC channel activator STIM1 binds and inhibits L-type voltage-gated calcium channels. *Science* 330: 101–105.
61. Badou A, Basavappa S, Desai R, Peng YQ, Matza D, Mehal WZ, Kaczmarek LK, Boulpaep EL, Flavell RA. 2005. Requirement of voltage-gated calcium channel beta4 subunit for T lymphocyte functions. *Science* 307: 117–121.
62. Badou A, Jha MK, Matza D, Mehal WZ, Freichel M, Flockerzi V, Flavell RA. 2006. Critical role for the beta regulatory subunits of Cav channels in T lymphocyte function. *Proc Natl Acad Sci USA* 103: 15529–15534.
63. Jha MK, Badou A, Meissner M, McRory JE, Freichel M, Flockerzi V, Flavell RA. 2009. Defective survival of naive CD8+ T lymphocytes in the absence of the beta3 regulatory subunit of voltage-gated calcium channels. *Nat Immunol* 10: 1275–1282.

64. Omilusik K, Priatel JJ, Chen X, Wang YT, Xu H, Choi KB, Gopaul R et al. 2011. The Ca(v)1.4 calcium channel is a critical regulator of T cell receptor signaling and naive T cell homeostasis. *Immunity* 35: 349–360.
65. Zeng W, Yuan JP, Kim MS, Choi YJ, Huang GN, Worley PF, Muallem S. 2008. STIM1 gates TRPC channels, but not Orai1, by electrostatic interaction. *Mol Cell* 32: 439–448.
66. Huang GN, Zeng W, Kim JY, Yuan JP, Han L, Muallem S, Worley PF. 2006. STIM1 carboxyl-terminus activates native SOC, I(crac) and TRPC1 channels. *Nat Cell Biol* 8: 1003–1010.
67. Worley PF, Zeng W, Huang GN, Yuan JP, Kim JY, Lee MG, Muallem S. 2007. TRPC channels as STIM1-regulated store-operated channels. *Cell Calcium* 42: 205–211.
68. Ong HL, Cheng KT, Liu X, Bandyopadhyay BC, Paria BC, Soboloff J, Pani B et al. 2007. Dynamic assembly of TRPC1-STIM1-Orai1 ternary complex is involved in store-operated calcium influx. Evidence for similarities in store-operated and calcium release-activated calcium channel components. *J Biol Chem* 282: 9105–9116.
69. DeHaven WI, Jones BF, Petranka JG, Smyth JT, Tomita T, Bird GS, Putney JW, Jr. 2009. TRPC channels function independently of STIM1 and Orai1. *J Physiol* 587: 2275–2298.
70. Wenning AS, Neblung K, Strauss B, Wolfs MJ, Sappok A, Hoth M, Schwarz EC. 2011. TRP expression pattern and the functional importance of TRPC3 in primary human T-cells. *Biochim Biophys Acta* 1813: 412–423.
71. Quintana A, Pasche M, Junker C, Al-Ansary D, Rieger H, Kummerow C, Nunez L et al. 2011. Calcium microdomains at the immunological synapse: How ORAI channels, mitochondria and calcium pumps generate local calcium signals for efficient T-cell activation. *EMBO J* 30: 3895–3912.
72. Ritchie MF, Samakai E, Soboloff J. 2012. STIM1 is required for attenuation of PMCA-mediated Ca^{2+} clearance during T-cell activation. *EMBO J* 31: 1123–1133.
73. Thompson JL, Mignen O, Shuttleworth TJ. 2013. The ARC channel—An endogenous store-independent Orai channel. *Curr Top Membr* 71: 125–148.
74. Grigoriev I, Gouveia SM, van der Vaart B, Demmers J, Smyth JT, Honnappa S, Splinter D et al. 2008. STIM1 is a MT-plus-end-tracking protein involved in remodeling of the ER. *Curr Biol* 18: 177–182.
75. Honnappa S, Gouveia SM, Weisbrich A, Damberger FF, Bhavesh NS, Jawhari H, Grigoriev I et al. 2009. An EB1-binding motif acts as a microtubule tip localization signal. *Cell* 138: 366–376.
76. Putney JW, Jr. 1986. A model for receptor-regulated calcium entry. *Cell Calcium* 7: 1–12.
77. Prins D, Groenendyk J, Touret N, Michalak M. 2011. Modulation of STIM1 and capacitative Ca^{2+} entry by the endoplasmic reticulum luminal oxidoreductase ERp57. *EMBO Rep* 12: 1182–1188.
78. Kim KD, Srikanth S, Tan YV, Yee MK, Jew M, Damoiseaux R, Jung ME et al. 2014. Calcium signaling via Orai1 is essential for induction of the nuclear orphan receptor pathway to drive Th17 differentiation. *J Immunol* 192: 110–122.
79. Kim KD, Srikanth S, Yee MK, Mock DC, Lawson GW, Gwack Y. 2011. ORAI1 deficiency impairs activated T cell death and enhances T cell survival. *J Immunol* 187: 3620–3630.
80. Piston DW, Kremers GJ. 2007. Fluorescent protein FRET: The good, the bad and the ugly. *Trends Biochem Sci* 32: 407–414.
81. Kerppola TK. 2008. Bimolecular fluorescence complementation (BiFC) analysis as a probe of protein interactions in living cells. *Annu Rev Biophys* 37: 465–487.
82. Parekh AB. 2008. Ca^{2+} microdomains near plasma membrane Ca^{2+} channels: Impact on cell function. *J Physiol* 586: 3043–3054.
83. Parekh AB. 2011. Decoding cytosolic Ca^{2+} oscillations. *Trends Biochem Sci* 36: 78–87.

84. Walsh CM, Doherty MK, Tepikin AV, Burgoyne RD. 2010. Evidence for an interaction between Golli and STIM1 in store-operated calcium entry. *Biochem J* 430: 453–460.
85. Woodward OM, Li Y, Yu S, Greenwell P, Wodarczyk C, Boletta A, Guggino WB, Qian F. 2010. Identification of a polycystin-1 cleavage product, P100, that regulates store operated Ca entry through interactions with STIM1. *PLoS One* 5: e12305.
86. Zeiger W, Ito D, Swetlik C, Oh-hora M, Villereal ML, Thinakaran G. 2011. Stanniocalcin 2 is a negative modulator of store-operated calcium entry. *Mol Cell Biol* 31: 3710–3722.
87. Krapivinsky G, Krapivinsky L, Stotz SC, Manasian Y, Clapham DE. 2011. POST, partner of stromal interaction molecule 1 (STIM1), targets STIM1 to multiple transporters. *Proc Natl Acad Sci USA* 108: 19234–19239.
88. Zhao G, Li T, Brochet DX, Rosenberg PB, Lederer WJ. 2015. STIM1 enhances SR Ca^{2+} content through binding phospholamban in rat ventricular myocytes. *Proc Natl Acad Sci USA* 112: E4792–E4801.

5 CRAC Channels and Ca^{2+}-Dependent Gene Expression

Yi-Chun Yeh and Anant B. Parekh

CONTENTS

5.1 INTRODUCTION

In order to respond to changes in the environment, cells need to switch genes on and off. One conserved mechanism that links events at the cell surface to gene expression in the nucleus employs intracellular Ca^{2+}. A cytoplasmic Ca^{2+} rise stimulates Ca^{2+}-dependent transcription factors, which then regulate gene activity. In neurons, Ca^{2+} entry through voltage-gated Ca^{2+} channels activates Ca^{2+}-calmodulin-dependent protein kinases, leading to the phosphorylation of the transcription factor CREB (cAMP response-element binding protein) [1,2]. Phosphorylated CREB is thought to be involved in long-term potentiation [3], a form of learning and memory in the nervous system. In immune cells, transcription factors such as c-fos and the nuclear factor of activated T cells (NFAT) are activated by Ca^{2+} influx through store-operated Ca^{2+}-release-activated Ca^{2+} (CRAC) channels [4,5]. These transcription factors often work in tandem to control the expression of chemokines and cytokines that are involved in both the innate and adaptive immune responses [6]. NFAT proteins are a family of five cytosolic proteins that migrate into the nucleus upon activation. For four of the members, translocation is triggered by a rise in cytoplasmic Ca^{2+} concentration [5,7]. NFATs are extensively phosphorylated at rest, resulting in the masking of a nuclear localization sequence. A rise in Ca^{2+} activates the enzyme calcineurin, the target of immunosuppressants cyclosporine A and tacrolimus. Dephosphorylation exposes nuclear localization sequences, enabling NFAT to migrate into the nucleus in a complex with tubulin alpha [8]. Activator protein-1 (AP-1) is a transcription complex composed of members of the fos and jun families that can associate to form a range of homo- and heterodimers [9]. AP-1 activity can be increased by the stimulation of preexisting protein or following enhanced expression and many extracellular signals including G-protein-coupled receptor agonists, cytokines, and growth factors target AP-1 activity. Here, the involvement of cytoplasmic Ca^{2+}, particularly localized Ca^{2+} signals or Ca^{2+} microdomains, in activating NFAT and c-fos will be described. The focus is on immune cells, where these processes have been studied in considerable detail.

5.2 Ca^{2+} ENTRY THROUGH CRAC CHANNELS ACTIVATES GENE EXPRESSION

In immune cells, it has been known for many years that a rise in cytoplasmic Ca^{2+} is essential for expression of various chemokines and cytokines that drive T cell proliferation and the subsequent immune response [10]. Although cells can raise Ca^{2+} through either Ca^{2+} release from internal stores or Ca^{2+} influx across the plasma membrane, Ca^{2+} entry was found to play the major role in immune cell activation. Stimulation in the absence of external Ca^{2+} often failed to evoke any detectable gene expression, whereas robust transcription occurred when external Ca^{2+} was present. For example, studies on the Jurkat human T cell line revealed that antibodies against the T lymphocyte antigen receptor complex evoked Ca^{2+} release from the stores but this failed to induce interleukin 2 (IL-2) expression [11]. In the murine T cell hybridoma B3Z expressing the lacZ reporter gene under the transcriptional control of NFAT, the fraction of lacZ-positive cells fell after stimulation when external Ca^{2+} was reduced [12]. Further evidence for the major role of Ca^{2+} influx in gene expression came from studies on T cells isolated from a 3-month-old immunodeficient boy [13]. Stimulation with anti-CD3 antibody evoked less IL-2 production than in the control and this was pinpointed to impaired Ca^{2+} influx. By contrast, Ca^{2+} release was unaffected. Similar conclusions were reached in rat bone marrow–derived mast cells and RBL-2H3 immortalized mast cell line [14]. Cross-linking of FCεRI receptors led to the activation of NFAT and this was suppressed by the removal of external Ca^{2+}. Collectively, these and other studies established the importance of Ca^{2+} influx as a key activator of NFAT and transcription of IL-2.

Cells express a plethora of Ca^{2+} entry channels [15], and each could potentially underlie the Ca^{2+} influx needed for gene expression in immune cells. Several studies, using different approaches, identified the CRAC channel as being the dominant plasmalemmal Ca^{2+} channel that coupled to NFAT activation and gene expression. First, noise analysis of the whole cell Ca^{2+} current in T cells revealed that the chord conductance of the Ca^{2+} channel activated by the T cell receptor was identical to that evoked by thapsigargin [16] (see Chapter 1). Hence, the CRAC channel and the Ca^{2+} channel activated by the T cell receptor, which leads to NFAT activation, are one and the same. Second, the elevation of external K^+ depolarizes the membrane potential and this was found to reduce Ca^{2+} influx and NFAT reporter gene expression [12] (see Figure 1.1). Because the open probability of voltage-gated Ca^{2+} channels increases as the membrane potential decreases, these results established that the Ca^{2+} influx pathway was not depolarization activated. Third, patch clamp studies on T cells from an immunodeficient infant, in which T cell proliferation was lost, revealed the complete absence of the non-voltage-activated CRAC current following stimulation of either the T cell receptor or after store depletion with thapsigargin [17]. Fourth, mutant cells selected from a population of gamma-irradiated Jurkat NZDipA cells showed defective CRAC channel activity to varying degrees, and the severity of the defect correlated well with the extent of compromised NFAT-dependent reporter gene expression [18]. Collectively, these findings constructed a body of evidence that suggested Ca^{2+} entry through the CRAC channel was required for NFAT activation followed by gene expression. Direct evidence in support of this came when the CRAC channel components were identified at a molecular level.

The first critical component to be identified was the stromal interaction molecule (STIM) family of proteins, STIM1 and STIM2 [19,20]. STIM proteins are found mainly spanning the endoplasmic reticulum, where they function as Ca^{2+} sensors to monitor the Ca^{2+} content of the store [21]. Upon store depletion, they form multimers that migrate to specialized regions of the endoplasmic reticulum located just beneath the plasma membrane [22,23]. Here, they bind to and activate CRAC channels in the plasma membrane.

Gene linkage analysis and siRNA knockdown strategies identified Orai1 as being essential for CRAC channel function [24–26] (see Chapters 2 and 3). Orai1 is a plasma membrane protein with four transmembrane domains. The mutation of the conserved acidic residue E106 in transmembrane domain reduced Ca^{2+} influx and changed ion selectivity, establishing Orai1 as the pore-forming subunit of the CRAC channel [27–29]. A point mutation, R91W, in Orai1 was identified in patients with one form of hereditary severe combined immunodeficiency [25]. This mutant channel was expressed in the plasma membrane but was unable to gate. Importantly, the expression of wild-type Orai1 rescued Ca^{2+} influx in the immunodeficient T cells. Consistent with a fundamental role for Orai1 in T cells, the expression of pore mutant E106Q Orai1 channels in T cells reduced cell proliferation and cytokine secretion [30]. Furthermore, the overexpression of the recombinant Orai1 channel in HEK cells accelerated NFAT activation and reporter gene expression, whereas the knockdown of endogenous Orai1 prevented NFAT activation and gene expression in response to thapsigargin [31]. Moreover, the knockdown of Orai1 suppressed NFAT-driven gene expression following stimulation of either FCεF1 or cysteinyl leukotriene type I G-protein-coupled receptors, demonstrating the physiological relevance of CRAC channels to this form of excitation–transcription coupling [32]. Finally, a relatively specific CRAC channel blocker, Synta66 [33], inhibited NFAT activation following the stimulation of native CRAC channels [31].

In addition to activating NFAT, CRAC channels also induce the expression of the immediate early gene c-fos [34], a component of the AP-1 transcription factor complex. AP-1 and NFAT are considered partners in transcriptional regulation because composite NFAT and AP-1 sites are found on numerous genes that are regulated during the immune response [6].

Stimulation with thapsigargin or leukotriene C_4 (LTC_4), an agonist of G-protein-coupled cysteinyl leukotriene type I receptors, which couple to phospholipase C to generate IP_3 and stimulate protein kinase C, increased c-fos transcription but only when external Ca^{2+} was present [35]. A pharmacological block of CRAC channels with Synta66 also suppressed c-fos induction [35], as did knockdown of Orai1 [36].

5.3 THE IMPORTANCE OF Ca^{2+} MICRODOMAINS NEAR OPEN CRAC CHANNELS IN THE REGULATION OF TRANSCRIPTION

Although Ca^{2+} entry through CRAC channels is important in activating NFAT and c-fos, several lines of evidence show that Ca^{2+} microdomains near the mouth of open channels are more effective in activating the transcription factors than a rise in bulk cytoplasmic Ca^{2+}. First, the Ca^{2+} response to thapsigargin in Ca^{2+}-free external solution is transient because plasma membrane Ca^{2+} clearance mechanisms remove cytoplasmic Ca^{2+}. In rat basophilic leukemia (RBL) and human embryonic kidney (HEK) cells, the inhibition of the plasma membrane Ca^{2+} ATPase (PMCA) pump with La^{3+} leads to a large and sustained rise in bulk Ca^{2+} following stimulation with thapsigargin in a Ca^{2+}-free solution. Although the bulk Ca^{2+} rise was larger than that seen when cells were exposed to thapsigargin in the presence of external Ca^{2+}, c-fos transcription was not stimulated [37]. Interestingly, NFAT1 migration into the nucleus did occur upon challenge with thapsigargin in Ca^{2+}-free solution supplemented with La^{3+}, suggesting a large, nonphysiological bulk cytoplasmic Ca^{2+} rise is sufficient to activate NFAT1 [31]. Second, one important factor that determines the size of Ca^{2+} microdomains is the unitary flux through each channel. This is dictated by the prevailing electrochemical gradient. The K_D for Ca^{2+} permeation through CRAC channels in RBL cells is ~0.7 mM [38]. Therefore, the unitary flux in 0.5 mM external Ca^{2+} should be considerably smaller than in 2 mM Ca^{2+}. When RBL cells were stimulated with thapsigargin in either 0.5 or 2 mM Ca^{2+}, bulk Ca^{2+} was similar but the transcription of c-fos only occurred in the higher external Ca^{2+} [37]. Similar findings were obtained for NFAT activation in HEK cells [31]. Conversely, a reduction of the membrane potential by the inhibition of inward rectifier K^+ channels in RBL cells had little effect on bulk Ca^{2+} but suppressed CRAC channel–driven c-fos expression [37]. Hence, manipulations that reduce the unitary flux and thus the size of the Ca^{2+} microdomain, without compromising the bulk Ca^{2+} rise, impair c-fos and NFAT activation. Third, loading the cytoplasm with the Ca^{2+} chelators EGTA or BAPTA both substantially reduce the rise in bulk cytoplasmic Ca^{2+} following stimulation with thapsigargin. However, EGTA is too slow to prevent the build-up of the Ca^{2+} microdomain and had no inhibitory effect on either c-fos transcription or NFAT activation in RBL or HEK cells [31,37,39]. By contrast, BAPTA has an on-rate for Ca^{2+} ~500-fold faster than EGTA and therefore reduces the size and lateral extent of the microdomain. In support of a central role for Ca^{2+} microdomains, BAPTA was found to impair both c-fos and NFAT activation following CRAC channel opening [31,37]. Consistent with these earlier studies, BAPTA but not EGTA was also found to suppress NFAT activation by CRAC channels in neural stem cells [40]. Finally, a comparison of the signaling capability of clustered Orai1 channels with dispersed constitutively active V102C mutant Orai1 channels showed that channel clustering was far more effective in activating both c-fos and NFAT than a similar number of active but diffuse ones [36]. Although the Ca^{2+} selectivity of the V102C Orai1 channels is slightly less than for the wild-type channel, the bulk Ca^{2+} was nevertheless increased to a similar extent. Clustering appears favored in signal transduction because the overlapping Ca^{2+} microdomains result in a much higher local Ca^{2+} signal that robustly activates downstream signaling pathways [36].

5.4 HOW LOCAL IS LOCAL?

An interesting question concerns how local the Ca^{2+} signal near Orai1 needs to be to activate NFAT. Is the Ca^{2+} signal spatially restricted to Orai1 or is a more general smearing of subplasmalemmal Ca^{2+} important? One way of addressing this is to compare the ability of two different plasma membrane Ca^{2+}-permeable channels that both raise bulk cytoplasmic Ca^{2+} to similar extent to activate NFAT. The activation of recombinant TRPC3 channels with 1-oleoyl-2-acetyl-sn-glycerol (OAG), a diacylglycerol analogue, led to a rise in cytoplasmic Ca^{2+} that was dependent entirely on Ca^{2+} entry [31]. The rise in cytoplasmic Ca^{2+} was similar to that evoked following CRAC channel opening. However, NFAT1 consistently failed to migrate into the nucleus in the presence of OAG [31]. Hence, a general rise in subplasmalemmal Ca^{2+} is not sufficient for NFAT1 activation; rather, it is the Ca^{2+} in the vicinity of CRAC channels that is important.

These findings should not be interpreted to mean that Ca^{2+} entry through TRPC3 is unable to activate NFAT1 in other systems [41]. TRPC3 can form heteromultimers with TRPC1 [42], and TRPC1 channels can be inserted into the plasma membrane following local Ca^{2+} entry through Orai1 [43], suggesting that TRPC and Orai1 channels might colocalize in the membrane. Nevertheless, in mast cells, the direct opening of recombinant TRPC3 channels with OAG is considerably less effective than CRAC channels in activating NFAT1.

5.5 SENSING LOCAL Ca^{2+} NEAR CRAC CHANNELS

For Ca^{2+} microdomains near CRAC channels to activate transcription factors, a Ca^{2+} sensor is required that relays local Ca^{2+} to a downstream signaling pathway. Interestingly, different signal transduction pathways are used to activate c-fos and NFAT [36,37].

In RBL-1 cells, Ca^{2+} microdomains near CRAC channels stimulate c-fos expression through recruitment of the nonreceptor tyrosine kinase Syk [33,37]. A pharmacological block of Syk and a reduction in protein levels using an siRNA-based approach both inhibited c-fos expression in response to thapsigargin [37] or cysteinyl leukotriene type I receptor stimulation [39]. Ca^{2+} entry through the channels was unaffected, suggesting that the involvement of Syk was downstream of CRAC channel activation. Syk is expressed mainly at the cell periphery and remains so after CRAC channel opening [37]. Co-immunoprecipitation studies have found association between Syk and Orai1 under nonstimulated conditions and this increases ~2-fold after store depletion with thapsigargin [36]. The signal transducers and activators of transcription (STAT) family of transcription factors are widely expressed in the immune system and are activated following phosphorylation by nonreceptor tyrosine kinases [44,45]. Phosphorylated STATs dimerize and then rapidly translocate into the nucleus, where they bind to enhancer elements to regulate gene expression. Following CRAC channel opening, STAT5 was phosphorylated within minutes and this was suppressed by the Syk inhibitor [37]. Hence, Ca^{2+} entry through CRAC channels activates c-fos through the Syk/STAT5 pathway. More recent work using phospho-specific antibodies for the highly related STAT5a and STAT5b proteins found that STAT5a phosphorylation was significantly increased by Ca^{2+} entry through CRAC channels [46]. Basal STAT5b phosphorylation was ~2-fold higher than for STAT5a, but did not change after stimulation.

NFAT activation is mediated through the Ca^{2+}-dependent stimulation of protein phosphatase 2B (calcineurin), the target of immunosuppressants cyclosporine A and tacrolimus. In neurons, the plasmalemmal scaffold protein AKAP79 binds both calcineurin and $Ca_v1.2$ Ca^{2+} channels, thereby bringing the phosphatase within the realm of the Ca^{2+} microdomain [47,48]. Biochemical pull-down studies revealed that Orai1 was associated with calcineurin after store depletion, and this was prevented following the knockdown of AKAP79 [49]. The knockdown of AKAP79 prevented CRAC channels from activating NFAT, an effect that was rescued by the overexpression of AKAP79 but not by a mutant AKAP protein that was unable to bind calcineurin. In addition to binding calcineurin, AKAP79 was also associated with a fraction of the total NFAT pool [50]. These results show that calcineurin and NFAT are brought close to the CRAC channel microdomain after store depletion, providing a mechanism to activate the enzyme selectively while providing a high local concentration of its target NFAT.

5.6 PARALLEL PROCESSING OF THE CRAC CHANNEL Ca^{2+} MICRODOMAIN

Pharmacological and siRNA-based approaches reveal that CRAC channel–gated Ca^{2+} microdomains activate Syk and calcineurin through distinct signaling pathways [36]. The inhibition or knockdown of Syk impaired c-fos expression following CRAC channel opening without affecting NFAT activation. Conversely, the inhibition of calcineurin with cyclosporine A suppressed NFAT activation without affecting c-fos expression. Ca^{2+} microdomains therefore recruit two different transcription factors through distinct signaling mechanisms.

In addition to enhancing c-fos expression, Syk also stimulates extracellular signal–regulated kinases (ERK) via protein kinase C [34,51]. This leads to the phosphorylation of both Ca^{2+}-dependent phospholipase A_2 ($cPLA_2$) and 5-lipoxygenase, increasing activities of both enzymes. $cPLA_2$ hydrolyzes arachidonyl phospholipids to release arachidonic acid, which is then metabolized by 5-lipoxygenase to produce leukotrienes such as LTC_4, a powerful pro-inflammatory signal (see Chapter 11). Although a pharmacological block of Syk suppresses both c-fos transcription and LTC_4 production, ERK inhibition impacted only on LTC_4 production [34]. The parallel processing of the Ca^{2+} microdomain by Syk through two distinct signaling pathways therefore constitutes an effective mechanism to evoke spatially and temporally different cellular responses.

5.7 CAVEOLIN-1 DIFFERENTIALLY REGULATES NFAT AND c-FOS ACTIVITIES

AP-1 and NFAT often interact cooperatively to control gene expression. For example, genes that encode IL-2–4 and granulocyte-macrophage colony-stimulating factor are transcribed when both NFAT and c-fos are present [52]. By contrast, the transcription of some other genes such as interleukin-13 and tumor necrosis factor-α is induced by NFAT alone [52]. Different gene expression programs can therefore be activated depending on whether NFAT and c-fos operate together or in isolation. This raises an interesting question: If the same signal, namely, Ca^{2+} microdomains near open CRAC channels, activates both NFAT and c-fos, then how can one be activated and not the other? A mechanism must exist that is capable of tunneling the Ca^{2+} microdomain to recruit one transcription factor and not the other. Because Ca^{2+} microdomains activate NFAT and c-fos indirectly, via AKAP79 and Syk, respectively, one possibility is that a membrane scaffolding complex that interacts with numerous signaling pathways determines which transcription factor is activated by the Ca^{2+} microdomain. One candidate for this role is caveolin-1, a protein found in the plasma membrane. Caveolin-1 forms a large oligomeric complex [53], interacts with multiple ion channels [54,55], and helps form signalosomes at the cell periphery [56]. Co-immunoprecipitation studies have found that caveolin-1 associates with Orai1 [57], and two potential interaction sites are on the cytoplasmic N terminus between amino acids 52 and 60 [57] and on transmembrane domain 4, between amino acids 250 and 258 [21].

The overexpression of caveolin-1 in RBL mast cells resulted in enhanced store-operated Ca^{2+} entry [46], which seemed to arise through the stabilization of the STIM1–Orai1 interaction. This action required the scaffolding domain of caveolin-1 because mutations within the critical core motif that is required for association with signaling molecules [58,59] abolished the potentiation of Ca^{2+} entry. Because of the increase in Ca^{2+} flux through CRAC channels, a simple expectation would be that both NFAT and c-fos activities should increase in the presence of caveolin-1. However, this was not the case. Although NFAT reporter gene expression increased after the activation of CRAC channels in cells overexpressing caveolin-1, c-fos expression was actually inhibited [46]. The phosphorylation of STAT5 following Ca^{2+} flux through CRAC channels was prevented by caveolin-1 [46]. Therefore, caveolin-1 differentially regulates the activation of the two transcription factors in response to the same local Ca^{2+} stimulus.

5.8 MODULAR REGULATION BY CAVEOLIN-1

One possible explanation for the opposing effects of caveolin-1 on NFAT and c-fos activities relates to the increased Ca^{2+} influx through CRAC channels. c-Fos activity could have a bell-shaped dependence on local Ca^{2+}; in this scenario, Ca^{2+} would initially stimulate c-fos but a further rise would then inhibit expression. By contrast, NFAT activity increases quasi-monotonically with local Ca^{2+} until the pathway reaches its maximal capacity. Several lines of evidence argue against this possibility [46]. First, caveolin-1 expression still resulted in the inhibition of c-fos even when stimulus intensity was reduced. Second, reducing the rise in cytoplasmic Ca^{2+} by loading the cytoplasm with EGTA failed to rescue c-fos expression in the presence of caveolin-1. Third, the application of ionomycin after exposure to thapsigargin in wild-type cells raised cytoplasmic Ca^{2+} to high levels, and this led to a further increase in c-fos expression. This Ca^{2+} rise was greater than that seen in cells expressing caveolin-1 and stimulated with thapsigargin, conditions that failed to activate c-fos. In a model where c-fos exhibits a bell-shaped dependence on Ca^{2+} levels, ionomycin stimulation after thapsigargin should have reduced c-fos expression, not increased it. Finally, mutations within the scaffolding domain prevented the increase in Ca^{2+} influx following store depletion but c-fos activity remained suppressed.

Evidence for modular regulation by caveolin-1 of NFAT and c-fos activities has come from experiments with a tyrosine phosphorylation-resistant mutant [46]. Caveolin-1 is phosphorylated by Src family tyrosine kinases on cytosol-facing amino acid residue tyrosine 14 [60], which is thought to tether the Src SH2 domain to phosphorylated caveolin-1 and thus retain Src at specific plasma membrane regions such as focal adhesions [61]. In RBL cells, recombinant caveolin-1 was phosphorylated on tyrosine 14 under resting conditions. Following stimulation with thapsigargin, phosphorylation was only slightly reduced and c-fos expression was suppressed [46]. Expression of a caveolin-1 protein in which tyrosine 14 had been mutated to phenylalanine (caveolin-1-Y14F) fully rescued STAT5a phosphorylation and c-fos expression following CRAC channel opening [46]. This mutant, like wild-type caveolin-1, increased store-operated Ca^{2+} entry as well as NFAT reporter gene expression.

Tyrosine 14 on the cytoplasmic region of caveolin-1 is the locus for the inhibition of c-fos transcription. The phosphorylation of Tyr14 inhibited the ability of Ca^{2+} entry through CRAC channels to activate STAT5a and thus inhibited c-fos expression in RBL cells. By contrast, Tyr 14 phosphorylation of caveolin-1 had no inhibitory effect on NFAT activation, revealing that the phosphorylation status of this site helps determine whether the same local Ca^{2+} signal is capable of simultaneously activating the two transcription factors.

Caveolin-1 is expressed in CD4+ and CD8+ T lymphocytes [62]. CD8+ T cells deficient in caveolin-1 showed normal NFkB activity but an impaired NFAT pathway [62]. Caveolin-1 expression has been reported to inhibit HIV replication through defective NFkB signaling [63]. Hence, caveolin-1 seems to regulate NFAT and NFkB in T cells in a reciprocal manner in certain situations, reminiscent of the regulation of c-fos and NFAT in mast cells.

5.9 LARGE BULK Ca^{2+} RISES AND C-FOS GENE EXPRESSION

In many studies of Ca^{2+}-dependent regulation of gene expression, high doses of Ca^{2+} ionophores such as ionomycin have been used to raise bulk Ca^{2+} to high levels in order to activate NFAT or c-fos and thereby gene transcription. At low concentrations (sub μM), ionomycin first releases Ca^{2+} from the stores and this leads to the opening of CRAC channels [64]. However, at higher doses (μM), ionomycin additionally increases Ca^{2+} flux across the plasma membrane directly via its ionophoretic activity. In RBL cells, CRAC channel blockers fail to prevent the rise in bulk Ca^{2+} due to Ca^{2+} entry that is elicited by high concentrations of ionomycin [35]. These large increases in bulk Ca^{2+} may elevate Ca^{2+} sufficiently close to the plasma membrane to match the high local Ca^{2+} within the Ca^{2+} microdomain. However, interpretation is not straightforward. In RBL cells in which both CRAC channels and Ca^{2+} extrusion via the PMCA pumps were inhibited with La^{3+}, stimulation with a high dose of ionomycin (5 μM) led to a large increase in cytoplasmic Ca^{2+} and in c-fos transcription [35]. The inhibition of Syk or overexpression of caveolin-1, which both inhibit receptor and thapsigargin-evoked c-fos expression, failed to reduce c-fos activity to ionomycin under these conditions [46]. Hence, a large, grossly nonphysiological rise in bulk Ca^{2+} can activate c-fos gene expression through a pathway not engaged by more physiologically relevant stimuli.

5.10 CONCLUSION

Ca^{2+} microdomains near open Ca^{2+} channels confer several signaling advantages over a rise in bulk Ca^{2+}. These include specificity, speed, and high fidelity in the signal transduction process. Although local Ca^{2+} signals near voltage-operated Ca^{2+} channels have long been known to play a major role in neurotransmitter release at presynaptic active zones [65], the importance of Ca^{2+} signals confined to the vicinity of Ca^{2+} channels in non-excitable cells is now being appreciated. Ca^{2+} microdomains near CRAC channels signal effectively to the nucleus through the recruitment of Ca^{2+}-dependent transcription factors including NFATs and c-fos. In mast cells, the scaffolding protein caveolin-1 plays an important role in dictating whether both transcription factors will be activated by the Ca^{2+} microdomain (Figure 5.1). Whereas NFAT signaling is maintained in the presence of caveolin-1, c-fos activity is suppressed. The inhibition is linked to the phosphorylation of tyrosine 14 of caveolin-1, although how

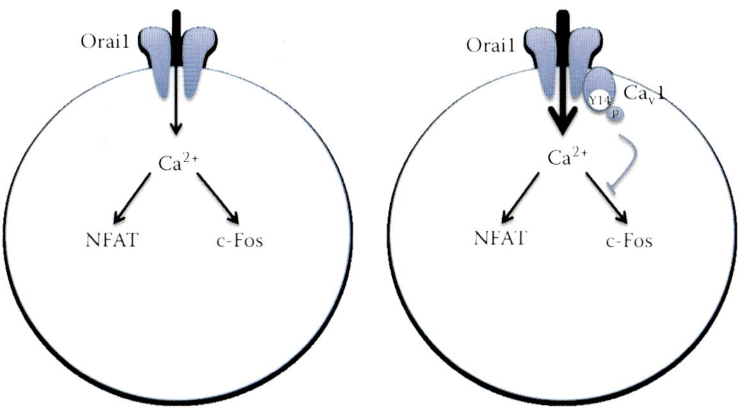

FIGURE 5.1 Cartoon summarizing modular regulation of NFAT and c-fos by caveolin-1 in mast cells. In the absence of caveolin-1, NFAT and c-fos both activate in response to Ca^{2+} microdomains near open CRAC channels. However, when caveolin-1 is phosphorylated on tyrosine 14, the c-fos pathway is suppressed. Note the increase in Ca^{2+} flux through CRAC channels when caveolin-1 is present.

this impairs the phosphorylation of STAT5, the transcription factor that couples Ca^{2+} microdomains near CRAC channels to c-fos expression, is currently not known.

Modular regulation by the phosphorylated state of a membrane scaffolding protein like caveolin-1 might be a general way to control selectively downstream signaling pathways in response to the same Ca^{2+} microdomain.

REFERENCES

1. Wheeler DG, Barrett CF, Groth RD, Safa P, and Tsien RW. CaMKII locally encodes L-type channel activity to signal to nuclear CREB in excitation-transcription coupling. *Journal of Cell Biology*. 2008;183:849–863.
2. Ma H, Groth RD, Cohen SM, Emery JF, Li B, Hoedt E, Zhang G, Neubert TA, and Tsien RW. γCaMKII shuttles Ca^{2+}/CaM to the nucleus to trigger CREB phosphorylation and gene expression. *Cell*. 2014;159:281–294.
3. Barco A, Alarcon JM, and Kandel ER. Expression of constitutively active CREB protein facilitates the late phase of long-term potentiation by enhancing synaptic capture. *Cell*. 2002;108:689–703.
4. Parekh AB. Store-operated CRAC channels: Function in health and disease. *Nature Reviews Drug Discovery*. 2010;9:399–410.
5. Hogan PG, Chen L, Nardone J, and Rao A. Transcriptional regulation by calcium, calcineurin, and NFAT. *Genes and Development*. 2003;17:2205–2232.
6. Macian F, Lopez-Rodriguez C, and Rao A. Partners in transcription: NFAT and AP-1. *Oncogene*. 2001;20:2476–2489.
7. Wu H, Peisley A, Graef IA, and Crabtree GR. NFAT signalling and the invention of vertebrates. *Trends in Cell Biology*. 2007;17:251–260.
8. Ishiguro K, Ando T, Maeda O, Watanabe O, and Goto H. Tubulin alpha functions as an adaptor in NFAT-importin beta interaction. *Journal of Immunology*. 2011;186:2710–2713.
9. Eferl R and Wagner EF. AP-1: A double-edged sword in tumorigenesis. *Nature Reviews Cancer*. 2003;3:859–868.
10. Rao A, Luo C, and Hogan PG. Transcription factors of the NFAT family: Regulation and function. *Annual Review of Immunology*. 1997;15:707–747.
11. Goldsmith MA and Weiss A. Early signal transduction by the antigen receptor without commitment to T cell activation. *Science*. 1988;240:1029–1031.
12. Negulescu PA, Shastri N, and Cahalan MD. Intracellular calcium dependence of gene expression in single T lymphocytes. *Proceedings of the National Academy of Sciences of the United States of America*. 1994;91:2873–2877.
13. Le Deist F, Hivroz C, Partiseti M, Thomas C, Buc HA, Oleastro M, Belohradsky B, Choquet D, and Fischer A. A primary T-cell immunodeficiency associated with defective transmembrane calcium influx. *Blood*. 1995;85:1053–1062.
14. Hutchinson LE and McCloskey MA. Fc epsilon RI-mediated induction of nuclear factor of activated T-cells. *Journal of Biological Chemistry*. 1995;270:16333–16338.
15. Clapham DE. Calcium signalling. *Cell*. 1995;80:259–268.
16. Zweifach A and Lewis RS. Mitogen-regulated Ca^{2+} current of T lymphocytes is activated by depletion of intracellular Ca^{2+} stores. *Proceedings of the National Academy of Sciences of the United States of America*. 1993;90:6295–6299.
17. Partiseti M, Ledeist F, Hivroz C, Fischer A, Kom H, and Choquet D. The calcium current activated by T cell receptor and store depletion in human lymphocytes is absent in a primary immunodeficiency. *Journal of Biological Chemistry*. 1994;269:32327–32335.
18. Fanger CM, Hoth M, Crabtree GR, and Lewis RS. Characterization of T cell mutants with defects in capacitative calcium entry: Genetic evidence for the physiological roles of CRAC channels. *Journal of Cell Biology*. 1995;131:655–667.

19. Liou J, Kim ML, Heo WD, Jones JT, Myers JW, Ferrell Jr. JE, and Meyer T. STIM is a calcium sensor essential for calcium-store-depletion-triggered calcium influx. *Current Biology*. 2005;15:1235–1241.
20. Roos J, DiGregorio PJ, Yeromin AV, Ohlsen K, Lioudyno M, Zhang S, Safrina O et al. STIM1, an essential and conserved component of store-operated Ca^{2+} channel function. *Journal of Cell Biology*. 2005;169:435–445.
21. Hogan PG, Lewis RS, and Rao A. Molecular basis of calcium signalling in lymphocytes: STIM and ORAI. *Annual Review of Immunology*. 2010;28:491–533.
22. Liou J, Fivaz M, Inoue T, and Meyer T. Live-cell imaging reveals sequential oligomerization and local plasma membrane targeting of stromal interaction molecule 1 after calcium store depletion. *Proceedings of the National Academy of Sciences of the United States of America*. 2007;104:9301–9306.
23. Wu MM, Buchanan J, Luik RM, and Lewis RS. Ca store depletion causes STIM1 to accumulate in ER regions closely associated with the plasma membrane. *Journal of Cell Biology*. 2006;174:803–813.
24. Zhang SL, Yeromin AV, Zhang XH-F, Yu Y, Safrina O, Penna A, Roos J, Stauderman KA, and Cahalan MD. Genome-wide RNAi screen of calcium influx identifies genes that regulate Ca release-activated Ca channel activity. *Proceedings of the National Academy of Sciences of the United States of America*. 2006;103:9357–9362.
25. Feske S, Gwack Y, Prakriya M, Srikanth S, Puppel S-V, Tanasa B, Hogan PG, Lewis RS, Daly M, and Rao A. A mutation in Orai1 causes immune deficiency by abrogating CRAC channel function. *Nature*. 2006;441:179–185.
26. Vig M, Peinelt C, Beck A, Koomoa DL, Rabah D, Koblan-Huberson M, Kraft S et al. CRACM1 is a plasma membrane protein essential for store-operated calcium entry. *Science*. 2006;312:1220–1223.
27. Prakriya M, Feske S, Gwack Y, Srikanth S, Rao A, and Hogan PG. Orai1 is an essential pore subunit of the CRAC channel. *Nature*. 2006;443:230–233.
28. Yeromin AV, Zhang SL, Jiang W, Yu Y, Safrina O, and Cahalan MD. Molecular identification of the CRAC channel by altered ion selectivity in a mutant of Orai. *Nature*. 2006;443:226–229.
29. Vig M, Beck A, Billingsley JM, Lis A, Parvez S, Peinelt C, Koomoa DL et al. CRACM1 multimers form the ion-selective pore of the CRAC channel. *Current Biology*. 2006;16:2073–2079.
30. Gwack Y, Srikanth S, Feske S, Cruz-Guilloty F, Oh-hora M, Neems D, Hogan PG, and Rao A. Biochemical and functional characterization of Orai proteins. *Journal of Biological Chemistry*. 2007;282:16232–16243.
31. Kar P, Nelson C, and Parekh AB. Selective activation of the transcription factor NFAT1 by calcium microdomains near Ca^{2+} release-activated Ca^{2+} (CRAC) channels. *Journal of Biological Chemistry*. 2011;286:14795–14803.
32. Kar P, Bakowski D, Di Capite J, Nelson C, and Parekh AB. Different agonists recruit different stromal interaction molecule proteins to support cytoplasmic Ca^{2+} oscillations and gene expression. *Proceedings of the National Academy of Sciences of the United States of America*. 2012;109:6969–6974.
33. Ng S-W, DiCapite JL, Singaravelu K, and Parekh AB. Sustained activation of the tyrosine kinase Syk by antigen in mast cells requires local Ca^{2+} influx through Ca^{2+} release-activated Ca^{2+} channels. *Journal of Biological Chemistry*. 2008;283:31348–31355.
34. Chang W-C, Nelson C, and Parekh AB. Ca^{2+} influx through CRAC channels activates cytosolic phospholipase A2, leukotriene C4 secretion and expression of c-fos through ERK-dependent and independent pathways in mast cells. *FASEB Journal*. 2006;20:1681–1693.
35. Di Capite J, Ng S-W, and Parekh AB. Decoding of cytoplasmic Ca^{2+} oscillations through the spatial signature drives gene expression. *Current Biology*. 2009;19:853–858.

36. Samanta K, Kar P, Mirams GR, and Parekh AB. Ca^{2+} channel re-localization to plasma-membrane microdomains strengthens activation of Ca^{2+}-dependent nuclear gene expression. *Cell Reports*. 2015;12:203–216.
37. Ng S-W, Nelson C, and Parekh AB. Coupling of Ca^{2+} microdomains to spatially and temporally distinct cellular responses by the tyrosine kinase Syk. *Journal of Biological Chemistry*. 2009;284:24767–24772.
38. Fierro L and Parekh AB. Substantial depletion of the intracellular Ca^{2+} stores is required for macroscopic activation of the Ca^{2+} release-activated Ca^{2+} current in rat basophilic leukaemia cells. *Journal of Physiology (London)*. 2000;522:247–257.
39. Ng SW, Bakowski D, Nelson C, Mehta R, Almeyda R, Bates G, and Parekh AB. Cysteinyl leukotriene type I receptor desensitization sustains Ca^{2+}-dependent gene expression. *Nature*. 2012;482:111–115.
40. Somasundaram S, Shum AK, McBride HJ, Kessler JA, Feske S, Miller RJ, and Prakriya M. Store-operated CRAC channels regulate gene expression and proliferation in neural progenitor cells. *Journal of Neuroscience*. 2014;34:9107–9123.
41. Poteser M, Schleifer H, Lichtenegger M, Schernthaner M, Stockner T, Kappe CO, Glasnov TN, Romanin C, and Groschner K. PKC-dependent coupling of calcium permeation through transient receptor potential canonical 3 (TRPC3) to calcineurin signaling in HL-1 myocytes. *Proceedings of the National Academy of Sciences of the United States of America*. 2011;108:10556–10661.
42. Yuan JP, Zeng W, Huang GN, Worley PF, and Muallem S. STIM1 heteromultimerizes TRPC channels to determine their function as store-operated channels. *Nature Cell Biology*. 2007;9:636–645.
43. Cheng KT, Liu X, Ong HL, Swaim W, and Ambudkar IS. Local Ca^{2+} entry via Orai1 regulates plasma membrane recruitment of TRPC1 and controls cytosolic Ca^{2+} signals required for specific functions. *PLoS Biology*. 2011;9:e10011025.
44. Darnell JEJ. STATs and gene regulation. *Science*. 1997;277:1630–1635.
45. Bromberg JF. Activation of STAT proteins and growth control. *Bioessays*. 2001; 23:161–169.
46. Yeh Y-C and Parekh AB. Distinct structural domains of caveolin-1 independently regulate Ca^{2+} release-activated Ca^{2+} channels and Ca^{2+} microdomain-dependent gene expression. *Molecular and Cellular Biology*. 2015;35:1341–1349.
47. Oliveria SF, Dell'Acqua ML, and Sather WA. AKAP79/150 anchoring of calcineurin controls neuronal L-type Ca^{2+} channel activity and nuclear signalling. *Neuron*. 2007;55:261–275.
48. Li H, Pink MD, Murphy JG, Stein A, Dell'Acqua ML, and Hogan PG. Balanced interactions of calcineurin with AKAP79 regulate Ca^{2+}-calcineurin-NFAT signaling. *Nature Structural & Molecular Biology*. 2012;19:337–345.
49. Kar P, Samanta K, Kramer H, Morris O, Bakowski D, and Parekh AB. Dynamic assembly of a membrane signaling complex enables selective activation of NFAT by orai1. *Current Biology*. 2014;24:1361–1368.
50. Kar P and Parekh AB. Distinct spatial Ca^{2+} signatures selectively activate different NFAT transcription factor isoforms. *Molecular Cell*. 2015;58:232–243.
51. Chang WC, Di Capite J, Singaravelu K, Nelson C, Halse V, and Parekh AB. Local calcium influx through calcium release-activated calcium (CRAC) channels stimulates production of an intracellular messenger and an intercellular pro-inflammatory signal. *Journal of Biological Chemistry*. 2008;283:4622–4631.
52. Macian F, Garcia-Rodriguez C, and Rao A. Gene expression elicited by NFAT in the presence or absence of cooperative recruitment of Fos and Jun. *EMBO Journal*. 2000;19:4783–4795.
53. Sargiacomo M, Scherer PE, Tang Z, Kubler E, Song KS, Sanders MC, and Lisanti MP. Oligomeric structure of caveolin: Implications for caveolae membrane organization.

Proceedings of the National Academy of Sciences of the United States of America. 1995;92:9407–9911.
54. Alioua A, Lu R, Kumar Y, Eghbali M, Kundu P, Toro L, and Stefani E. Slo1 caveolin-binding motif, a mechanism of caveolin-1-Slo1 interaction regulating Slo1 surface expression. *Journal of Biological Chemistry.* 2008;283:4808–4817.
55. Pani B, Ong HL, Brazer SW, Liu X, Rauser K, Singh BJ, and Ambudkar IS. Activation of TRPC1 by STIM1in ER-PM microdomains involves release of the channel from its scaffold caveolin-1. *Proceedings of the National Academy of Sciences of the United States of America.* 2009;106:20087–20092.
56. Ludwig A, Howard G, Mendoza-Topaz C, Deerinck T, Mackey M, Sandin S, Ellisman MH, and Nichols BJ. Molecular composition and ultrastructure of the caveolar coat complex. *PLoS Biology.* 2013;11:e1001640.
57. Yu F, Sun L, and Machaca K. Constitutive recycling of the store-operated Ca^{2+} channel Orai1 and its internalization during meiosis. *Journal of Cell Biology.* 2010;191:523–535.
58. Couet J, Li S, Okamoto T, Ikezu T, and Lisanti MP. Identification of peptide and protein ligands for the caveolin scaffolding domain. *Journal of Biological Chemistry.* 1997;272:6525–6533.
59. Li S, Okamoto T, Chun M, Sargiacomo M, Casanova JE, Hansen SH, Nishimoto I, and Lisanti MP. Evidence for a regulated interaction between heterotrimeric G proteins and caveolin. *Journal of Biological Chemistry.* 1995;270:15693–15701.
60. Lee H, Volonte D, Galbiati F, Iyengar P, Lublin DM, Bregman DB, Wilson MT et al. Constitutive and growth factor-regulated phosphorylation of caveolin-1 occurs at the same site (Tyr-14) in vivo: Identification of a c-src/Ca_v-1/Grb7 signaling cassette. *Molecular Endocrinology.* 2000;14:1750–1775.
61. Gottlieb-Abraham E, Shvartsman DE, Donaldson JC, Ehrlich M, Gutman O, Martin GS, and Henis YI. Src-mediated caveolin-1 phosphorylation affects the targeting of active Src to specific membrane sites. *Molecular Biology of the Cell.* 2013;24:3881–3895.
62. Tomassian T, Humphries LA, Liu SD, Silva O, Brooks DG, and Miceli MC. Caveolin-1 orchestrates TCR synaptic polarity, signal specificity, and function in CD8 T cells. *Journal of Immunology.* 2011;187:2992–3002.
63. Wang XM, Nadeau PE, Lin S, Abbott JR, and Mergia A. Caveolin 1 inhibits HIV replication by transcriptional repression mediated through NF-kB. *Journal of Virology.* 2011;85:5483–5493.
64. Morgan AJ and Jacob R. Ionomycin enhances Ca^{2+} influx by stimulating store-regulated cation entry and not by a direct action at the plasma membrane. *Biochemical Journal.* 1994;300:665–672.
65. Neher E. Vesicle pools and Ca^{2+} microdomains: New tools for understanding their roles in neurotransmitter release. *Neuron.* 1998;20:389–399.

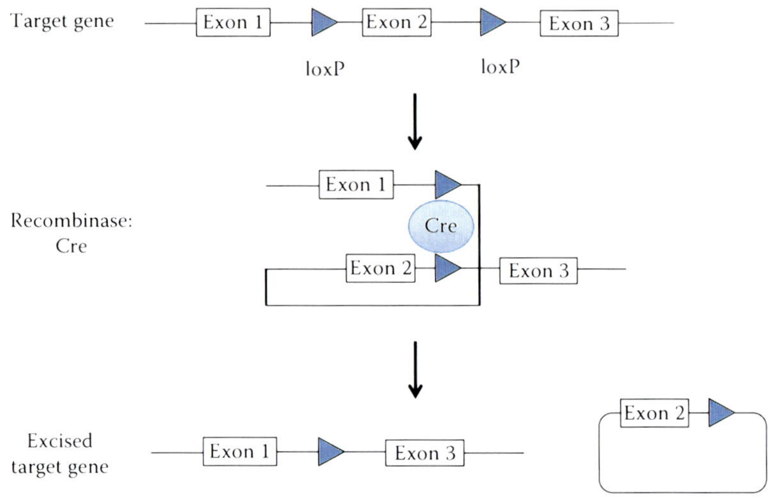

FIGURE 6.2 Principle of the Cre-*loxP* system as an example of conditional gene targeting. The target exon(s) of the gene is flanked by two loxP sites. Cre recognizes loxP sites in the genomic DNA and excises the target exon along with one loxP sequence off the gene while recombining the two ends of remained sequences, causing the permanent deletion of the target gene.

the gene in a specific cell type(s) in the living mice by crossing with transgenic (Tg) or knock-in mice expressing a recombinase under the promoter of a specific cell type(s) (Figure 6.2). These systems also can delete genes in a spatially and temporally restricted manner. Especially, a variety of Cre-expressing mice have been established so far and many of them are commercially available from providers such as the Jackson Laboratory and Taconic Biosciences.

6.2.3 Gene Trapping

Gene-trap strategy is used for making gene-deficient mice and detecting patterns of gene expression [19–22]. A reporter gene construct is randomly integrated into the genome and the reporter gene is expressed under the transcriptional regulation of an endogenous gene. A gene-trap vector in its most basic version contains a splice-acceptor sequence at the 5′ end, a reporter gene, a selectable marker gene, and a polyadenylation signal sequence at the 3′ end (Figure 6.3a). Since the *Escherichia coli* β-galactosidase (β-gal) protein is stable and easy to detect by X-Gal staining in cells, tissues, and embryos, β-galactosidase gene (*lacZ*) is most commonly utilized as the reporter gene. The gene-trap cassette is introduced into ES cells by electroporation or retroviral transduction method. The insertion of the vector construct in a genetic region typically results in complete inactivation of the "trapped" gene. However, in some cases, vector insertion can fail to inactivate the gene, lead to hypomorphic gene function, or result in a dominant negative phenotype [23]. To avoid these problems, it is desirable that the position of vector insertion is close to the 5′ end of a gene.

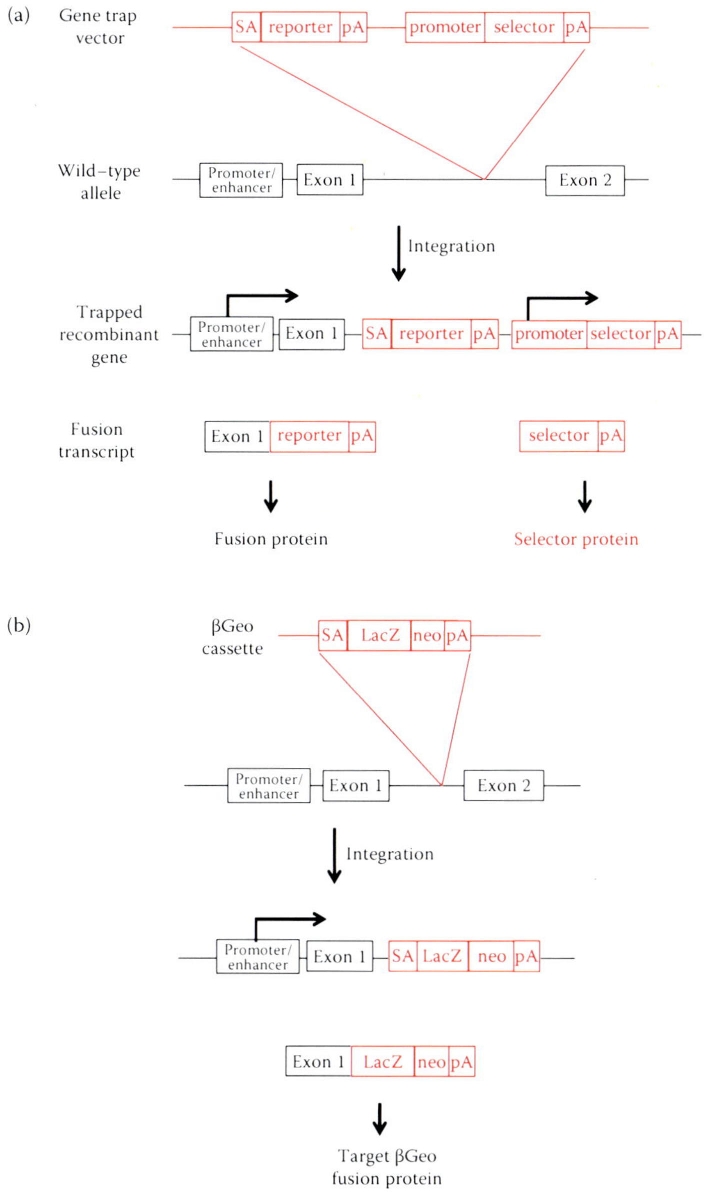

FIGURE 6.3 Principle of gene trapping. (a) Schematic diagram of a basic gene-trap strategy. A gene-trap vector is composed of a splice-acceptor (SA) sequence at the 5′ end, a reporter gene followed by a polyadenylation (pA) signal sequence, and a selector sequence with a strong promoter (e.g., PGK) and polyadenylation signal sequence at 3′ end. Once the gene-trap vector is integrated into an intron of target gene, a splice donor of the target gene recognizes the SA sequence of the vector, which leads to formation of a fusion of reporter protein instead of the target protein. On the other hand, the selector protein can be expressed independently. (b) Schematic diagram of βGeo method. A βGeo protein is a fusion protein of β-galactosidase and a neomycin resistance (*Neo*) gene. In this case, all neomycin-resistant ES cells can express β-galactosidase.

6.3 ESTABLISHMENT OF ORAI/STIM-DEFICIENT MICE AT THE WHOLE-BODY LEVEL

Patients who have nonfunctional mutations in the *ORAI1* or *STIM1* gene died mainly from immunodeficiency [24]. Therefore, knowledge obtained from these patients is limited. Many researchers, including us, established the Orai1, Stim1, and Stim2 knockout mice by using various gene-targeting technologies (Table 6.1). These mice have made significant contributions to the advancement of our understanding of the roles of these gene products *in vivo*. In this section, we will describe how the Orai1, Stim1, Stim2, and other mutant mice were established.

TABLE 6.1
List of Orai/Stim-Deficient Lines

Gene	Target Tissues/Cells	Targeting Method/ Cre-Deleter Line	References
Orai1	Whole body	Gene trap	[25]
	Whole body	Conventional	[26]
	T cells	Cd4-Cre, Lck-Cre	[39,41]
	Neural stem cells/progenitors	Nestin-Cre	[40]
	Mammary gland	MMTV-Cre	[42]
	Whole body, temporally	UBC-Cre/ERT2	[41]
Stim1	Whole body	Gene trap	[35,36]
	Whole body	CMV-Cre, CAG-Cre	[28,29]
	T cells	Cd4-Cre	[29]
	B cells	mb1-Cre	[43]
	Macrophage and neutrophils	LysM-Cre	[44]
	Skeletal muscle	Myogenin-Cre	[45]
	Smooth muscle	SM22α-Cre KI	[46]
	Purkinje cells	Grid2-Cre KI	[47]
	Forebrain	CaMKIIα-Cre	[48]
	Whole body, temporally	UBC-Cre/ERT2	[41]
Stim2	Whole body	Conventional	[37]
	Whole body	CMV-Cre	[29]
	T cells	Cd4-Cre	[29]
	B cells	mb1-Cre	[43]
	Smooth muscle	SM22α-Cre KI	[46]
	Forebrain	CaMKIIα-Cre	[48]
Stim1 and Stim2	Whole body	CMV-Cre	[50]
	T cells	Cd4-Cre, Lck-Cre	[29,50]
	B cells	mb1-Cre	[43]
	Hematopoietic cells	Vav-iCre	[50]
	Smooth muscle	SM22α-Cre KI	[46]
	Forebrain	CaMKIIα-Cre	[48]
	Hematopoietic cells, temporally	Mx1-Cre	[50]

6.3.1 Orai1 Knockout Mouse

Orai1 protein, encoded by the *Orai1* gene located on chromosome 5 in the mouse, is a small and widely expressed tetraspanning membrane glycoprotein, whose N- and C-termini are both located in the cytoplasm [10–12]. Orai1 belongs to a novel class of channel proteins with no sequence homology to other known ion channels except its paralogues Orai2 and Orai3 (also known as CRACM2 and CRACM3, respectively).

Initially the Orai1 KO mouse was established by the gene trapping method using ES cells derived from 129P2/OraHsd [24]. The gene-trap cassette was composed of a splice-acceptor site followed by a "βGeo" cassette encoding a fusion of β-galactosidase and a neomycin resistance (*Neo*) gene (Figure 6.3b). The insertion site for this gene-trap cassette is in intron 1 of *Orai1*. As a result, the Orai1-βGeo fusion protein contains only a part of the first transmembrane region of Orai1. Thus, Orai1 gene-trap mutant mouse can be considered an Orai1 KO mouse.

Orai1 KO mice using the conventional gene targeting method was generated by us [26]. Exon 1 of the *Orai1* gene encoding the 5′ UTR, the N-terminal cytoplasmic tail region, and part of the first transmembrane region of Orai1 was deleted by replacing with a *Neo* gene. To save the time required for breeding and promoter interference, the Cre recombinase sequence and the testis-specific ACE promoter sequence were introduced into the targeting vector. The first *loxP* site was inserted upstream of the ACE promoter sequence and the second *loxP* site was inserted downstream of the Neo cassette. In this system, the *Neo* gene is constitutively transcribed, while the Cre recombinase is expressed under the control of the testis-specific ACE promoter; thus, the Neo cassette is deleted during spermatogenesis in chimeric mice, not in ES cells. Gene targeting of the *Orai1* gene was performed by using homologous recombination in B6/3 ES cells derived from C57BL/6 mice.

6.3.2 Stim1 Knockout Mouse

Stim1 protein, encoding the *Stim1* gene located on chromosome 7 in the mouse, is a type I transmembrane protein predominantly localized in the ER membrane and expressed broadly [27–29]. Stim1 protein is composed of several functional domains including a Ca^{2+}-binding EF hand motif in the ER lumen portion and the CRAC activation domain (CAD)/the STIM Orai activating region (SOAR)/the Orai1 activating small fragment (OASF) domain in the cytoplasm [27,30–33]. The C-terminal cytosolic portion of Stim1 contains a polybasic lysine-rich region, which is required for attachment to the plasma membrane [31,34]. Stim1 KO mouse was generated by gene trapping [35,36] and conditional gene targeting technology [28,29].

As for the Orai1 gene-trap mutant mouse, the same gene-trap cassette was utilized to make mutant ES cells. Two groups obtained the same Stim1 gene-trap mutant ES cell clone (129P2/OraHsd background) from Baygenomics, then established Stim1 gene-trap mutant mouse independently [35,36]. The insertion site is 600 bp downstream of exon 7, into intron 7 of the *Stim1* gene, which results in the production of a fusion protein containing the EF hand motif, SAM domain, and transmembrane domain, but not most of the cytoplasmic domain of Stim1. Because of the absence of the CAD/SOAR/OASF domain, the fusion protein cannot bind or activate Orai1. Therefore, Stim1 gene-trap mutant mice can be considered Stim1 KO mice.

Even though the Stim1 KO mouse by the conventional gene targeting technology has not been established so far, a mouse lacking Stim1 at the whole-body level (*Stim1*$^{(\Delta/\Delta)}$ mouse) was generated by intercrossing the progeny of founder *Stim*$^{flox/+}$ mice (see the following text) after breeding with *CMV*-Cre Tg mice [28,29].

6.3.3 Stim2 Knockout Mouse

Stim2, encoding the *Stim2* gene located on chromosome 5 in the mouse, is a paralogue of Stim1 and shares its overall protein domain architecture with Stim1 [27]. The conventional Stim2 KO mouse was generated by Nieswandt and colleagues [37]. The targeting vector was designed to delete most of exons 4–7 of the Stim2 gene. The corresponding locus was replaced with neomycin resistance and LacZ cassettes. In addition, the *Stim2*$^{(\Delta/\Delta)}$ mouse was established in the same way as *Stim1*$^{(\Delta/\Delta)}$ mice [29].

6.3.4 Orai1 (R93W) Knock-In (KI) Mouse

The Orai1 (R93W) KI mouse was generated by us [38]. R93W, corresponding to R91W in a human [10], results in the loss of function of Orai1. We replaced codon 93 (CGG encoding R93) in exon 1 of the Orai1 gene with TGG (encoding W93) and introduced the genomic fragments containing the mutant exon 1 in the same vector used for the Orai1 KO mouse. The replacement of the *Orai1* gene was performed by using homologous recombination in B6/3 ES cells derived from C57BL/6 mice.

6.4 ESTABLISHMENT OF TISSUE-SPECIFIC ORAI/STIM-DEFICIENT MOUSE LINES

In contrast to the conventional KO mouse and gene-trap mouse, mice generated by the conditional gene targeting method may enable us to analyze a cell-type-specific function *in vivo* and avoid potential lethality during development. In this section, we will describe the strategies for making tissue-specific Orai1, Stim1, and Stim2 KO mouse.

6.4.1 Orai1 Conditional KO (cKO) Mouse

The Orai1 cKO mouse was generated by two groups [39,40]. In both cases, the first *loxP* site was inserted into intron 1 at around 1–2 kb upstream of exon 2 and the second *loxP* site was inserted at around 0.3–0.8 kb downstream of exon 2, which contains the polyA signal sequence and 3′ UTR. Although Prakriya and colleagues described the presence of exon 3 of the *Orai1* gene [40], it seems to be identical to a part of exon 2 of the *Orai1* gene available in public databases. Both groups utilized the *Neo* gene as a positive selection cassette flanked by *FRT* recombination sites, although it was inserted into a different position in the *Orai1* gene.

To establish T cell–specific Orai1 KO mouse model, Orai1-floxed mice were crossed with mice expressing Cre under the control of the *Cd4* or *Lck* promoter/enhancer [39,41].

Although these lines express Cre recombinase in T cells, the expression of Cre recombinase is induced at different developmental stages in the thymus, that is, at the early stage of CD4 and CD8-double positive immature thymocytes in *Cd4*-Cre Tg mice and at the double negative stage in *Lck*-Cre Tg mice. In the case of other tissues, *nestin*-Cre and *mouse mammary tumor virus (MMTV)*-Cre line D were utilized for generating mice to delete the *Orai1* gene in neural progenitor cells and mammary glands, respectively [40,42]. A mouse model in which the Orai1 gene can be deleted at any time was established by crossing Orai1-floxed mice with Cre-ERT2 Tg mice expressing tamoxifen-inducible Cre recombinase [41].

6.4.2 Stim1 Conditional KO (cKO) Mouse

The Stim1 cKO mouse was generated by two groups, including us [28,29]. We designed the targeting vector to delete exon 2, encoding the EF hand motif of the *Stim1* gene, and whose deletion is predicted to result in a frameshift, which generates a premature stop codon in the next exon [29]. Exon 2 was flanked by *loxP* recombination sites and the *Neo* gene flanked by *FRT* recombination sites was inserted upstream of exon 2. We did not detect any mRNA and protein in cells when we bred with Cre-expressing mice. We performed gene targeting of the *Stim1* gene using homologous recombination in Bruce-4 ES cells derived from C57BL/6 mice. To generate the conditional *Stim1*$^{flox/+}$ alleles, we bred founder *Stim1*$^{neo/+}$ chimeric mice to Flpe Tg "female" mice to remove the neomycin resistance cassette from the targeted *Stim1* alleles.

In another line, the deletion of exon 6, encoding the transmembrane region, of the *Stim1* gene creates a premature stop codon in the next exon [28]. Although this line is capable of creating the predicted truncated form of Stim1, it was not detected by Western blotting, suggesting that its dominant negative effect can be excluded. Therefore, both lines have properties identical to Stim1 cKO mice.

Using Stim1-floxed mice, a variety of tissue-specific Stim1 KO mice have been established so far. In the immune system, T cell–specific Stim1 KO mice were first generated by crossing with *Cd4*-Cre Tg mice to analyze Stim1 function in peripheral T cells [29]. B cell–specific Stim1 KO mice were made by using *mb1*-Cre Tg mice [43]. LysM-Cre mice were used for generating macrophage and neutrophil-specific Stim1 KO mice [44]. In the muscular system, each muscle-specific Stim1 KO mice was generated by breeding with *Myogenin*-Cre Tg mice for skeletal muscle or *SM22*α-Cre knock-in mice for smooth muscle cells, respectively [45,46]. In the nervous system, Purkinje neuron-specific Stim1 KO mice were generated by crossing Stim1-floxed mice with *Grid2*-Cre knock-in mice expressing Cre under the control of the GluD2 promoter δ2 [47]. *CaMKII*α-Cre Tg mice were utilized to generate forebrain-specific Stim1 KO mice [48]. As in the case of Orai1 KO, Cre-ERT2-mediated inducible Stim1 KO mouse model was established [41].

6.4.3 Stim2 Conditional KO (cKO) Mouse

We also generated Stim2 cKO mice. We designed the targeting vector to delete exon 3, encoding a part of the EF hand motif of the *Stim2* gene, and whose deletion is predicted to result in a frameshift, which generates a premature stop codon in the next exon [29]. Exon 3 was flanked by *loxP* recombination sites and the *Neo* gene flanked by *FRT* recombination sites was inserted upstream of exon 3. As with Stim1 cKO mice, no mRNA and protein were detected in cells when we bred with Cre-expressing mice. Bruce-4 ES cells were utilized to make *Stim2*$^{neo/+}$ ES cell clones. Mice bearing the conditional *Stim2*$^{flox/+}$ alleles were obtained by breeding with Flpe Tg "female" mice. *Stim2*$^{(\Delta/\Delta)}$ mice were established in the same way as *Stim1*$^{(\Delta/\Delta)}$ mice were generated.

In addition to Stim1 cKO mice, Kurosaki's group also generated another line of Stim2 cKO mice [43]. They designed the targeting vector to delete exons 6 and 7, which encode the transmembrane region and a part of the first coiled-coil domain of the *Stim2* gene. In this line, no predicted truncated form of Stim2 was detected.

As in the case of Orai1 and Stim1 cKO mice, Stim2-floxed mice were crossed with *Cd4*-Cre Tg, *mb1*-Cre Tg, *SM22*α-Cre KI, and *CaMKII*α-Cre Tg mice for T cell-, B cell-, smooth muscle-, and forebrain-specific Stim2 KO mice, respectively [29,43,46,48]. In addition to crossing with Cre-expressing mouse lines, a specific gene deletion is achieved by introducing Cre recombinase into desired tissues and cells both *in vitro* and *in vivo* with viral vectors, such as lentiviral and adeno-associated vectors. Using a lentiviral vector encoding nuclear-targeted Cre (NLS-cre), Stim2-deficient hippocampal neurons were established [49].

6.4.4 Stim1 and Stim2 Double Conditional KO Mouse

The Stim1 and Stim2 doubly floxed mouse was developed by crossing Stim1 floxed mice with Stim2 floxed mice [29,43]. All tissue-specific Stim1 and Stim2 doubly conditional KO mouse models were derived from one of these lines.

Stim1 and Stim2 doubly floxed mice were crossed with *Cd4*-Cre Tg, *Lck*-Cre Tg, *mb1*-Cre Tg, *CaMKII*α-Cre Tg, and *SM22*α-Cre KI mice to generate conventional T cell-, B cell-, forebrain-, and smooth muscle-specific Stim1 and Stim2-double KO (dKO) mice, respectively [29,43,46,48,50]. In addition to the tissue-specific model mentioned earlier, a hematopoietic cell–specific dKO mouse model was generated by using *Vav*-iCre mice [50]. Inducible dKO mice were also established by crossing with Mx1-cre mice in which the expression of Cre recombinase in hematopoietic cells is induced by treatment with polyinosinic-polycytidylic acid (Poly I:C), which is recognized by Toll-like receptor 3 (TLR3) and induces type I interferon production [50].

6.5 METHODS TO ESTABLISH MURINE EMBRYONIC FIBROBLAST LINES FROM KO MICE

Cells isolated from genetically manipulated mouse models are indispensable tools to study the molecular and cellular mechanisms and regulatory gene networks involving the mutated gene but also the cell-specific function of the gene. Among various cells, MEFs are common and very useful. MEF cells can be established even if gene mutations are lethal in a midgestation mouse embryo. In the case of Orai/Stim genes, systemically deleting each gene in the mouse results in lethality. Orai1 KO,

Orai1(R93W) KI, *Stim1*$^{(\Delta/\Delta)}$, and Stim1 gene-trap mice displayed perinatal lethality [28,29,35,36]. Sudden death was observed in both *Stim2*$^{(\Delta/\Delta)}$ mice and Stim2 KO mice in young adulthood [29,37].

Here we describe a standard protocol to establish MEF cells:

1. Take out day 13.5–15.5 embryo and place them in cold phosphate buffered saline (PBS) on appropriate dish/plate.
2. Remove placenta and other maternal tissues.
3. Transfer the embryo to a new plate and cut a piece of their tail and put in a 1.5 mL tube for genotyping.
4. Remove the head, heart, liver, and gut.
5. Add 1 mL trypsin and cut the embryo into small pieces with a razor blade or scissors.
6. Add another 1 mL trypsin and then incubate them at 37°C for 20 min.
7. Add 4 mL MEF media (DMEM supplemented with 10% FCS and antibiotics).
8. Pipet up and down to disaggregate the tissue chunks.
9. Transfer to a 15 mL tube.
10. Let the undigested tissues settle to the bottom of the tube.
11. Transfer the suspension including fibroblasts to a new tube.
12. Add 4 mL of new medium to the tissues and repeat steps 8–11 twice.
13. Combine the suspension.
14. Centrifuge cells at 1200 rpm, 4°C for 5 min.
15. Aspirate off medium and resuspend cells in 10 mL of MEF media.
16. Plate cells on a 10 cm dish and allow cells to grow to confluency (2–3 days).
17. Aspirate off medium and rinse with 10 mL PBS.
18. Add 3 mL trypsin and incubate at 37°C for 5 min.
19. Inactivate trypsin by adding 3 mL MEF medium.
20. Transfer cells to a 15 mL tube and spin them down at 1200 rpm, 4°C for 5 min.
21. Freeze down some cells, plate the rest of cells, and allow to grow to confluency.

Obtained MEF cells should be frozen down as early as possible, ideally on the first or second time of cell passage. If necessary, MEF cells can be immortalized by introducing SV40 large T antigen, hTERT, or EVB. In the case of Stim1- or Stim2-deficient MEF cells, we established immortalized MEF cells using SV40 large T antigen [29].

6.6 FUNCTION OF ORAI/STIM PROTEINS IN THE IMMUNE SYSTEM

The absence of SOC entry through Orai1 critically influences the immune system. Indeed, SOC entry is known to regulate the development and function of various immune cells including T lymphocytes and mast cells. This section focuses on the functions of Orai/Stim proteins in immunity, focusing on findings in mice mentioned earlier.

6.6.1 T Lymphocytes

In T cells, SOC entry is induced after the engagement of TCR with complexes of peptide and the major histocompatibility complex (pMHC). Patients who have non-functional mutations in the *ORAI1* or the *STIM1* gene developed a form of severe combined immunodeficiency (SCID), clearly indicating that Orai1 and Stim1 are essential for T cell activation in the periphery. Indeed, knockout mice lacking the *Stim1* or *Orai1* gene in T cells displayed similar but not identical phenotypes [26,29,41,51]. Sitm1-deficient T cells abrogated SOC entry and impaired the production of effector cytokines such as IL-2, IFNγ, and IL-17, regardless of naïve or effector T cells [29,51]. Consistently, Stim1-deficient T cells had a lower capacity to induce T cell–mediated immune response as judged by graft-versus-host disease [52]. Unexpectedly, in the mouse, Orai1 predominantly functions in effector T cells rather than naïve T cells [26,41,51]. This may partially explain the reason why Orai1-deficient patients have normal numbers of T cells in the periphery. In contrast to these genes, the function of Stim2 protein was elusive at the time and the generation of the Stim2 knockout mouse contributed to the elucidation of its function. Analyses of Stim2-deficient mice and their MEF cells revealed that the Stim2 protein positively regulates SOC entry depending on the cell type, governs the maintenance of long-term SOC entry in activated T cells, and is essential for individual survival [29].

One of the most important contributions of Orai/Stim-deficient mice is the elucidation of the role of these proteins in T cell development. Ca^{2+} signaling has long been thought to be crucial for positive selection during T cell development in the thymus by studies using the pharmacological or genetic inactivation of calcineurin activity [53–56]. Given that it is the chief mechanism of Ca^{2+} entry in T cells, SOC entry has been predicted to be crucial for T cell development; however, this assumption has not been directly demonstrated. Mice with transplanted bone marrow cells from Stim1 gene-trap mutant mice showed normal conventional T cell development, suggesting that Stim2 may still support the SOC entry necessary for T cell development [52]. To clarify the role of SOC entry in T cell development in the thymus, mice in which both Stim1 and Stim2 are deleted systemically or specifically in T cells or hematopoietic cells using *CMV*-Cre Tg, *Lck*-Cre Tg, or *Vav*-iCre Tg mice, respectively, were generated [50]. Contrary to previous assumptions, SOC entry is not essential for the positive selection of conventional TCRαβ$^+$ T cells, consistent with the fact that ORAI1- or STIM1-deficient patients have a normal number of T cells in the periphery. However, only mice lacking both Stim1 and Stim2 specifically in T cells (T-dKO mice) unexpectedly developed a severe autoimmune disease [29]. T-dKO mice develop severe primary Sjögren's syndrome with high levels of autoantibodies, such as antinuclear, SSA/Ro, and SSB/La antibodies, in sera [57]. The structure of their salivary and submandibular glands was progressively destroyed by T cell infiltration, which resulted in significant decrease of saliva secretion. Further analyses of T-dKO mice revealed that Stim-mediated SOC entry is essential for the development and function of regulatory T (Treg) cells, which express Forkhead box P3 (Foxp3) as a master transcription factor. Although Treg precursor cells normally exist in the thymus of T-dKO mice, there are virtually no mature CD4$^+$ CD25$^+$ Foxp3$^+$ Treg cells. This is partly due to impaired proliferation since Stim-deficient precursor cells were capable of differentiating into mature $Foxp3^+$ Treg cells by the treatment with IL-2 both *in vitro* and *in vivo*. However, these recovered Stim-deficient Treg cells lost the suppressive activity due to decreased expressions of inhibitory surface molecules such as CTLA-4 [58] and TIGIT [59]. In addition to $CD4^+$ $CD25^+$ $Foxp3^+$ Treg cells, invariant natural killer T (iNKT) cells and $CD8\alpha\alpha^+$ $TCR\alpha\beta^+$ intestinal intraepithelial lymphocytes (IELs) were absent and their differentiation was also blocked at the transition of their precursor cells to mature cells in T-dKO mice. These cells are classified into agonist-selected T cells [60,61]. In contrast to the indispensable role of Stim1 and Stim2 in Treg cells, Orai1 seems to contribute less to the development and function of Treg cells [39,41]. It is speculated that Treg cells may utilize other SOC channels such as Orai2, Orai3, or TRP channels.

A series of analyses of KO mice clearly demonstrated for the first time that Stim-mediated SOC entry is essential for the development of agonist-selected T cells but not conventional $TCR\alpha\beta^+$ T cells.

6.6.2 B Lymphocytes

SOC entry is also present in B cells. Although B cells that lack Orai/Stim genes singly or doubly showed a profound defect of SOC entry and proliferation by the engagement of B cell receptor (BCR) [26,43], B cell development and antibody responses were normal [43]. As in the case of T cells, mice lacking both Stim proteins in B cells (B-dKO mice) revealed that SOC entry contributes to an unexpected function of B cells. B cells from B-dKO mice failed to produce an anti-inflammatory cytokine, IL-10 [43]. Due to the lack of these cells, experimental autoimmune encephalomyelitis, a mouse model for multiple sclerosis, was exacerbated in B-dKO mice. These findings demonstrate that SOC entry is a key signaling event for B cell regulatory function [43].

6.6.3 Mast Cells

Myeloid lineage cells, such as mast cells, macrophages, and dendritic cells (DCs), express Fc receptors (FcRs), which activates phospholipase C and subsequently induces Ca^{2+} depletion from the ER. It is well known that the CRAC current was originally defined in T cells and mast cells [6,7]. Mast cells from both *Stim1*-deficient mice and mice with a gene-trap insertion in the *Orai1* gene were defective in degranulation owing to a severe decrease in SOC entry [25,28]. Passive cutaneous anaphylaxis, a reaction that measures mast degranulation *in vivo*, was substantially impaired in the *Orai1* gene-trap mutant mice and in fact was decreased even in heterozygous (haploinsufficient) $Stim1^{+/-}$ mice. These data emphasize that SOC entry through Stim1/Orai1 is essential for mast cell function.

6.7 FUNCTION OF ORAI/STIM PROTEINS IN THE MUSCLE

Patients bearing nonfunctional mutations of the *Orai1* gene or the *Stim1* gene showed muscular hypotonia [62–64], suggesting that SOC entry is highly likely to regulate the function of muscle cells. Interestingly, Stim1 deficiency leads to the decreased activation of multiple signal transduction cascades associated with growth signaling (calcineurin, AKT, and ERK1/2), which results in impaired neonatal muscle growth and differentiation [45]. As a result, both Stim1 gene-trap mice and skeletal muscle–specific Stim1 KO mice displayed severe delay in the growth of skeletal muscle and died of skeletal myopathy by 5 weeks after birth [36,45]. Intriguingly, various gain-of-function mutations in each gene have been recently identified in patients suffering from either tubular aggregate myopathy (TAM) or Stormorken syndrome [64].

Smooth muscle cells are also known to express Orai1 and Stim1 proteins [65]. Smooth muscle-specific Stim1 KO mice also died, but their mortality is milder than that observed in skeletal muscle–specific Stim1 KO mice, as half of them survive longer than 20 weeks after birth [46]. Although smooth muscle-specific Stim2 KO mice had no detectable phenotypes, the deletion of both genes in the smooth muscle resulted in perinatal lethality, suggesting that Stim2 may compensate for an essential role of Stim1. The absence of Stim1 substantially reduced the Ca^{2+} store–refilling rate in smooth muscle cells and abolished platelet-derived growth factor (PDGF)-induced Ca^{2+} entry, which leads to impaired nuclear factor of activated T cells (NFAT) activation.

These findings demonstrate that strict control of SOC entry is required for the muscular homeostasis.

6.8 FUNCTION OF ORAI/STIM PROTEINS IN THE NERVOUS SYSTEM

Ca^{2+} signals regulate various events during brain development including neurogenesis [66], neurotransmitter specification [67], and axonal outgrowth [68]. Although no signs of phenotypes have been reported so far in SOC entry-deficient patients, recent studies indicate that SOC entry through Orai/Stim proteins is involved in the regulation of neuronal functions, including neurotransmitter release, synaptic plasticity, and Ca^{2+} oscillations. To elucidate the role of SOC entry in the nervous system, several mouse models have been analyzed.

Mice in which the Orai1 gene is deleted at neural stem/progenitor cells by *Nestin*-Cre Tg mice showed significant decrease of proliferation of neuronal progenitor cells; however, no abnormal structure in the brain was reported [40], suggesting that SOC entry may not be essential for neuronal development. In contrast, SOC entry seems to be involved in neuronal functions. Stim2 was first reported to regulate neuronal cell death by ischemia but not neuronal activities [37]; however, later Stim2-mediated neuronal SOC entry was reported to stabilize mushroom spines through the persistent activation of Ca^{2+}/calmodulin-dependent protein kinase II, suggesting that Stim2 may be involved in the maintenance of memory [49]. In another mouse model, the absence of Stim2 in the forebrain did not influence learning and memory, while Stim1-deficiency resulted in a mild learning delay [48]. However, the deletion of both *Stim* genes in the forebrain resulted in a pronounced impairment in spatial learning and memory, suggesting that Stim2 may compensate for Stim1 function in the underlying neural circuits [48]. Surprisingly, Stim1 and Stim2-double KO mice exhibited enhanced long-term potentiation and PKA signaling in the hippocampus. Another group showed that Stim1 is essential for cerebellar motor control through the activation of TRPC3 downstream of metabotropic glutamate receptor type1 [47].

Although these findings demonstrate that Orai/Stim proteins are involved in neuronal functions, their roles in the nervous system remain largely unknown. Intriguingly, the expression of Stim2 in mushroom spines is also decreased in sporadic Alzheimer's disease patients [49]. It is tempting to speculate that abnormal control of SOC entry may cause neuronal diseases. Further studies will be required to elucidate the roles of Orai/Stim proteins in the nervous system.

6.9 CONCLUDING REMARKS

The genetically manipulated mouse model is one of the most powerful tools to elucidate roles of specific genes *in vivo*. KO mouse studies revealed that each Stim available protein has redundant roles but also distinct cell-type-specific roles. Although a variety of mouse models have been developed by many researchers, the roles of SOC entry through Orai/Stim proteins have not yet been completely understood. Neither Orai2 nor Orai3 KO mice have been reported; such KO mice would be helpful for clarifying the function of each Orai protein. It is critical to keep in mind that functional redundancy among Orai/Stim family proteins or a compensatory mechanism by other Ca^{2+} channels may mask phenotypes. Indeed, some examples indicated here demonstrate that double knockout mice show more profound phenotypes, as is the case for Stim1/Stim2. Although it takes a long time to establish mice lacking multiple genes, recent advances of technology called genome editing, such as CRISPR/Cas9 and TAREN method, enables us to make mouse models faster and easier than the classical method. Another advantage of genome editing is that neighboring genes can also be inactivated. For example, it would be easy to generate mice doubly deficient in Orai1 and Orai2, which locate on the same chromosome in mouse.

ACKNOWLEDGMENTS

We wish to thank many of our colleagues in the Harvard Immune Disease Institute, Tokyo Medical and Dental University, and Kyushu University for their contributions to establishing Orai/Stim KO mice. This work was supported in part by a Grant-in-Aid for Scientific Research (B) and Grant-in-Aid for Challenging Exploratory Research from the Japan Society for the Promotion of Science and grants from the Takeda Science Foundation and Daiichi Sankyo Foundation of Life Science.

REFERENCES

1. Feske, S., Skolnik, E. Y., and Prakriya, M. (2012) Ion channels and transporters in lymphocyte function and immunity. *Nat. Rev. Immunol.* 12, 532–547.
2. Putney, J. W., Jr. (1986) A model for receptor-regulated calcium entry. *Cell Calcium* 7, 1–12.
3. Oh-Hora, M. and Rao, A. (2008) Calcium signaling in lymphocytes. *Curr. Opin. Immunol.* 20, 250–258.

4. Mogami, H., Nakano, K., Tepikin, A. V., and Petersen, O. H. (1997) Ca^{2+} flow via tunnels in polarized cells: Recharging of apical Ca^{2+} stores by focal Ca^{2+} entry through basal membrane patch. *Cell* 88, 49–55.
5. Vazquez, G., de Boland, A. R., and Boland, R. (1997) Stimulation of Ca^{2+} release-activated Ca^{2+} channels as a potential mechanism involved in non-genomic 1,25$(OH)_2$-vitamin D_3-induced Ca^{2+} entry in skeletal muscle cells. *Biochem. Biophys. Res. Commun.* 239, 562–565.
6. Hoth, M. and Penner, R. (1992) Depletion of intracellular calcium stores activates a calcium current in mast cells. *Nature* 355, 353–356.
7. Zweifach, A. and Lewis, R. S. (1993) Mitogen-regulated Ca^{2+} current of T lymphocytes is activated by depletion of intracellular Ca^{2+} stores. *Proc. Natl. Acad. Sci. USA* 90, 6295–6299.
8. Liou, J., Kim, M. L., Heo, W. D., Jones, J. T., Myers, J. W., Ferrell, J. E., Jr., and Meyer, T. (2005) STIM is a Ca^{2+} sensor essential for Ca^{2+}-store-depletion-triggered Ca^{2+} influx. *Curr. Biol.* 15, 1235–1241.
9. Roos, J., DiGregorio, P. J., Yeromin, A. V., Ohlsen, K., Lioudyno, M., Zhang, S., Safrina, O. et al. (2005) STIM1, an essential and conserved component of store-operated Ca^{2+} channel function. *J. Cell Biol.* 169, 435–445.
10. Feske, S., Gwack, Y., Prakriya, M., Srikanth, S., Puppel, S. H., Tanasa, B., Hogan, P. G., Lewis, R. S., Daly, M., and Rao, A. (2006) A mutation in Orai1 causes immune deficiency by abrogating CRAC channel function. *Nature* 441, 179–185.
11. Vig, M., Peinelt, C., Beck, A., Koomoa, D. L., Rabah, D., Koblan-Huberson, M., Kraft, S. et al. (2006) CRACM1 is a plasma membrane protein essential for store-operated Ca^{2+} entry. *Science* 312, 1220–1223.
12. Zhang, S. L., Yeromin, A. V., Zhang, X. H., Yu, Y., Safrina, O., Penna, A., Roos, J., Stauderman, K. A., and Cahalan, M. D. (2006) Genome-wide RNAi screen of Ca^{2+} influx identifies genes that regulate Ca^{2+} release-activated Ca^{2+} channel activity. *Proc. Natl. Acad. Sci. USA* 103, 9357–9362.
13. Joyner, A. L. (2000) *Gene Targeting: A Practical Approach.* Oxford University Press, Oxford, U.K.
14. Muller, U. (1999) Ten years of gene targeting: Targeted mouse mutants, from vector design to phenotype analysis. *Mech. Dev.* 82, 3–21.
15. Lakso, M., Sauer, B., Mosinger, B., Jr., Lee, E. J., Manning, R. W., Yu, S. H., Mulder, K. L., and Westphal, H. (1992) Targeted oncogene activation by site-specific recombination in transgenic mice. *Proc. Natl. Acad. Sci. USA* 89, 6232–6236.
16. Orban, P. C., Chui, D., and Marth, J. D. (1992) Tissue- and site-specific DNA recombination in transgenic mice. *Proc. Natl. Acad. Sci. USA* 89, 6861–6865.
17. Gu, H., Marth, J. D., Orban, P. C., Mossmann, H., and Rajewsky, K. (1994) Deletion of a DNA polymerase beta gene segment in T cells using cell type-specific gene targeting. *Science* 265, 103–106.
18. Dymecki, S. M. (1996) Flp recombinase promotes site-specific DNA recombination in embryonic stem cells and transgenic mice. *Proc. Natl. Acad. Sci. USA* 93, 6191–6196.
19. Gossler, A., Joyner, A. L., Rossant, J., and Skarnes, W. C. (1989) Mouse embryonic stem cells and reporter constructs to detect developmentally regulated genes. *Science* 244, 463–465.
20. Friedrich, G. and Soriano, P. (1991) Promoter traps in embryonic stem cells: A genetic screen to identify and mutate developmental genes in mice. *Genes Dev.* 5, 1513–1523.
21. Skarnes, W. C., Auerbach, B. A., and Joyner, A. L. (1992) A gene trap approach in mouse embryonic stem cells: The lacZ reported is activated by splicing, reflects endogenous gene expression, and is mutagenic in mice. *Genes Dev.* 6, 903–918.

22. Wurst, W., Rossant, J., Prideaux, V., Kownacka, M., Joyner, A., Hill, D. P., Guillemot, F. et al. (1995) A large-scale gene-trap screen for insertional mutations in developmentally regulated genes in mice. *Genetics* 139, 889–899.
23. Mitchell, K. J., Pinson, K. I., Kelly, O. G., Brennan, J., Zupicich, J., Scherz, P., Leighton, P. A. et al. (2001) Functional analysis of secreted and transmembrane proteins critical to mouse development. *Nat. Genet.* 28, 241–249.
24. Feske, S., Muller, J. M., Graf, D., Kroczek, R. A., Drager, R., Niemeyer, C., Baeuerle, P. A., Peter, H. H., and Schlesier, M. (1996) Severe combined immunodeficiency due to defective binding of the nuclear factor of activated T cells in T lymphocytes of two male siblings. *Eur. J. Immunol.* 26, 2119–2126.
25. Vig, M., DeHaven, W. I., Bird, G. S., Billingsley, J. M., Wang, H., Rao, P. E., Hutchings, A. B., Jouvin, M. H., Putney, J. W., and Kinet, J. P. (2008) Defective mast cell effector functions in mice lacking the CRACM1 pore subunit of store-operated calcium release-activated calcium channels. *Nat. Immunol.* 9, 89–96.
26. Gwack, Y., Srikanth, S., Oh-Hora, M., Hogan, P. G., Lamperti, E. D., Yamashita, M., Gelinas, C. et al. (2008) Hair loss and defective T- and B-cell function in mice lacking ORAI1. *Mol. Cell Biol.* 28, 5209–5222.
27. Williams, R. T., Manji, S. S., Parker, N. J., Hancock, M. S., Van Stekelenburg, L., Eid, J. P., Senior, P. V. et al. (2001) Identification and characterization of the STIM (stromal interaction molecule) gene family: Coding for a novel class of transmembrane proteins. *Biochem. J.* 357, 673–685.
28. Baba, Y., Nishida, K., Fujii, Y., Hirano, T., Hikida, M., and Kurosaki, T. (2008) Essential function for the calcium sensor STIM1 in mast cell activation and anaphylactic responses. *Nat. Immunol.* 9, 81–88.
29. Oh-Hora, M., Yamashita, M., Hogan, P. G., Sharma, S., Lamperti, E., Chung, W., Prakriya, M., Feske, S., and Rao, A. (2008) Dual functions for the endoplasmic reticulum calcium sensors STIM1 and STIM2 in T cell activation and tolerance. *Nat. Immunol.* 9, 432–443.
30. Park, C. Y., Hoover, P. J., Mullins, F. M., Bachhawat, P., Covington, E. D., Raunser, S., Walz, T., Garcia, K. C., Dolmetsch, R. E., and Lewis, R. S. (2009) STIM1 clusters and activates CRAC channels via direct binding of a cytosolic domain to Orai1. *Cell* 136, 876–890.
31. Yuan, J. P., Zeng, W., Dorwart, M. R., Choi, Y. J., Worley, P. F., and Muallem, S. (2009) SOAR and the polybasic STIM1 domains gate and regulate Orai channels. *Nat. Cell Biol.* 11, 337–343.
32. Muik, M., Fahrner, M., Derler, I., Schindl, R., Bergsmann, J., Frischauf, I., Groschner, K., and Romanin, C. (2009) A cytosolic homomerization and a modulatory domain within STIM1 C terminus determine coupling to ORAI1 channels. *J. Biol. Chem.* 284, 8421–8426.
33. Kawasaki, T., Lange, I., and Feske, S. (2009) A minimal regulatory domain in the C terminus of STIM1 binds to and activates ORAI1 CRAC channels. *Biochem. Biophys. Res. Commun.* 385, 49–54.
34. Jardin, I., Dionisio, N., Frischauf, I., Berna-Erro, A., Woodard, G. E., Lopez, J. J., Salido, G. M., and Rosado, J. A. (2013) The polybasic lysine-rich domain of plasma membrane-resident STIM1 is essential for the modulation of store-operated divalent cation entry by extracellular calcium. *Cell Signal.* 25, 1328–1337.
35. Varga-Szabo, D., Braun, A., Kleinschnitz, C., Bender, M., Pleines, I., Pham, M., Renne, T., Stoll, G., and Nieswandt, B. (2008) The calcium sensor STIM1 is an essential mediator of arterial thrombosis and ischemic brain infarction. *J. Exp. Med.* 205, 1583–1591.
36. Stiber, J., Hawkins, A., Zhang, Z. S., Wang, S., Burch, J., Graham, V., Ward, C. C. et al. (2008) STIM1 signalling controls store-operated calcium entry required for development and contractile function in skeletal muscle. *Nat. Cell Biol.* 10, 688–697.

37. Berna-Erro, A., Braun, A., Kraft, R., Kleinschnitz, C., Schuhmann, M. K., Stegner, D., Wultsch, T. et al. (2009) STIM2 regulates capacitive Ca^{2+} entry in neurons and plays a key role in hypoxic neuronal cell death. *Sci. Signal.* 2, ra67.
38. Bergmeier, W., Oh-Hora, M., McCarl, C. A., Roden, R. C., Bray, P. F., and Feske, S. (2009) R93W mutation in Orai1 causes impaired calcium influx in platelets. *Blood* 113, 675–678.
39. Jin, S., Chin, J., Kitson, C., Woods, J., Majmudar, R., Carvajal, V., Allard, J., Demartino, J., Narula, S., and Thomas-Karyat, D. A. (2013) Natural regulatory T cells are resistant to calcium release-activated calcium (CRAC/ORAI) channel inhibition. *Int. Immunol.* 25, 497–506.
40. Somasundaram, A., Shum, A. K., McBride, H. J., Kessler, J. A., Feske, S., Miller, R. J., and Prakriya, M. (2014) Store-operated CRAC channels regulate gene expression and proliferation in neural progenitor cells. *J. Neurosci.* 34, 9107–9123.
41. Kaufmann, U., Shaw, P. J., Kozhaya, L., Subramanian, R., Gaida, K., Unutmaz, D., McBride, H. J., and Feske, S. (2016) Selective ORAI1 inhibition ameliorates autoimmune central nervous system inflammation by suppressing effector but not regulatory T cell function. *J. Immunol.* 196, 573–585.
42. Davis, F. M., Janoshazi, A., Janardhan, K. S., Steinckwich, N., D'Agostin, D. M., Petranka, J. G., Desai, P. N. et al. (2015) Essential role of Orai1 store-operated calcium channels in lactation. *Proc. Natl. Acad. Sci. USA* 112, 5827–5832.
43. Matsumoto, M., Fujii, Y., Baba, A., Hikida, M., Kurosaki, T., and Baba, Y. (2011) The calcium sensors STIM1 and STIM2 control B cell regulatory function through interleukin-10 production. *Immunity* 34, 703–714.
44. Steinckwich, N., Myers, P., Janardhan, K. S., Flagler, N. D., King, D., Petranka, J. G., and Putney, J. W. (2015) Role of the store-operated calcium entry protein, STIM1, in neutrophil chemotaxis and infiltration into a murine model of psoriasis-inflamed skin. *FASEB J.* 29, 3003–3013.
45. Li, T., Finch, E. A., Graham, V., Zhang, Z. S., Ding, J. D., Burch, J., Oh-Hora, M., and Rosenberg, P. (2012) STIM1-Ca^{2+} signaling is required for the hypertrophic growth of skeletal muscle in mice. *Mol. Cell Biol.* 32, 3009–3017.
46. Mancarella, S., Potireddy, S., Wang, Y., Gao, H., Gandhirajan, R. K., Autieri, M., Scalia, R. et al. (2013) Targeted STIM deletion impairs calcium homeostasis, NFAT activation, and growth of smooth muscle. *FASEB J.* 27, 893–906.
47. Hartmann, J., Karl, R. M., Alexander, R. P., Adelsberger, H., Brill, M. S., Ruhlmann, C., Ansel, A. et al. (2014) STIM1 controls neuronal Ca^{2+} signaling, mGluR1-dependent synaptic transmission, and cerebellar motor behavior. *Neuron* 82, 635–644.
48. Garcia-Alvarez, G., Shetty, M. S., Lu, B., Yap, K. A., Oh-Hora, M., Sajikumar, S., Bichler, Z., and Fivaz, M. (2015) Impaired spatial memory and enhanced long-term potentiation in mice with forebrain-specific ablation of the Stim genes. *Front. Behav. Neurosci.* 9, 180.
49. Sun, S., Zhang, H., Liu, J., Popugaeva, E., Xu, N. J., Feske, S., White, C. L., 3rd, and Bezprozvanny, I. (2014) Reduced synaptic STIM2 expression and impaired store-operated calcium entry cause destabilization of mature spines in mutant presenilin mice. *Neuron* 82, 79–93.
50. Oh-Hora, M., Komatsu, N., Pishyareh, M., Feske, S., Hori, S., Taniguchi, M., Rao, A., and Takayanagi, H. (2013) Agonist-selected T cell development requires strong T cell receptor signaling and store-operated calcium entry. *Immunity* 38, 881–895.
51. McCarl, C. A., Khalil, S., Ma, J., Oh-Hora, M., Yamashita, M., Roether, J., Kawasaki, T. et al. (2010) Store-operated Ca^{2+} entry through ORAI1 is critical for T cell-mediated autoimmunity and allograft rejection. *J. Immunol.* 185, 5845–5858.

52. Beyersdorf, N., Braun, A., Vogtle, T., Varga-Szabo, D., Galdos, R. R., Kissler, S., Kerkau, T., and Nieswandt, B. (2009) STIM1-independent T cell development and effector function in vivo. *J. Immunol.* 182, 3390–3397.
53. Jenkins, M. K., Schwartz, R. H., and Pardoll, D. M. (1988) Effects of cyclosporine A on T cell development and clonal deletion. *Science* 241, 1655–1658.
54. Gao, E. K., Lo, D., Cheney, R., Kanagawa, O., and Sprent, J. (1988) Abnormal differentiation of thymocytes in mice treated with cyclosporin A. *Nature* 336, 176–179.
55. Neilson, J. R., Winslow, M. M., Hur, E. M., and Crabtree, G. R. (2004) Calcineurin B1 is essential for positive but not negative selection during thymocyte development. *Immunity* 20, 255–266.
56. Bueno, O. F., Brandt, E. B., Rothenberg, M. E., and Molkentin, J. D. (2002) Defective T cell development and function in calcineurin Aβ-deficient mice. *Proc. Natl. Acad. Sci. USA* 99, 9398–9403.
57. Cheng, K. T., Alevizos, I., Liu, X., Swaim, W. D., Yin, H., Feske, S., Oh-Hora, M., and Ambudkar, I. S. (2012) STIM1 and STIM2 protein deficiency in T lymphocytes underlies development of the exocrine gland autoimmune disease, Sjögren's syndrome. *Proc. Natl. Acad. Sci. USA* 109, 14544–14549.
58. Takahashi, T., Tagami, T., Yamazaki, S., Uede, T., Shimizu, J., Sakaguchi, N., Mak, T. W., and Sakaguchi, S. (2000) Immunologic self-tolerance maintained by $CD25^{+}CD4^{+}$ regulatory T cells constitutively expressing cytotoxic T lymphocyte-associated antigen 4. *J. Exp. Med.* 192, 303–310.
59. Yu, X., Harden, K., Gonzalez, L. C., Francesco, M., Chiang, E., Irving, B., Tom, I. et al. (2009) The surface protein TIGIT suppresses T cell activation by promoting the generation of mature immunoregulatory dendritic cells. *Nat. Immunol.* 10, 48–57.
60. Stritesky, G. L., Jameson, S. C., and Hogquist, K. A. (2012) Selection of self-reactive T cells in the thymus. *Annu. Rev. Immunol.* 30, 95–114.
61. Baldwin, T. A., Hogquist, K. A., and Jameson, S. C. (2004) The fourth way? Harnessing aggressive tendencies in the thymus. *J. Immunol.* 173, 6515–6520.
62. McCarl, C. A., Picard, C., Khalil, S., Kawasaki, T., Rother, J., Papolos, A., Kutok, J. et al. (2009) ORAI1 deficiency and lack of store-operated Ca^{2+} entry cause immunodeficiency, myopathy, and ectodermal dysplasia. *J. Allergy Clin. Immunol.* 124, 1311e7–1318e7.
63. Picard, C., McCarl, C. A., Papolos, A., Khalil, S., Luthy, K., Hivroz, C., LeDeist, F. et al. (2009) STIM1 mutation associated with a syndrome of immunodeficiency and autoimmunity. *N. Engl. J. Med.* 360, 1971–1980.
64. Lacruz, R. S. and Feske, S. (2015) Diseases caused by mutations in ORAI1 and STIM1. *Ann. N. Y. Acad. Sci.* 1356, 45–79.
65. Potier, M., Gonzalez, J. C., Motiani, R. K., Abdullaev, I. F., Bisaillon, J. M., Singer, H. A., and Trebak, M. (2009) Evidence for STIM1- and Orai1-dependent store-operated calcium influx through I_{CRAC} in vascular smooth muscle cells: Role in proliferation and migration. *FASEB J.* 23, 2425–2437.
66. Nakanishi, S. and Okazawa, M. (2006) Membrane potential-regulated Ca^{2+} signalling in development and maturation of mammalian cerebellar granule cells. *J. Physiol.* 575, 389–395.
67. Spitzer, N. C., Root, C. M., and Borodinsky, L. N. (2004) Orchestrating neuronal differentiation: Patterns of Ca^{2+} spikes specify transmitter choice. *Trends Neurosci.* 27, 415–421.
68. Leclerc, C., Webb, S. E., Daguzan, C., Moreau, M., and Miller, A. L. (2000) Imaging patterns of calcium transients during neural induction in *Xenopus laevis* embryos. *J. Cell Sci.* 113(Pt 19), 3519–3529.

7 Assessing the Molecular Nature of the STIM1/Orai1 Coupling Interface Using FRET Approaches

Yandong Zhou, Youjun Wang, and Donald L. Gill

CONTENTS

7.1 INTRODUCTION

Ca^{2+} signals control a vast array of cellular processes and are mediated by the concerted effort of a spectrum of Ca^{2+} channels, transporters, and pumps present in the plasma membrane (PM) and endoplasmic reticulum (ER) membrane [1,2]. In nonexcitable cell types, store-operated channels (SOCs) are the major means through which extracellular Ca^{2+} enters cells to generate Ca^{2+} signals. The two major components of SOCs, STIM1 and Orai1, were identified a decade ago [3–8], and extensive studies have focused on the mechanisms of how STIM1 becomes activated in response to store depletion, how Orai1 subunits assemble to form the channel, and how the STIM1 molecule interacts with Orai1 to achieve channel gating [1,9–13] (discussed in Chapters 2 and 3). The physical interaction between STIM1 and Orai1 has been one of the most important parameters in order to understand the stoichiometry and gating mechanism of the STIM1/Orai1 complex. The physical interaction between STIM and Orai1 has been studied extensively using Förster (fluorescence) resonance energy transfer (FRET) imaging technology [13–16]. FRET allows the visualization and quantification of macromolecular interactions in living cells by measuring light energy transfer between closely associated fluorescently tagged proteins [17]. Since STIM1 and Orai1 are in different membranes and interact at discrete ER–PM junctions, FRET is a highly effective means for assessing this interaction. FRET occurs in a short range (5–10 nm) across which energy is transferred from an excited donor to acceptor fluorophore [18]. The efficiency of energy transfer is inversely proportional to the sixth power of the distance between donor and acceptor fluorophores; hence, FRET measurements give extremely sensitive information on the distance separating the pair. The mechanism and uses of FRET are well described in other reviews [19–23].

Despite intense study, the molecular nature of the coupling interaction between the activated STIM1 protein and the Orai1 channel remains elusive. Our approach to studying this interaction is to use a fragment of the STIM1 protein that itself is able to mediate full activation of the Orai1 channels. This 100-amino acid fragment is known as SOAR (STIM–Orai-activating region) [24,25] (see Chapter 2). In a previous report, Shen et al. [26] purified and crystallized this fragment from STIM1 and revealed that it can exist as a dimer. Indeed, the SOAR fragment appears to be an important "core" structure within the STIM1 protein contributing to dimerization of the whole STIM1 protein [12]. We have been able to express the SOAR protein as a concatenated dimer construct and therefore genetically manipulate the exact dimeric composition of SOAR expressed in cells [13]. We revealed that a single point mutation (F394H) in the Orai1 binding site of the SOAR fragment from STIM1 can completely prevent STIM1 binding to and activation of Orai1 channels [16]. Importantly, since the SOAR dimer contains two of these sites, we can modify either one or both of these sites to study the requirements for interaction with the Orai1 channel. Using a set of concatemer–dimers of SOAR containing one or two F394H mutations, we are able to study how the SOAR dimer interacts with Orai1 and whether each SOAR unit within a dimer is equivalent. The aim of this chapter is to provide a detailed protocol to assess the STIM/Orai interaction by FRET. We describe how to ensure the FRET assay is reliable and consistent and how to analyze and how to interpret the FRET data.

7.2 STRATEGY FOR QUANTITATIVE FRET MEASUREMENT

We first introduce the term "E-FRET," which is a fundamental "instrument-independent" measurement of FRET, representing the "FRET efficiency" or what percentage of donor emission is quenched by the acceptor [27]. Any microscope that can simultaneously collect 3-channel images is usable for such FRET experiments.

Some imaging software programs have already been embedded with E-FRET formulation capability. It is also possible to use either MATLAB® or Excel to manually calculate the E-FRET values. The current chapter will focus on E-FRET analysis methodology since it is the most widely used in our field and, importantly, the values of E-FRET are directly comparable between different labs and imaging systems [27]. In this procedure, 3-channel image collection is undertaken for FRET measurements using a traditional epifluorescence microscope. It is important to correct for cross-channel "bleed-through" of fluorescence. The following formula is used to calculate the corrected FRET ($FRET_C$):

$$FRET_C = I_{DA} - a * I_{AA} - d * I_{DD} \tag{7.1}$$

in which

I_{DD}, I_{AA}, and I_{DA} represent the intensity of the background-subtracted CFP, YFP, and FRET images, respectively (these are defined further later)
$FRET_C$ represents the corrected energy transfer
a represents the measured coefficient of the acceptor (YFP) bleed-through the FRET filter cube
d is the measured coefficient of the donor (CFP) bleed-through the FRET filter cube

Under this condition, any excess CFP donor or YFP acceptor molecules will also be counted during FRET analysis even though there is no FRET occurring for these excess molecules, as shown in Figure 7.1. For example, in an experiment comparing the FRET between protein "X" (CFP labeled) and protein "Y" (YFP labeled) under two different conditions (i.e., either excess CFP donor or excess YFP acceptor), it would be easy to come to the wrong conclusion when comparing the FRET between Figure 7.1a (excess CFP donor) and Figure 7.1b (excess YFP acceptor), even though the real energy transfer should be the same. Thus, the expression level of donor and the ratio of donor to acceptor must be kept similar between the groups when undertaking quantitative FRET. For this, one of the two tagged proteins should be stably transfected to maximize consistency. The other tagged protein in the FRET pair can be transiently expressed in the stable line. Cells selected for analysis should have a narrow range of transiently expressed protein to assure that the ratio of fluorescence for the two proteins does not significantly vary. This is discussed further later. Another case is shown in Figure 7.1c and d, in which we are comparing FRET between protein X and another protein "Z" with two binding sites for X, or a mutated protein Z* in which one binding site is eliminated. As shown in the diagram, it would be easy to obtain a different result under conditions of altered ratio of X and Z proteins. With an excess of acceptor protein Z (Figure 7.1c), the E-FRET level might not be so different for Z versus Z* even though Z has twice as many X binding sites as Z*. However, with an excess of donor X protein (Figure 7.1d), the binding to Z or Z* will be saturated and there will be a quantitative difference in the amount of FRET depending on the number

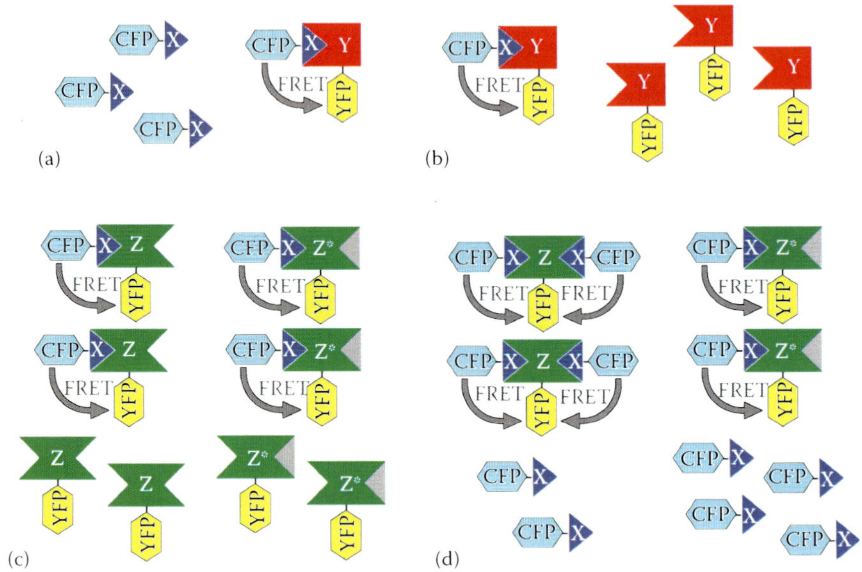

FIGURE 7.1 FRET measurements and the relationship with donor and acceptor expression levels. (a and b) Assessing the interaction between protein X and protein Y by FRET. X is labeled with CFP (donor) and Y is labeled with YFP (acceptor). The ratio of the expression of donor to acceptor is different in two versions of the experiment. In (a), CFP-labeled protein X is in excess over YFP-labeled protein Y. In (b), YFP-labeled protein Y is in excess over CFP-labeled protein X. The E-FRET value obtained from these two experiments (a and b) will not be consistent—the value from (a) will be smaller than that from (b). Thus, even if the amount of X–Y complex was the same in each experiment (i.e., the theoretical FRET level would the same), the calculated E-FRET result obtained from Equation 7.6 would vary because the amount of free unbound donor or acceptor would be different. Therefore, for this experiment, in order to obtain consistent and comparable results, it is important that the ratio of YFP and CFP is maintained within and between experiments. (c and d) A different experiment aimed at determining the interaction between CFP-labeled protein X and a different protein Z that contains two binding sites for X. The aim is to determine if a mutation that blocks binding of X to one site (protein Z*) will affect binding of X to the other site. Both proteins Z and Z* are labeled with acceptor YFP. In (c), YFP-Z (or YFP-Z*) is expressed in excess over CFP-X, whereas in (d) CFP-X is expressed in excess of YFP-Z (or YFP-Z*). In this case, any defect in protein Z* in binding X will only be detected by FRET under the condition in (d) and not under the condition in (c). Thus, only under conditions where there is an excess of donor and thus CFP-X can saturate the YFP-Z protein sites (i.e., the condition in (d)) will it be possible to assess the difference in binding caused by the Z* mutation. Conversely, under the conditions in (c) in which there are no saturating levels of CFP-X, the difference in binding to Z as opposed to Z* would be much smaller and difficult to determine.

of available binding sites for X. Taken together, it is crucial to strictly control the expression level of donor and acceptor and the ratio of donor to acceptor in FRET analyses to obtain accurate and consistent data. In our studies aiming to quantitate FRET between SOAR and Orai1, we therefore utilize stable cell lines in which the expression of at least one of the FRET pairs remains constant.

7.3 FRET MEASUREMENTS TO QUANTITATE SOAR–ORAI1 INTERACTIONS

7.3.1 Generation of Stable Cell Lines

We recently constructed a number of concatemer–dimers of SOAR as described in Figure 7.2a in order to study the STIM1–Orai1 coupling interaction. As described in our recent study [13], we hypothesize that the wild-type SOAR homodimer (YFP-S-S) has two Orai1 binding sites, while the F394H-mutated SOAR heterodimers (YFP-SH-S or YFP-S-SH) have only one Orai1 binding site. When both SOAR monomers are mutated to F394H, there is no interaction with Orai1. One important aspect of these studies was to assess FRET between each concatemer and Orai1 channels. As described in Figure 7.1c and d, it is important to ensure an excess of donor is expressed in cells (e.g., Orai1–CFP or a CFP-tagged STIM1-binding fragment of Orai1, as described later). In addition, it is necessary that cell lines are generated that

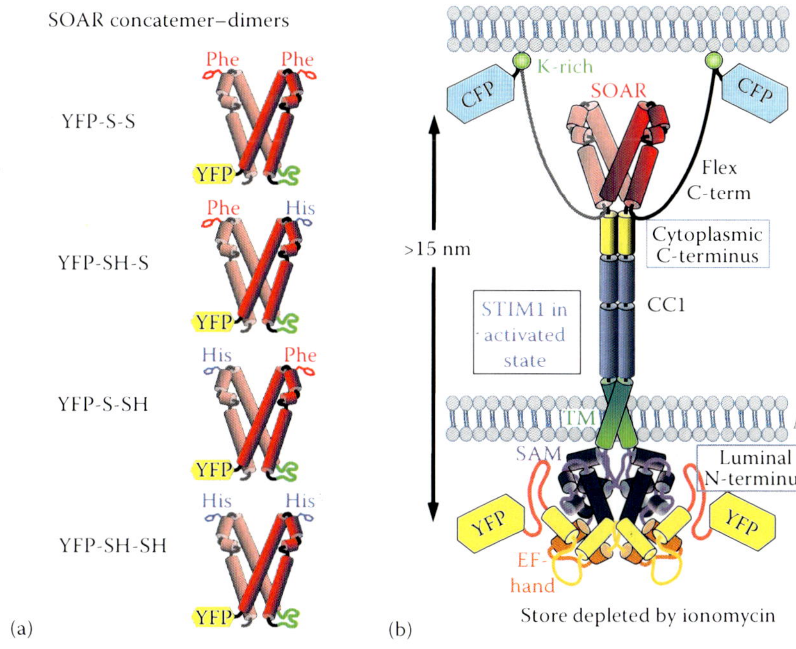

FIGURE 7.2 (a) Schematic diagram of the four SOAR concatemers. YFP-S-S is the homodimer concatemer of wild-type SOAR concatemer–dimer, YFP-SH-S is the YFP-tagged heterodimer concatemer of SOAR containing one SOAR-F394H unit and one wild-type SOAR unit, YFP-S-SH is the YFP-tagged heterodimer concatemer of SOAR containing one wild-type SOAR unit and one SOAR-F394H unit, and YFP-SH-SH is the YFP-tagged homodimer concatemer of SOAR containing two SOAR-F394H units. (b) YFP–STIM1–CFP construct: This probe was used to determine the fluorescence intensity of equimolar CFP and YFP while assuring minimal intramolecular CFP–YFP FRET since under conditions of STIM1 activation (emptying of stores with ionomycin), the two fluorophores are maximally separated (>15 nm) at each end of the STIM1 molecule.

of view with a 20× objective, they are ready for expanding into 6-well plates. Cells should not become confluent (usually they reach confluence in 4 days). Contaminated wells should be emptied by aspiration. After confluence is reached, aspirate the old medium out, add 1.5 mL fresh medium, detach the cells by pipetting up and down to separate cells (9–12 times is sufficient; avoid bubble formation), and transfer all the cells into a 10 cm dish containing 10 mL medium. Let the cells grow until they reach confluence, trypsinize and resuspend cells, and utilize some for functional tests (e.g., Fura-2 Ca^{2+} imaging or electrophysiology). If three or four clones test positive for desired Ca^{2+} signaling and/or patch-clamp results, these clones can be grown further and the rest discarded. For selected clones, cells are grown in 10 cm dishes and frozen in approximately 10 vials.

7.3.2 Calibration of FRET Imaging System

We use a Leica DMI6000B automatic turret-equipped microscope and a 40× oil objective (N.A.1.35; Leica) with the following filter cubes: CFP (438Ex/483Em), YFP (500Ex/542Em), and FRET (438Ex/542Em). We collect images using a Hamamatsu ORCA-Flash4.0 camera controlled by SlideBook 6.0 software (Intelligent Imaging Innovations; Denver, CO) as described in detail elsewhere [13]. We use a xenon lamp and Lambda 10-3 shutter wheel (Sutter), although the shutter wheel is not used for FRET measurements. Before any FRET imaging measurements are taken, the instrument must be calibrated to determine the following parameters:

1. The ratio of fluorescence intensity for equimolar donor and acceptor used
2. Bleed-through coefficients for donor and acceptor
3. The *G* constant for E-FRET analysis

These are explained below. Our FRET imaging experiments are undertaken with cells in modified Ringer's solution [13] containing (in mM) 107 NaCl, 7.2 KCl, 1.2 $MgCl_2$, 1 $CaCl_2$, 11.5 glucose, and 20 HEPES, with pH 7.2.

7.3.2.1 Assessment of Equimolar Fluorescence Intensity of Donor and Acceptor

The fluorescence intensities of YFP and CFP tags are not the same even when the expression level of YFP and CFP are equal. These intensities also vary with different platforms and settings. It is necessary to calculate the ratio of the fluorescence intensity of the same molarity of YFP to CFP. This information is particularly important in assessing whether there is excess donor or acceptor in the region of interest for FRET analysis. To accomplish this goal, we have designed a specific STIM1-derived construct as shown in Figure 7.2b. We added a YFP tag to the N-terminus of the STIM1 molecule that is located in the ER lumen and a CFP tag to the STIM1 C-terminus located in the cytosol. When this construct is expressed, it will ensure that in the same cell there is an identical level of expression of YFP and CFP. With resting STIM1, although the N- and C-termini are separated across the ER membrane, their distance is unknown, and there may be some FRET between them. However, after store depletion, the protein unfolds and stretches out particularly in the ER–PM junctions, so that the distance between N- and C-termini is about 20 nm. The maximal distance for FRET between fluorophores to obtain measurable FRET is approximately 10 nm [27]. Since we measure fluorescence exclusively at the cell periphery, and since all the STIM1 molecules in this region are adjacent to one another in ER–PM junctions, all the signal comes from YFP and CFP molecules that are undergoing minimal FRET. Thus, to avoid FRET signals, we express YFP–STIM1–CFP in cells and treat them with ionomycin to empty stores and assure that STIM1 molecules move into ER–PM junctions. Ionomycin can be used up to 2.5 μM to selectively release Ca^{2+} from stores but not directly move Ca^{2+} into the cell across the PM. Alternatively, the ER Ca^{2+} pump inhibitor, thapsigargin, can be applied to cells to empty stores that cause slightly slower Ca^{2+} release. The CFP tag is brought close to PM in junctions, but the intermembrane distance of the junctions is 12–17 nm [29,30], and the CFP–YFP distance will be larger than the 10 nm maximum for FRET signals (Figure 7.2b).

At the beginning of a FRET imaging experiment, all microscope parameters are fixed, including exposure times for YFP, CFP, and FRET channels. We randomly select different fields and collect the images with the YFP or CFP channels. Under our normal settings, using the YFP–STIM1–CFP probe with equimolar YFP and CFP, the ratio of fluorescence intensity is about 1.33 ± 0.04. In our FRET experiments, if the ratio is smaller than 1.33, this indicates there is excess of CFP donor–tagged protein. All the parameters of light source output, microscope settings, and camera settings in the software should be kept exactly the same for the calibration and data collection as follows.

7.3.2.2 Calibration of Bleed-Through Coefficients

As shown in formula (7.1), the accuracy of the corrected energy transfer, $FRET_C$, depends on determination of the bleed-through parameters. The bleed-through parameters need to be recalibrated if the xenon light bulb is changed or anything in the light pathway is replaced. Although only two bleed-through parameters (*d* and *a*) were considered in formula (7.1), in fact there are four bleed-through coefficients:

$$d = \frac{I_{DA}}{I_{DD}} \tag{7.2}$$

$$a = \frac{I_{DA}}{I_{AA}} \tag{7.3}$$

$$b = \frac{I_{AA}}{I_{DD}} \tag{7.4}$$

$$c = \frac{I_{DD}}{I_{AA}} \tag{7.5}$$

The definitions of I_{DD}, I_{AA}, and I_{DA} were mentioned earlier and are explained further in Figure 7.3a. Thus, I_{DD} is the intensity of fluorescence collected using the donor excitation and emission wavelengths, I_{AA} is the fluorescence intensity collected using

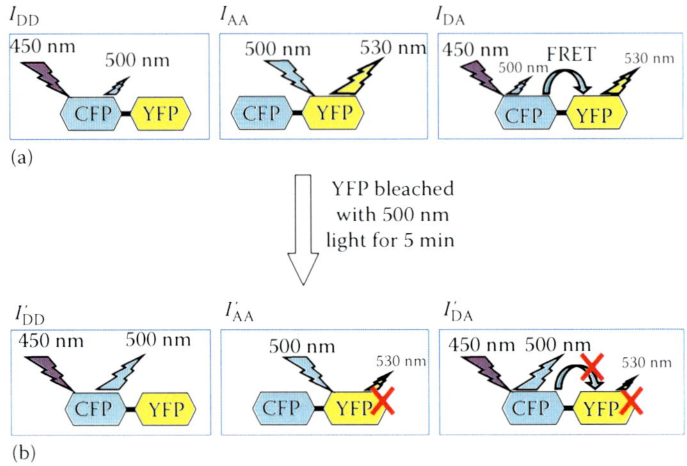

FIGURE 7.3 Three-channel demonstration of the partial receptor YFP-photobleaching method to determine the value of the G parameter using the YFP–CFP construct. Left, I_{DD} is the CFP fluorescence intensity from images collected through the CFP filter cube; middle, I_{AA} is the YFP fluorescence intensity from images collected through the YFP filter cube; and right, I_{DA} is the FRET intensity from images collected through the FRET filter cube. (a) Before YFP photobleaching, the emission of CFP will be quenched by YFP that causes FRET to occur. (b) After YFP photobleaching (with application of 500 nm light), the emission of CFP will no longer be quenched by YFP.

the acceptor excitation and emission wavelengths, and I_{DA} is the fluorescence intensity collected using the donor excitation and acceptor emission wavelengths. In our experiments, bleed-through coefficient (7.2) and (7.4) represent the bleed-through of donor (CFP) through the FRET filter cube and YFP filter cube, respectively. Bleed-through coefficients (7.3) and (7.5) represent the bleed-through of acceptor (YFP) through the FRET filter cube and CFP filter cube, respectively. However, with the advanced technology for YFP or CFP filter manufacture, the narrower bandwidth of exciter and emitter has avoided the YFP fluorescence bleed-through CFP filter cube and vice versa. Thus, in our system, $b = c = 0$. Therefore, we need to obtain accurate values for a and d. However, if you are not sure about the spectra and bandwidth of your filter sets, it is recommended to perform the calibration for all four coefficients.

For these calibrations, it is important to use the protein of interest tagged with CFP or YFP, but not the CFP or YFP empty vector constructs since these would be expressed at much higher levels. The following is a detailed protocol for obtaining the bleed-through coefficients in our experiments on Orai1–CFP and the YFP–SOAR dimers:

1. HEK cells are transfected with the Orai1–CFP construct alone or the YFP-S-S construct alone for 24 h.
2. Using the settings that were optimized in Section 7.3.2.1, two sets of images (CFP and FRET) are collected from different fields at room temperature

using HEK cells transiently expressing Orai1–CFP. Another two sets of images (YFP and FRET) are collected from different fields using HEK cells expressing YFP-S-S.
3. Formulas (7.2) and (7.3) are used to calculate the bleed-through coefficients using Excel. We use SlideBook software to automatically calculate the bleed-through coefficients after the images have been collected.

7.3.3 Determination of *G* Constant Number

As described earlier, *G* is a constant parameter for a specified imaging system and a pair of FRET fluorophores. It is described in more detail elsewhere [27]. *G* is pivotal for the calculation of FRET efficiency (E_{app}):

$$E_{app} = \frac{\text{FRET}_C}{\text{FRET}_C + G * I_{DD}} \tag{7.6}$$

G is defined as the ratio of the sensitized emission FRET_c to the corresponding amount of recovery of donor emission in the I_{DD} channel after photobleaching the acceptor (Figure 7.3). This gives the following formula:

$$G = \frac{\text{FRET}_C}{\left(I'_{DD} - I_{DD}\right)} \tag{7.7}$$

As described later, photobleaching of the YFP acceptor can be undertaken by applying 500 nm light for 5 min. The acceptor photobleaching is usually not complete, just partial. The formula is therefore modified as follows:

$$G = \frac{\text{FRET}_C - \text{FRET}'_C}{I'_{DD} - I_{DD}} \tag{7.8}$$

in which FRET_C represents the corrected energy transfer before acceptor photobleaching (see formula (7.1)), and FRET'_C is the corrected energy transfer after acceptor photobleaching. I_{DD} (defined earlier) represents the intensity of the background-subtracted CFP before acceptor photobleaching, while I'_{DD} represents the intensity of the background-subtracted CFP after acceptor photobleaching.

We made a specific construct designed for the determination of the *G* constant. This is the YFP–CFP construct that we described in a recent paper [13]. CFP was flanked with HindIII and BamHI restriction sites and subcloned into the pEYFP vector. The expressed YFP–CFP proteins will result in maximum FRET between CFP and YFP fluorophores as shown in Figure 7.3.

The following is our protocol for determining the *G* constant:

1. HEK cells are transfected with the pEYFP–CFP construct by electroporation.
2. Using the settings that were optimized in Section 7.3.2.1, three sets of images (CFP, YFP, and FRET) are collected at room temperature for HEK

cells transiently expressing the YFP–CFP protein. This step is called pre-photobleaching, and $FRET_C$ and I_{DD} are obtained from this step, as shown in Figure 7.3a.
3. The photobleaching step can vary in effectiveness and how confined the area of bleaching is, depending on the microscope/software platforms. The process described here should apply to most platforms. To bleach the YFP fluorophore in this field, the output of the light source is set to 100% and exposure time for YFP is set to 5 min (this time can increase if 5 min is insufficient).
4. After photobleaching, it is important to restore the settings back to step 2.
5. A further three sets of images (CFP, YFP, and FRET) are collected in the same field at room temperature for HEK cells transiently expressing the YFP–CFP protein. This step is called post-photobleaching, and $FRET'_C$ and I'_{DD} are obtained from this step, as shown in Figure 7.3b. At this step, the YFP image should be dim compared to that obtained in step 2. If the difference in intensity is hard to tell, repeat step 3 and increase the exposure time to 15 min.
6. The value of the G constant that was determined using formula (7.8) for the data in Figure 7.4 was 1.9 ± 0.1 (n = 32 cells) using formula (7.8).

7.3.4 Data Collection

A useful example of our measurement of E-FRET is given in Figure 7.4 in which we investigate whether the two SOAR units within the SOAR dimer are equivalent. Thus, an important question was whether the active SOAR dimer unit (the active site within the dimeric STIM1 molecule) functions in a bimolecular manner in which each of the two SOAR units in the dimer together associate with each of the two adjacent Orai1 subunits within the hexameric Orai1 channel, as shown in Figure 7.4a. Alternatively, each SOAR unit in the SOAR dimer may interact independently with a single Orai1 subunit in a unimolecular manner, as shown in Figure 7.1b.

To investigate this further, we designed a specific Orai1-derived construct that expresses only the strong STIM1-binding C-terminus of Orai1 (aa 267–301). This construct, named PM–CFP–Orai1CT, includes a PM-directed transmembrane helix, which assures that it is expressed only in the PM [13] (Figure 7.4a). We generated a stable HEK cell line expressing the PM–CFP–Orai1CT construct comparable to the Orai1–CFP-expressing line described earlier. HEK–PM–CFP–Orai1CT stable cell lines were transfected with YFP-S-S, YFP-SH-S, YFP-S-SH, YFP-SH-SH, or YFP vector, 24 h prior to experiments. Transfected cells are seeded on 25 mm round glass coverslips.

1. To ensure consistent data collection, images for each coverslip are obtained under identical conditions. During experiments, unused coverslips are kept in 6-well plates in the incubator until utilized and assembled in the chamber.
2. Fields are selected with more YFP-positive cells, using the settings that were optimized in Section 7.3.2.1. Sets of images (CFP, YFP, and FRET) are collected at room temperature. Further sets of images (CFP, YFP, and FRET)

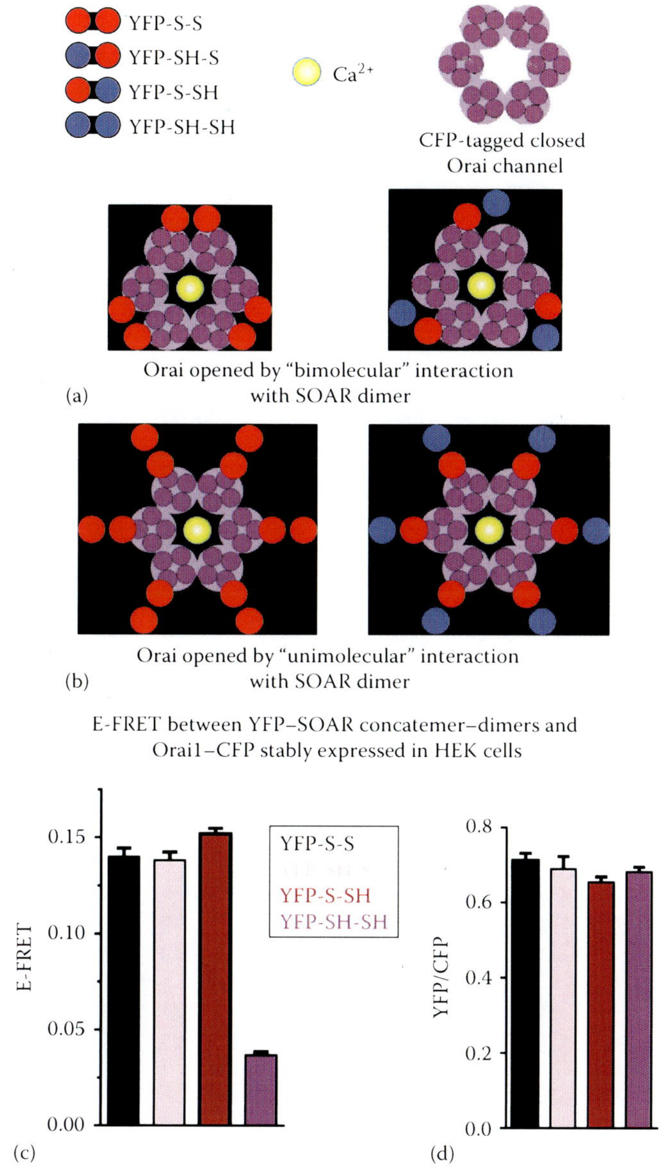

FIGURE 7.4 Evidence for a "unimolecular" interaction between the SOAR dimer and the Orai1 subunit within hexamer. Two possible models for the SOAR–Orai1 interaction: (a) The bimolecular binding of a homomeric SOAR dimer (red) to two adjacent Orai1 subunits (purple). (b) The unimolecular binding of a single monomer within the SOAR homodimer (red) to a single Orai1 subunit (purple). Changes in the association of heterodimers of SOAR comprising one WT and one F394H monomer (red/blue) occurring in the bimolecular (a) and unimolecular (b) models are shown on the right side of (a) and (b), respectively. (c) Near-PM values of E-FRET between Orai1–CFP in HEK–Orai1–CFP stable cells and transiently expressed YFP-S-S (black), YFP-SH-S (green), YFP-S-SH (red), or YFP-SH-SH (blue). (d) YFP/CFP ratios of the near-PM-expressed Orai1–CFP and YFP dimers in the cells used for E-FRET in (c). *(Continued)*

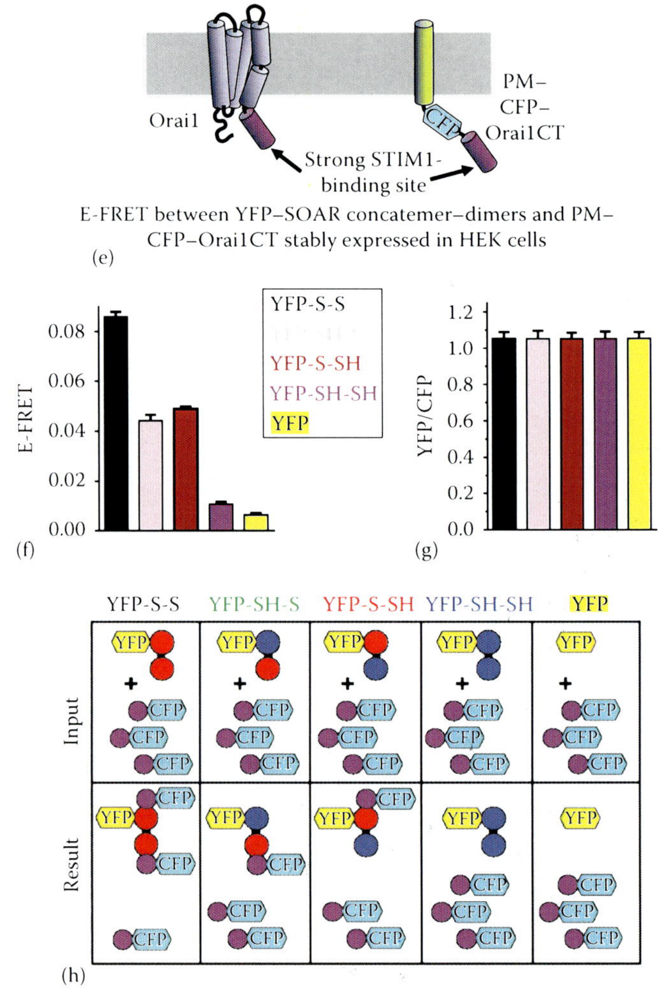

FIGURE 7.4 (*Continued*) (e) Cartoon of Orai1 protein (left) and PM–CFP–Orai1CT construct (right) comprising the C-terminus of Orai1 (residues 267–301) attached to CFP and a single PM-directed transmembrane-spanning helix. (f) Near-PM E-FRET between PM–CFP–Orai1CT stably expressed in HEK cells and transiently expressed SOAR concatemer–dimers: YFP-S-S (black), YFP-SH-S (green), YFP-S-SH (red), YFP-SH-SH (blue), and YFP alone (yellow). (g) YFP/CFP ratios of near-PM-expressed Orai1–CFP and YFP dimers in the cells used in (h). (h) Model interpreting the interactions between CFP-tagged Orai1CT (purple) and each of the four YFP-tagged SOAR dimer–concatemers (or YFP alone) used in the E-FRET experiments shown in (f). Top row shows input reactants for each condition; bottom row shows resulting interactions. (Modified from Fig. 5, Zhou, Y. et al., *Nat. Commun.*, 6, 8395, 2015, with permission.)

are collected again from different fields. The number of views will depend on cell density. If transfected cells are sparse, further views are needed until the total cell count is 30–40 cells or more.

3. After imaging, glass coverslips are discarded and the chamber is washed, and the subsequent coverslip readied for imaging. It is important to try to ensure that each coverslip is treated as consistently as possible.

7.3.5 Data Analysis

To analyze FRET images and export the data, we use SlideBook 6.0 (Intelligent Imaging Innovations). The FRET module comes with this software. For other platforms, the detailed process will be different in exporting the raw data. If the software being used does not have the module for E-FRET calculation, then raw data for I_{DD}, I_{AA}, and I_{DA} are exported manually and calculations for E-FRET are made in Excel using formulas (7.1) and (7.6) with the given bleed-through coefficients (*a* and *d*) and *G* parameter. After obtaining all the raw data for all groups, it is important to exclude those cells that have abnormal expression of YFP-tagged concatemer. As mentioned in Sections 7.2 and 7.3.2.1, only cells that have similar expression levels of YFP-tagged concatemer and also have YFP/CFP ratios that are smaller than 1.33 should be analyzed. To do this in an unbiased manner:

1. Export the data into an Excel file.
2. For each cell, calculate the YFP/CFP ratio using the values of I_{AA}/I_{DD}.
3. Each cell will have two important parameters: E-FRET and YFP/CFP. Sort the Excel sheet based on the value of YFP/CFP and then exclude all the cells with the value larger than 1.33. To ensure there is an excess of donor, we only select cells that have YFP/CFP around 1 or smaller as shown in Figure 7.4d and g. Thus, the E-FRET values are analyzed from cells with a narrow range of YFP/CFP ratios.

As shown in Figure 7.4c and d, the E-FRET between Orai1–CFP and YFP-S-S is similar to the E-FRET between Orai1–CFP and YFP-SH-S or YFP-S-SH. These results indicate that the single mutation (F394H) in the SOAR dimer has no effect on dimer binding to the Orai1 hexameric channel. This suggests that the bimolecular interaction shown in Figure 7.4a cannot explain how the SOAR dimer interacts with Orai1. Instead, we invoke a unimolecular hypothesis in which only one SOAR unit in the SOAR dimer is binding to Orai1 (Figure 7.4b). This explains the E-FRET results in which mutation of one of the two SOAR subunits in the SOAR dimer will not affect its interaction with Orai1 channel.

The strong binding site for STIM1 lies in the helical cytosolic Orai1 C-terminus (amino acids 267–301). Therefore, we made a new construct (PM–CFP–Orai1CT) in which we expressed this strong binding site attached via CFP to a PM-targeted transmembrane helix (Figure 7.4e). We used this probe to determine whether the SOAR dimer molecule could simultaneously bind to two Orai1 binding sites for STIM1. We examined this by measuring E-FRET. As shown in Figure 7.4f and g, the E-FRET between PM–CFP–Orai1CT and YFP-S-S is almost double that between

PM–CFP–Orai1CT and either YFP-SH-S or YFP-S-SH. In contrast, the E-FRET level with YFP-SH-SH was very close to background nonspecific FRET observed with cells expressing YFP (Figure 7.4f and g). We interpret these data in light of the models shown in Figures 7.1d and 7.4h. Thus, each YFP-S-S dimer is able to bind two PM–CFP–Orai1CT molecules, whereas YFP-SH-S or YFP-S-SH is only able to bind one PM–CFP–Orai1CT. This is almost twice as much CFP donor that can be quenched by the YFP acceptor. These data suggest that the two SOAR units within the SOAR dimer are equivalent and can each independently bind to one Orai1 subunit within the hexameric channel. This provides important new information on the molecular binding stoichiometry of the STIM1–Orai1 coupling interface, as we recently described [13].

7.4 CONCLUSIONS

FRET is one of the most powerful techniques for studying protein interactions. We detail here that meaningful quantitative analysis of protein–protein interactions can be achieved; however, careful awareness of the expression levels and ratio of donor and acceptor are crucial to such determination. The STIM and Orai coupling interaction is particularly suited to analysis by FRET. Thus, each protein resides in a separate, distinct membrane (ER or PM), and the proteins only interact within a "FRETable" distance (<10 nm) when they enter ER–PM junctions. Thus, there is little, if any, STIM1–Orai1 FRET in resting cells with full Ca^{2+} stores in which STIM has not been activated and has not been trapped within ER–PM junctions. For full-length STIM1, although FRET only occurs in the ER–PM region, the excess of Orai1 or STIM1 outside of the ER–PM region will affect the FRET analysis as described in Figure 7.1.

The use of SOAR instead of the full-length STIM1 molecule provides several strong advantages to study the coupling interaction with the Orai1 channel. SOAR is cytoplasmic and therefore is not restricted to the ER; hence, it can bind to any Orai1 in the PM where FRET will occur. By using a series of SOAR concatemers with either one or two F394H mutations expressed in the stable HEK–Orai1–CFP cell line, we were able to show that only one site in the SOAR dimer was required for activation of the Orai1 channel. Our FRET analysis was important in revealing that the interaction of the SOAR dimer with Orai1 required only one of the two Orai1 binding sites within the dimer. This revealed our "unimolecular coupling" model [13] that argues against the bimolecular coupling model that had been predicted earlier (see Chapter 2).

Using HEK–PM–CFP–Orai1CT stable cell line that expresses the simple PM–CFP–Orai1CT construct (as opposed to the full-length CFP–Orai1 hexamer), our quantitative FRET approach allowed us to reveal that one YFP–SOAR dimer could simultaneously interact with two PM–CFP–Orai1CT molecules. However, this analysis required that the donor PM–CFP–Orai1CT be expressed in excess of the acceptor YFP–SOAR dimer. This result was important in drawing the conclusion that the active SOAR dimer within the STIM1 protein could undergo binding to two Orai1 channel subunits and suggests that STIM1 can actively cluster Orai1 channels by cross-linking between them [13]. This provides an important new understanding of the STIM1–Orai1 interface.

REFERENCES

1. Soboloff, J., Rothberg, B. S., Madesh, M., and Gill, D. L. (2012) STIM proteins: Dynamic calcium signal transducers. *Nat. Rev. Mol. Cell Biol.* **13**, 549–565.
2. Kar, P., Nelson, C., and Parekh, A. B. (2012) CRAC channels drive digital activation and provide analog control and synergy to Ca^{2+}-dependent gene regulation. *Curr. Biol.* **22**, 242–247.
3. Liou, J., Kim, M. L., Heo, W. D., Jones, J. T., Myers, J. W., Ferrell, J. E., Jr., and Meyer, T. (2005) STIM is a Ca^{2+} sensor essential for Ca^{2+}-store-depletion-triggered Ca^{2+} influx. *Curr. Biol.* **15**, 1235–1241.
4. Roos, J., DiGregorio, P. J., Yeromin, A. V., Ohlsen, K., Lioudyno, M., Zhang, S., Safrina, O. et al. (2005) STIM1, an essential and conserved component of store-operated Ca^{2+} channel function. *J. Cell Biol.* **169**, 435–445.
5. Zhang, S. L., Yu, Y., Roos, J., Kozak, J. A., Deerinck, T. J., Ellisman, M. H., Stauderman, K. A., and Cahalan, M. D. (2005) STIM1 is a Ca^{2+} sensor that activates CRAC channels and migrates from the Ca^{2+} store to the plasma membrane. *Nature* **437**, 902–905.
6. Feske, S., Gwack, Y., Prakriya, M., Srikanth, S., Puppel, S. H., Tanasa, B., Hogan, P. G., Lewis, R. S., Daly, M., and Rao, A. (2006) A mutation in Orai1 causes immune deficiency by abrogating CRAC channel function. *Nature* **441**, 179–185.
7. Yeromin, A. V., Zhang, S. L., Jiang, W., Yu, Y., Safrina, O., and Cahalan, M. D. (2006) Molecular identification of the CRAC channel by altered ion selectivity in a mutant of Orai. *Nature* **443**, 226–229.
8. Prakriya, M., Feske, S., Gwack, Y., Srikanth, S., Rao, A., and Hogan, P. G. (2006) Orai1 is an essential pore subunit of the CRAC channel. *Nature* **443**, 230–233.
9. McNally, B. A. and Prakriya, M. (2012) Permeation, selectivity and gating in store-operated CRAC channels. *J. Physiol.* **590**, 4179–4191.
10. Zhou, Y., Srinivasan, P., Razavi, S., Seymour, S., Meraner, P., Gudlur, A., Stathopulos, P. B., Ikura, M., Rao, A., and Hogan, P. G. (2013) Initial activation of STIM1, the regulator of store-operated calcium entry. *Nat. Struct. Mol. Biol.* **20**, 973–981.
11. Korzeniowski, M. K., Manjarres, I. M., Varnai, P., and Balla, T. (2010) Activation of STIM1-Orai1 involves an intramolecular switching mechanism. *Sci. Signal.* **3**, ra82.
12. Covington, E. D., Wu, M. M., and Lewis, R. S. (2010) Essential role for the CRAC activation domain in store-dependent oligomerization of STIM1. *Mol. Biol. Cell* **21**, 1897–1907.
13. Zhou, Y., Wang, X., Wang, X., Loktionova, N. A., Cai, X., Nwokonko, R. M., Vrana, E., Wang, Y., Rothberg, B. S., and Gill, D. L. (2015) STIM1 dimers undergo unimolecular coupling to activate Orai1 channels. *Nat. Commun.* **6**, 8395.
14. Wang, Y., Deng, X., Zhou, Y., Hendron, E., Mancarella, S., Ritchie, M. F., Tang, X. D. et al. (2009) STIM protein coupling in the activation of Orai channels. *Proc. Natl. Acad. Sci. USA* **106**, 7391–7396.
15. Navarro-Borelly, L., Somasundaram, A., Yamashita, M., Ren, D., Miller, R. J., and Prakriya, M. (2008) STIM1-ORAI1 interactions and ORAI1 conformational changes revealed by live-cell FRET microscopy. *J. Physiol.* **586**, 5383–5401.
16. Wang, X., Wang, Y., Zhou, Y., Hendron, E., Mancarella, S., Andrake, M. D., Rothberg, B. S., Soboloff, J., and Gill, D. L. (2014) Distinct Orai-coupling domains in STIM1 and STIM2 define the Orai-activating site. *Nat. Commun.* **5**, 3183.
17. Gandhi, C. S. and Isacoff, E. Y. (2005) Shedding light on membrane proteins. *Trends Neurosci.* **28**, 472–479.
18. Clegg, R. M. (1995) Fluorescence resonance energy transfer. *Curr. Opin. Biotechnol.* **6**, 103–110.
19. Jares-Erijman, E. A. and Jovin, T. M. (2006) Imaging molecular interactions in living cells by FRET microscopy. *Curr. Opin. Chem. Biol.* **10**, 409–416.

20. Padilla-Parra, S. and Tramier, M. (2012) FRET microscopy in the living cell: Different approaches, strengths and weaknesses. *Bioessays* **34**, 369–376.
21. Arai, Y. and Nagai, T. (2013) Extensive use of FRET in biological imaging. *Microscopy (Oxf)* **62**, 419–428.
22. Heim, R. and Tsien, R. Y. (1996) Engineering green fluorescent protein for improved brightness, longer wavelengths and fluorescence resonance energy transfer. *Curr. Biol.* **6**, 178–182.
23. Periasamy, A. (2001) Fluorescence resonance energy transfer microscopy: A mini review. *J. Biomed. Opt.* **6**, 287–291.
24. Yuan, J. P., Zeng, W., Dorwart, M. R., Choi, Y. J., Worley, P. F., and Muallem, S. (2009) SOAR and the polybasic STIM1 domains gate and regulate Orai channels. *Nat. Cell Biol.* **11**, 337–343.
25. Park, C. Y., Hoover, P. J., Mullins, F. M., Bachhawat, P., Covington, E. D., Raunser, S., Walz, T., Garcia, K. C., Dolmetsch, R. E., and Lewis, R. S. (2009) STIM1 clusters and activates CRAC channels via direct binding of a cytosolic domain to Orai1. *Cell* **136**, 876–890.
26. Yang, X., Jin, H., Cai, X., Li, S., and Shen, Y. (2012) Structural and mechanistic insights into the activation of Stromal interaction molecule 1 (STIM1). *Proc. Natl. Acad. Sci. USA* **109**, 5657–5662.
27. Zal, T. and Gascoigne, N. R. (2004) Photobleaching-corrected FRET efficiency imaging of live cells. *Biophys. J.* **86**, 3923–3939.
28. Miller, M., Wu, M., Xu, J., Weaver, D., Li, M., and Zhu, M. X. (2011) High-throughput screening of TRPC channel ligands using cell-based assays. In *TRP Channels* (Zhu, M. X. ed.), Boca Raton, FL: CRC Press/Taylor & Francis.
29. Lur, G., Haynes, L. P., Prior, I. A., Gerasimenko, O. V., Feske, S., Petersen, O. H., Burgoyne, R. D., and Tepikin, A. V. (2009) Ribosome-free terminals of rough ER allow formation of STIM1 puncta and segregation of STIM1 from IP(3) receptors. *Curr. Biol.* **19**, 1648–1653.
30. Wu, M. M., Buchanan, J., Luik, R. M., and Lewis, R. S. (2006) Ca^{2+} store depletion causes STIM1 to accumulate in ER regions closely associated with the plasma membrane. *J. Cell Biol.* **174**, 803–813.

8 Optogenetic Approaches to Control Calcium Entry in Non-Excitable Cells

Lian He, Qian Zhang, Yubin Zhou, and Yun Huang

CONTENTS

8.1 INTRODUCTION

The calcium ion serves as a versatile and universal second messenger to control a myriad of biological processes, including muscle contraction, neurotransmission, hormone secretion, immune cell activation, cell motility, and apoptosis [1–3]. The calcium release–activated calcium (CRAC) channel mediated by ORAI and stromal interaction molecule (STIM) constitutes one of the primary Ca^{2+} entry routes in non-excitable cells [4–6] (see Chapter 2). The CRAC channel is regarded as a prototypical example of store-operated Ca^{2+} entry (SOCE), in which the depletion of Ca^{2+} store in the endoplasmic reticulum (ER) induces calcium influx from the extracellular space. Over the past decade, tremendous progress has been made with respect to

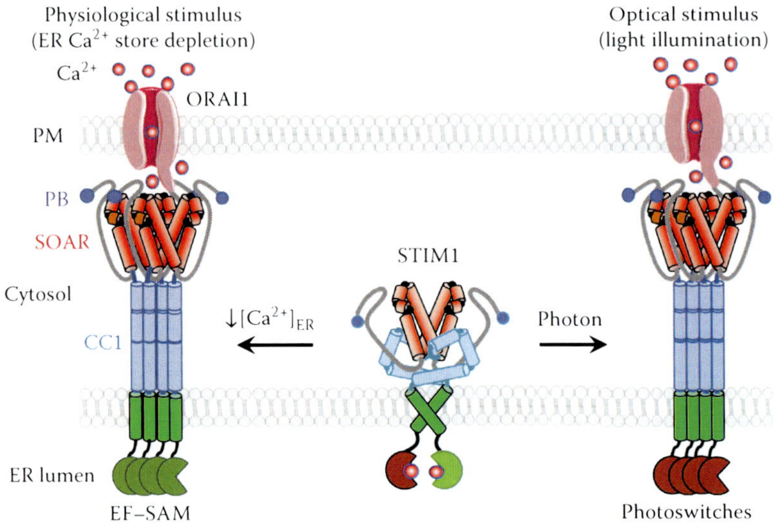

FIGURE 8.1 Converting store-operated calcium entry to light-operated calcium entry. Under physiological conditions, depletion of calcium store in the ER lumen initiates the oligomerization of the STIM1 EF–SAM domain to cause structural rearrangements in the STIM1 transmembrane domain (TM). Reorganized STIM1 TM further prompts a conformational change in the juxtamembrane coiled-coil region (CC1). As a consequence, CC1 adopts an extended conformation and assumes more helical content presumably by forming a stabilized coiled coil, thus masking the SOAR-docking sites in CC1 to overcome STIM1 intramolecular trapping. STIM1 then translocates toward the PM to engage and activate ORAI1 channels. Light sensitivity can be added to STIM1, either through photoinducible apposition of the N-termini of the cytoplasmic domain of STIM1 or light-triggered uncaging of SOAR, to enable optical control of calcium entry.

the critical steps required to activate SOCE (Figure 8.1) [5–7]. Under resting conditions, the STIM1 luminal EF–SAM domain is loaded with Ca^{2+} and exists largely as a monomer [8]. The STIM1 cytoplasmic domain (STIM1ct), consisting of a coiled-coil region (CC1), a minimal ORAI1-activating region (SOAR or CAD), and a C-terminal polybasic tail (PB), adopts a folded-back configuration that keeps itself inactive through autoinhibition [9,10]. Upon ER Ca^{2+} store depletion, dissociation of Ca^{2+} from the EF–SAM domain initiates a destabilization-coupled oligomerization process in the ER lumen [8]. Conformational changes in the canonical EF-hand Ca^{2+}-binding motif disrupt the intramolecular interaction between the EF-hands and SAM domains, thereby causing aggregation of the luminal EF–SAM domains. The luminal domain oligomerization further triggers conformational changes that propagate throughout STIM1ct. STIM1ct redeploys itself and adopts a more extended conformation by exposing the SOAR/CAD domain, as well as the polybasic C-tail. Next, the activated STIM1 multimerizes and moves toward the ER–plasma membrane (PM) junctional sites, where it recruits and gates ORAI1 channels through direct physical association with ORAI1. This process is also facilitated by the interaction between its polybasic C-tail and the negatively charged phosphoinositides in the inner leaflet of the plasma membrane (see Chapters 2 and 3). Sustained Ca^{2+} influx through ORAI1 channels activates downstream effectors such as calcineurin, a Ca^{2+}-dependent phosphatase that dephosphorylates the nuclear factor of activated T cells (NFAT) and triggers the nuclear translocation of NFAT to regulate gene expression during lymphocyte activation [4] (see Chapter 5).

Two critical steps during SOCE activation, oligomerization of STIM1 luminal domain and conformational switch within STIM1ct, can be mimicked by photosensitive domains that undergo light-inducible oligomerization or allosteric regulation to devise photoactivatable CRAC channels (termed Opto-CRAC), thereby enabling remote and optical control of calcium signaling [11–13]. Compared to the existing microbial opsin-based optogenetic tools, Opto-CRAC has several distinctive features such as the following:

- Unlike the widely used channelrhodopsin (ChR)-based tools [14] that exhibit less stringent ion selectivity and tend to perturb intracellular pH due to their high proton permeability, Opto-CRAC is engineered from a *bona fide* Ca^{2+} channel that is regarded as one of the most Ca^{2+}-selective ion channels (see Chapter 1).
- The Opto-CRAC tool is among the smallest optogenetic tools (<1 kb, compared to >2.2 kb of ChR) and is thus compatible with viral vectors used for *in vivo* gene delivery.
- Its tunable and relatively slow kinetics make Opto-CRAC most suitable for generating customized Ca^{2+} oscillation patterns in non-excitable cell types, such as cells of the immune and hematopoietic systems.

In this chapter, we present a brief overview of the design principles of Opto-CRAC constructs based on two different photoresponsive domains (CRY2 and LOV2) and illustrate how to use them to remotely control Ca^{2+} influx with at high spatiotemporal precision and subsequent nuclear translocation of a master transcriptional factor, the nuclear factor of activated T cells (NFAT), to fine-tune NFAT-dependent gene expression and control the function of immune cells.

8.2 DESIGN OF OPTO-CRAC CONSTRUCTS

8.2.1 CRY2-Based Strategy

Cryptochrome 2 (CRY2) from *Arabidopsis thaliana* is a blue light photoreceptor mediating the regulation of flowering induction. As a member of the flavoprotein superfamily, CRY2 contains an N-terminal photolyase homology (PHR) domain with flavin adenine dinucleotide (FAD) as its cofactor [15–17]. Blue light illumination in the range of 405–488 nm induces an intramolecular redox reaction that involves FAD and conserved tryptophan residues and subsequently leads to conformational changes to induce CRY2 oligomerization within seconds [16,17]. In the dark, CRY2 homo-oligomers return to monomeric state in approximately 5–10 min. It has been shown that CRY2PHR has improved expression over full-length CRY2 [18]. Two short truncated variants of CRY2 (residues 1–515 or 1–535) display improved dynamic ranges and reduced self-association when compared with CRY2PHR [16]. Mutations can also be carefully introduced into CRY2PHR to fine-tune its photochemical properties. For example, E490G at the predicted C-terminal helix of CRY2PHR (called CRY2olig; Figure 8.2a) remarkably can promote the efficiency of forming oligomers [16]. Taslimi and colleagues identified two adjacent mutations that pose opposite effects on the photocycle kinetics [16]. The L348F mutant has a prolonged half-life time of approximately 24 min, whereas W349R shortens the half-life time to 2.5 min.

Close apposition of STIM1 through chemical cross-linking at residue 233—the position where the transmembrane domain ends and the cytosolic portion of STIM1

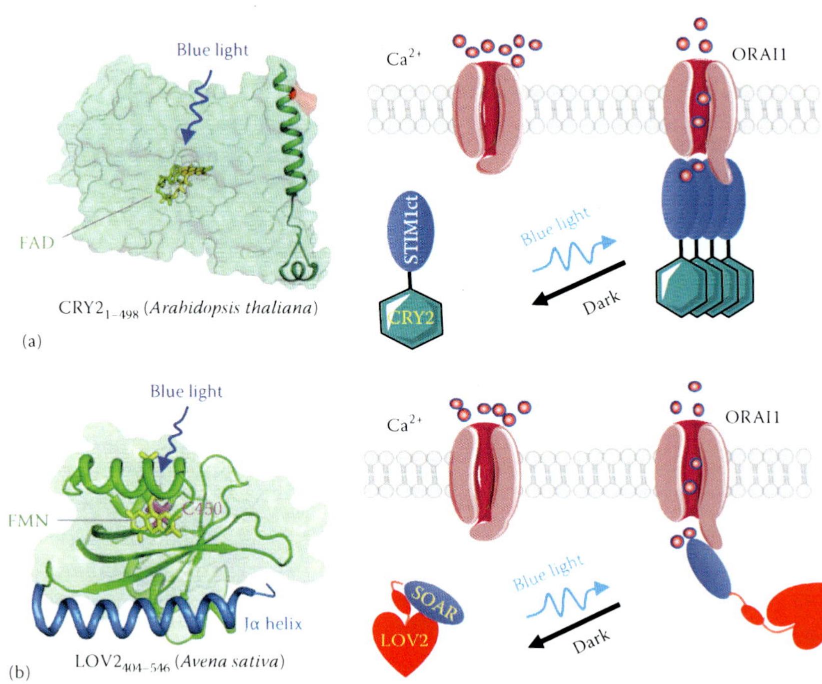

FIGURE 8.2 Photoactivatable calcium entry with CRY2- or LOV2-based optogenetic constructs. (a) CRY2PHR domain (residues 1–498) from *Arabidopsis thaliana* is fused to the cytoplasmic domain of STIM1 (STIM1ct; residues 233–685). In the dark, the fusion protein remains inactive. Following blue light stimulation, CRY2 undergoes oligomerization to bring the N-termini of STIM1ct into close apposition, much like calcium depletion–induced EF–SAM multimerization, to trigger activation of STIM1ct and subsequent calcium influx through ORAI1 channels. (b) LOV2 domain (residues 404–546) from *Avena sativa* phototropin is fused to STIM1ct fragments (e.g., residues 336–486). When expressed alone, these STIM1ct fragments elicit constitutive Ca^{2+} entry from the extracellular space to the cytosol. In the dark, the C-terminal Jα helix docks to the LOV2 domain and keeps STIM1ct fragments quiescent. Upon blue light illumination, photoexcitation generates a covalent adduct between Cys450 and the cofactor FMN, thereby promoting the undocking and unwinding of the Jα helix to expose the STIM1ct fragments. Unleashed STIM1ct fragments further move toward the plasma membrane to directly engage and activate ORAI1 channels. *(Continued)*

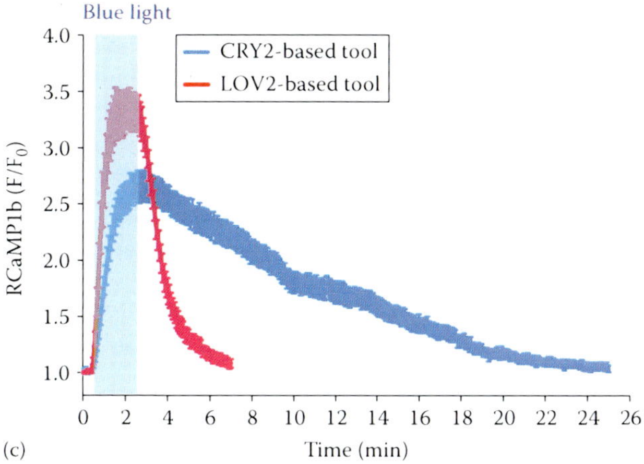

(c)

FIGURE 8.2 (*Continued*) Photoactivatable calcium entry with CRY2- or LOV2-based optogenetic constructs. (c) Light-inducible Ca^{2+} influx and recovery monitored with a GECI RCaMP1b. n = 8–10 cells from three independent experiments. Error bars denote SEM.

emerges from the ER membrane—could switch STIM1 from a folded-back to a more extended configuration [7] (see Chapter 3). The functional consequence of such conformational changes is to release the SOAR domain that is initially docked toward CC1 to overcome intramolecular trapping, thereby making the SOAR domain poised to engage and open the ORAI1 channels in the plasma membrane. To phenocopy this process with light-sensitive domains, we fused STIM1ct (residues 233–685) to the C-terminus of CRY2PHR. Following blue light stimulation, the chimeric construct was able to elicit robust calcium influx, as reported by a red-emitting genetically encoded calcium indicator (GECI), RCaMP1b, within 30 s (Figure 8.2c). After turning off the blue light, it took up to 20 min for the calcium signal to return to their basal level. CRY2PHR fused to shorter fragments of STIM1ct (238–685, 240–685, 245–685, or 250–685) can be similarly used to photoinduce calcium influx. Depending on the desired application and demand on the duration of calcium signaling, these STIM1ct fragments can be further paired with the aforementioned CRY2 mutants to generate calcium signals with varying kinetics. Nonetheless, the off kinetics of CRY2-based optogenetic tools is in the timescale of minutes, thus making it less compatible with physiological scenarios that occur within seconds. We overcame this major weakness by resorting to an alternative strategy based on the LOV2 photoswitchable domain.

8.2.2 LOV2-Based Strategy

Light–oxygen–voltage domain 2 (LOV2) derived from *Avena sativa* phototropin 1 contains a core per-arnt-sim (PAS) domain with a central β-scaffold and a helical interface that hold photoreactive flavin. LOV2 in the dark state contains FMN as cofactor that maximally absorbs light at 450 nm with the Jα helix tightly docked to the core body (Figure 8.2b). STIM1ct fragments fused immediately after the Jα helix may have a high chance to be trapped in an inactive state, thus mimicking the STIM1ct autoinhibition mediated by CC1 and SOAR interaction [9,10]. After blue light stimulation, the FMN cofactor forms a covalent bond with C450 to trigger conformational changes that ultimately result in the unwinding and unfolding of Jα helix, with subsequent release of STIM1ct fragments to activate ORAI1 calcium channels in the plasma membrane.

To explore this strategy, we generated a series of Opto-CRAC constructs by varying the length of STIM1ct fragments, introducing mutations into the LOV2 domain and optimizing the linker between them. After screening over 100 combinations by using Ca^{2+} influx and NFAT nuclear translocation as independent readouts, the chimeric construct LOV2-STIM$1_{336-486}$ (designated as "LOVSoc") stood out as the best candidate because it exhibited low dark activity but high dynamic range in terms of eliciting light-inducible Ca^{2+} influx (Figure 8.2c). The degree of Ca^{2+} influx can be modulated by changing the light power density. The *on* or *off* phase of photoinduced calcium influx takes approximately 30 s (Figure 8.3a). LOVSoc has been shown to induce calcium influx in a dozen of cell types derived from various non-excitable tissues at endogenous ORAI1 levels [19].

Electrophysiological measurements of photoactivated currents by whole-cell recording in HEK293-ORAI1 cells revealed a typical inwardly rectifying current characteristic of the CRAC channel (Figure 8.3b) (see Chapters 1 and 9). Replacing the extracellular Na^+ with a large impermeant cation NMDG did not alter the overall shape of the CRAC current (Figure 8.3b, red curve), further attesting to its high Ca^{2+} selectivity. Compared with CRY2-based Opto-CRAC constructs, LOVSoc showed more uniform responses in transfected cells with faster kinetics.

8.3 EXAMPLES OF OPTO-CRAC APPLICATIONS

8.3.1 Light-Operated Calcium Entry

Opto-CRAC-induced calcium influx in living cells can be monitored with both small molecule calcium indicators (e.g., Fura-2 AM) and GECIs (e.g., GCaMP, RCaMP, or R-GECO series) [20,21]. These calcium indicators bind calcium with dissociation constants in the range of 150–300 nM, at a level comparable to the resting cytosolic concentration, and are thus very sensitive to cytosolic calcium changes. The most popular GECIs are made of three parts: calmodulin, calmodulin-binding peptide M13, and engineered fluorescent proteins. Upon binding Ca^{2+}, calmodulin undergoes a substantial conformational change to wrap around its binding target M13 and thus induces changes in the fluorescence intensity [20]. Owing to its high sensitivity, precise targeting, and compatibility with prolonged imaging, GCaMP and its derivatives are gaining wide popularity in both *in vitro* and *in vivo* applications [22]. We describe herein the use of Fura-2 AM and GECIs to measure photoactivatable intracellular calcium elevations in HeLa or HEK293T cells (see Chapter 16). One caveat to be kept in mind is that the emission filter needs to be optimized to avoid undesired photoexcitation of CRY2 or LOV2 in the range of 400–490 nm. GECIs (e.g., RCaMP or R-GECO) or dyes (Fura-2 AM) with excitation peaks below

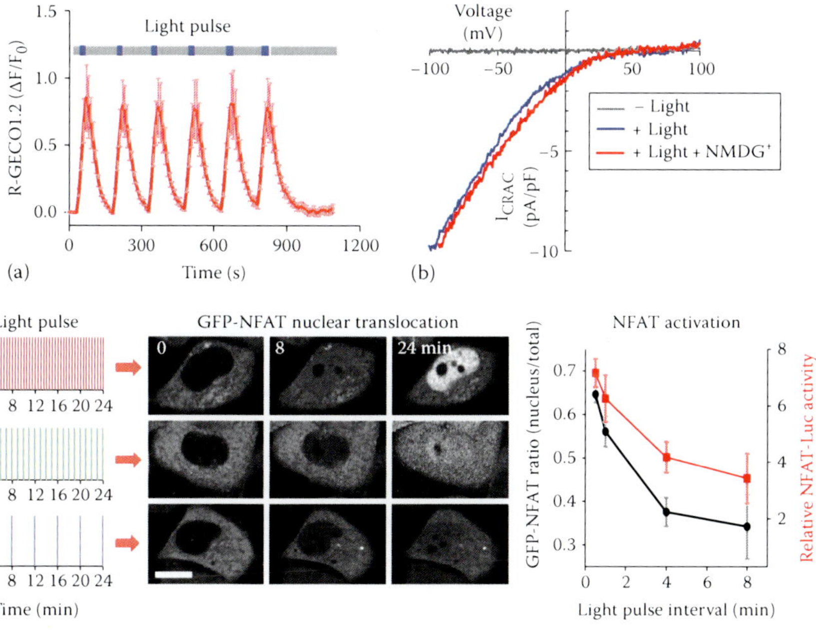

FIGURE 8.3 Optical control of the calcium/NFAT pathway in non-excitable cells. (a) Ca^{2+} oscillation induced by repeated light–dark cycles (30 s *on* and 120 s *off*; blue bar indicates light stimulation with a power density of 40 μW/mm^2 at 470 nm) in HeLa cells. Ca^{2+} fluctuation was reported by R-GECO1.2. n = 8 cells from three independent experiments. Error bars denote SEM. (b) Light-induced CRAC current–voltage relationships. HEK293 cells were cotransfected with ORAI1 and mCherry-tagged Opto-CRAC. Whole-cell patch-clamp recording utilizing a ramp protocol ranging from −100 to 100 mV was performed under the following conditions in mCherry-positive cells: dark (gray), blue light (blue), and blue light with the extracellular Na^+ replaced by $NMDG^+$ (red), a nonpermeant ion used to assess ion selectivity by examining the contribution of Na^+. Substitution of the most abundant extracellular cation Na^+ by $NMDG^+$ did not alter the amplitude or overall shape of the CRAC current, implying that Na^+ has negligible contribution to Opto-CRAC-mediated currents, which primarily conduct Ca^{2+}. (c) Light-tunable nuclear translocation of green fluorescent protein (GFP)-NFAT1 and NFAT-dependent luciferase (NFAT-Luc) gene expression. *Left panel*: blue light pulses (0.5 min *on*) with varying interpulse intervals (0.5, 1, or 4 min) were applied to HeLa cells stably expressing GFP-$NFAT_{1-460}$. *Middle panel*: representative images showing light-induced GFP-$NFAT_{1-460}$ nuclear translocation. *Right panel*: NFAT nuclear translocation (black) and NFAT-dependent luciferase activity (red) were plotted against different interpulse intervals. n = 15–20 cells from three independent experiments. Error bars denote SEM. Scale bar, 10 μm. (d–f) Near-infrared radiation (NIR) light-triggered Ca^{2+} influx and NFAT nuclear translocation in Opto-CRAC HeLa cells incubated with UCNPs, which act as nanoilluminators to convert NIR light into blue light and activate the LOV2 photoswitch. *(Continued)*

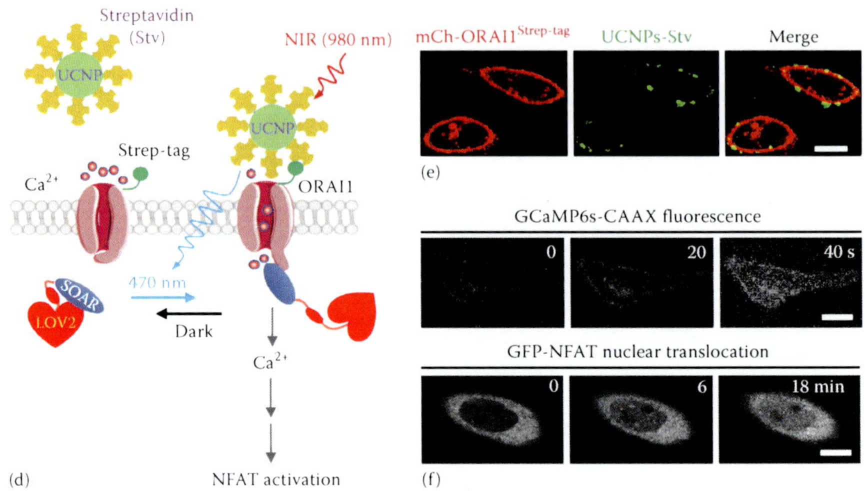

FIGURE 8.3 (*Continued*) Optical control of the calcium/NFAT pathway in non-excitable cells. (d) Schematic showing the engineered ORAI1 channel with a streptavidin-binding tag (Strep-tag) designed to recruit streptavidin-coated upconversion nanoparticles (UCNPs-Stv) to the cell surface. (e) Confocal images showing the accumulation of UCNPs-Stv (green, λ_{ex} = 980 nm, λ_{em} = 470 nm) on the plasma membrane of engineered HeLa cells. (f) Ca^{2+} influx (reported by GCaMP6s, upper panel) and GFP-NFAT nuclear translocation (lower panel) were triggered by NIR light (980 nm) in HeLa cells expressing mCh-ORAI1$^{Strep\text{-}tag}$ and Opto-CRAC in the presence of UCNPs-Stv (20 μg/μL). Scale bar, 10 μm.

400 nm or above 550 nm are most compatible with real-time imaging of repetitive calcium influx (Figure 8.3a).

8.3.1.1 Calcium Imaging Using Fura-2 AM

Fura-2 was invented in the 1980s as a ratiometric fluorescent dye that binds to free intracellular calcium [23]. Fura-2 can be excited at 340 and 380 nm with a maximal fluorescence emission at 510 nm. The ratio of the emitted light intensity (when excited at 340 and 380 nm, respectively) is proportional to the amount of bound calcium. Fura-2 acetoxymethyl ester (abbreviated as Fura-2 AM) is a membrane-permeable variant of Fura-2. Once inside the cells, cellular esterases remove the acetoxymethyl groups to retain Fura-2 within the cell. Because there is no transfection involved in this process, Fura-2 can be used for calcium imaging with both adherent and suspension cells, including difficult-to-transfect cardiac and neuronal cells [24–27], as well as cells of the immune and hematopoietic systems.

HeLa or HEK293T cells are cultured in Dulbecco's modified Eagle's medium (DMEM) supplemented with 10 mM HEPES and 10% heat-inactivated fetal bovine serum. Cultured cells are seeded on 35 mm glass-bottom dishes at a density of 0.2×10^6 in 1.5–2 mL culture medium. The cells are grown to 50%–70% confluence after overnight culture. The Opto-CRAC plasmid (200 ng) is transfected using Lipofectamine 3000 (Thermo Fisher Scientific). Twenty-four to thirty-six hours post-transfection, the culture medium is replaced with 1 × imaging bath solution (107 mM NaCl, 7.2 mM KCl, 1.2 mM $MgCl_2$, 11.5 mM glucose, 20 mM HEPES–NaOH, 1 mM $CaCl_2$, 0.1% [m/v] BSA, pH 7.2) in the presence of 2 μM Fura-2 AM (Life Technologies). Cells are incubated at 37°C for 30–45 min to allow the uptake of Fura-2 AM and then washed twice. Cells are kept in the imaging bath solution afterward.

Calcium imaging was carried out using a Zeiss observer-A1 microscope equipped with a Lambda DG4 light source (Sutter), BrightLine FURA2-C-000 filter set (Semrock). First, mCherry-positive transfected cells (mCh-LOVSoc) are identified in the imaging field, and care is taken that the cells are not exposed to blue or green light to avoid activation of CRY2 or LOV2 photosensory domains. Next, Fura-2 fluorescence at 510 nm generated by 340 nm excitation light (F_{340}) and 380 nm light (F_{380}) is measured every 5 s for 5 min. To photoactivate Opto-CRAC, cells are exposed to an external blue LED (470 nm at a power intensity of 40 μW/mm^2; ThorLabs). Alternatively, a light source commonly used for GFP excitation in fluorescence microscopy can also be repurposed to activate Opto-CRAC and induce calcium entry. Three to five fields are imaged to collect enough cells for statistical analysis. The resulting data are analyzed by MetaFluor software, and the intracellular Ca^{2+} levels are indicated as the ratio of F_{340}/F_{380}. The free calcium concentration in cells can be estimated by using Fura-2 calcium imaging calibration kit (Thermo).

8.3.1.2 Calcium Imaging with GECIs

GCaMP [28] contains a circularly permutated EGFP (cpEGFP) with a M13 fragment of myosin light chain kinase fused to the N-terminus of cpEGFP and engineered calmodulin connected to the C-terminal end. Ca^{2+}-dependent binding of M13 to calmodulin induces a large increase in the fluorescence intensity of EGFP. Similar strategies have been applied to engineer calcium-responsive fluorescent proteins that emit blue or red lights [29–32]. The Opto-CRAC system is compatible with both green (GCaMP6 [20]) and red color GECIs (R-GECO1 [33] and RCaMP [32]). GCaMP6, owing to the spectral overlap with blue light used to photoexcite CRY2 or LOV2, can only be used to monitor the *on* phase of light-inducible calcium influx. By contrast, R-GECO or RCaMP can be conveniently used to monitor both the *on* and *off* phases of cytosolic calcium fluctuations. However, the latter two GECIs have a much lower dynamic range.

HEK293T or HeLa cells are cultured and seeded as described earlier. The Opto-CRAC plasmids (200 ng; either mCh-LOVSoc or mCh-CRY2-$STIM1_{233-685}$) are cotransfected with 200 ng of GECIs (GCaMP6s for cytosolic calcium or membrane-tethered GCaMP6s-CAAX for more sensitive recording of calcium change near the plasma membrane) using Lipofectamine 3000. Calcium imaging is performed 24–36 h post-transfection. Since the excitation wavelength used to acquire the GCaMP6s (~488 nm) signals can photoexcite Opto-CRAC to elicit Ca^{2+} influx, it is important to avoid the use of the GFP channel before identifying mCherry-positive cells. Time-lapse imaging was performed by recording signals from both GFP (excited by a 488 nm laser) and mCherry (excited at 561 nm) channels with intervals of 1–5 s for 3 min. The GCaMP6 and Opto-CRAC cotransfected cells were selected for statistical analysis. The green GCaMP6 fluorescence intensity was exported, and the ratio of

F_t/F_0 (fluorescence at time t and time 0) was calculated to reflect the dynamic change of intracellular calcium. For imaging with red GECIs, 200 ng R-GECO1 or RCaMP was cotransfected with 200 ng Opto-CRAC. An external blue LED (ThorLabs, 470 nm with a power intensity of 40 μW/mm^2) was used to photoactivate Opto-CRAC without perturbing the red fluorescence used to monitor the expression of target proteins. Single-channel time-lapse imaging of mCherry signal was carried out by switching the blue light on and off with user-defined pulses. The ratio of F_t/F_0 was acquired and plotted using Prism 5 software. The half-life times of the *on* and *off* phases can be calculated by fitting the calcium response curve with a single exponential decay function. Compared with Fura-2 imaging, the use of GECIs is a more affordable choice and obviates the need for multiple incubation and washing steps. The development of a stable cell line coexpressing a red GECI and Opto-CRAC will likely provide a convenient all optical platform to aid the screening of CRAC channel modulators.

8.3.2 Spatial Control of Calcium Signals

Depending on the position and range of the light source, Opto-CRAC can be used to confer spatial control of calcium signals in living cells. A proof of concept experiment can be performed by using the fluorescence recovery after photobleaching module in a typical confocal microscope, such as an inverted Nikon Eclipse Ti-E microscope customized with Nikon A1R+ confocal laser sources (405/488/561/640 nm). Membrane-tethered GECIs like GCaMP6s-CAAX were recommended for this purpose to increase the detection sensitivity. Seeded HeLa or HEK293T cells on a 35 mm glass-bottom dish are mounted on the imaging platform. A 60X oil objective is recommended to identify mCherry-positive cells. Areas of interests with user-defined shapes and sizes can be selected for setting up local photoillumination with the GFP channel excitation at 488 nm. The power density should be reduced to less than 5% of the full laser power to minimize photobleaching. If the instrument allows, nanopatterned blue light source can be applied to generate calcium microdomains, thus opening new opportunities to dissect the functional roles of localized calcium signals in mammalian cells [34,35].

8.3.3 Temporal Control of Calcium Signals

Given its relatively faster kinetics, LOVSoc can be further used to generate calcium oscillatory patterns with pulsed light illumination (Figure 8.3a). Calcium oscillations shape the efficiency and specificity of gene expression [36] (see Chapter 5). Previously, customized calcium oscillations were made possible by adding ligands to their cognate receptors to activate phospholipase C and its downstream effectors or by using the "calcium clamp" method, in which Ca^{2+} oscillatory patterns were generated by alternately exposing cells to 0 or 1.5 mM calcium after store depletion [36]. With LOVSoc, we can conveniently generate oscillatory calcium patterns by simply exposing cells to repeated light–dark cycles (Figure 8.3a) [19]. We tested if we could manipulate the light pulse to generate diverse temporal patterns of Ca^{2+} signals to tune the degree of NFAT activation, which would be reflected in the efficiency of NFAT nuclear translocation and NFAT-dependent gene expression (Figure 8.3c). We applied a fixed light pulse of 30 s to HeLa cells transfected with mCh-LOVSoc and GFP-NFAT while varying the interpulse intervals from 0.5 to 8 min. We found that a prolonged interpulse interval is generally associated with a decrease in the nuclear accumulation of NFAT, as well as NFAT-dependent luciferase expression (Figure 8.3c).

8.3.4 Phototunable Ca^{2+}-Dependent Gene Expression in T Cells

8.3.4.1 Mouse Primary T Cell Isolation and Culture *In Vitro*

In T lymphocytes, NFAT cooperates with activator protein 1 (AP1) to regulate diverse inducible gene expressions (e.g., IL-2 and IFN-γ) to mount an efficient immune response [37]. Opto-CRAC, when coupled with phorbol 12-myristate 13-acetate (PMA) to activate the AP1 costimulatory pathway, can be used to phenocopy this physiological response by harnessing the power of light. Spleen and lymph nodes are isolated from 6- to 8-week-old female C57BL/6 mice and smashed against sterile nylon mesh by using the thumb side of a syringe plunger (1–3 mL) in 1% FBS/PBS to obtain single-cell suspensions of splenocytes and lymph node cells on Day 1. Isolated splenocytes are pooled together with lymph node cells for naïve $CD4^+$ T cells sorting by labeling cells with antibodies against CD4, CD25, CD62L, and CD44 to sort $CD4^+CD25^-CD62L^+CD44^-$ populations. Purified naïve $CD4^+$ T cells are then cultured in DMEM supplemented with 10% heat-inactivated fetal bovine serum, 2 mM L-glutamine, penicillin–streptomycin, nonessential amino acids, sodium pyruvate, vitamins, 10 mM HEPES, and 50 μM 2-mercaptoethanol. Cells are plated at a density of ~10^6 cells per mL in 6-well plates coated with anti-CD3 (clone 2C11, BioLegend) and anti-CD28 (clone 37.51, BioLegend) (1 μg/mL each) by precoating with 100 μg/mL goat anti-hamster IgG (MP Biomedicals). After 48 h (on Day 3), cells are removed and recultured at a concentration of 5×10^5 cells/well in an uncoated 6-well plate in T cell media supplemented with 20 U/mL recombinant human IL-2 (rhIL-2) and are ready for retroviral transduction on Day 4.

8.3.4.2 Retroviral Packaging and Transduction

Murine stem cell virus (MSCV)–based vector pMIG (MSCV-IRES-GFP) (Addgene#9044) is used for producing a retrovirus encoding mCherry-LOVSoc. The synthetic gene was inserted between XhoI and EcoRI sites of pMIG. For viral packaging, Platinum-E (Plat-E) retroviral packaging cells (Cell Biolabs) are seeded at a density of 8×10^6 cells in a 15 cm culture dish on Day 0 and cultured in DMEM supplemented with 10% FBS, 1% penicillin/streptomycin, and glutamine. After 16–24 h (Day 1), cells reaching 70%–80% confluency were transfected with pMIG-mCherry-LOVSoc and pCL-Eco by using Lipofectamine 3000. Forty-eight hours post-transfection (Day 3), the first supernatant is harvested and stored at 4°C and 20 mL of fresh culture medium is added to the Plat-E cells. The second supernatant is harvested on Day 4 and pooled together with the first collection. Cell debris in the supernatant is removed by centrifugation at 2000 rpm for 5 min. Next, retrovirus-containing supernatants are centrifuged at 20,000 rpm for 2 h at 4°C in a Beckman

SW28 swinging bucket rotor lined with an Open-Top polyclear centrifuge tube (Seton). The retroviral pellets are gently resuspended in 400 μL DMEM and added to isolated mouse naïve CD4 T cells (described in Section 8.3.4.1) in a 6-well plate in the presence of 8 μg/mL polybrene (EMD Millipore, Merck KGaA). The plate is sealed with a plastic wrap and subjected to centrifugation at 2000 rpm for 90 min at 37°C. The plate is incubated at 37°C with 5% CO_2 overnight and fresh T cell culture medium supplemented with rhIL-2. Forty-eight hours post-transduction, the percentage of infected cells is determined by flow cytometric analysis (FACS) of EGFP expression.

8.3.4.3 Induction of IL2 and IFN-γ Expression with Blue Light

GFP^+ CD4 T cells are treated with both PMA and blue light pulse (30 s pulse for every 1 min, 10–40 μW/mm^2) for 6–8 h. Expression and production of cytokines can be measured by enzyme-linked immunosorbent assay (ELISA), quantitative real-time polymerase chain reaction (PCR), and intracellular staining as described later.

1. *Quantitative RT-PCR* (*qRT-PCR*): Total RNA is isolated from isolated and transduced T cells, and the first-strand cDNA is generated from total RNA using oligo-dT primers and reverse transcriptase II (Invitrogen). Real-time PCR is performed using specific primers and the ABI Prism 7000 analyzer (Applied Biosystems) with SYBR Green Super Mix Universal (Life Technologies). Target gene expression values are normalized to mGapdh. The following primers for mGapdh and mIL2 are used: mGapdh; forward, 5′-TTGTCTCCTGCGACTTCAACAG-3′, mGapdh; reverse, 5′-GGTCTGGGATGGAAATTGTGAG-3′, mIL-2; forward, 5′-TGAGC AGGATGGAGAATTACAGG-3′, mIL-2; reverse, 5′-GTCCAAGTTCATC TTCTAGGCAC-3′, mIfn-γ; forward, 5′-ATGAACGCTACACACTGCA TC-3′, mIfn-γ; reverse, 5′-CCATCCTTTTGCCAGTTCCTC-3′.
2. *ELISA*: Supernatants are collected and diluted at optical concentrations suitable for the ELISA assay. In brief, a 96-well plate is precoated with the capture antibody (1:500 in coating buffer) at 4°C overnight. On the next day, the plate is washed with PBS/0.1% Tween 20 and blocked with 1% BSA/PBS or ELISA/ELISPOT. Diluted supernatants and cytokine standards are then applied to the plate and incubated for 2 h at room temperature (RT). The plate is then washed and incubated with biotin-conjugated detection antibody (1:1000 in 1% BSA/PBS or ELISA/ELISPOT [eBioscience] diluent buffer) for 1 h at RT. Next, the plate is washed and incubated with poly-HRP streptavidin (1:5000 in diluent buffer, Thermo Scientific) for 30 min. The plate is finally washed and incubated with the tetramethylbenzidine substrate solution (Sigma-Aldrich), and the reaction is stopped with 2 M H_2SO_4. Absorbance of each well is recorded at 450 nm. The absorbance of the standard sample can be used to construct a standard curve.
3. *Intracellular IL-2 and IFN-γ staining*: GFP^+ CD4 T cells are treated and grouped as described earlier in the presence of 2 μM GolgiStop (BD Biosciences) for 6–8 h. Cells are then stained with surface marker using PerCP-Cy5.5-CD4 antibody for 15 min on ice and permeabilized with

Cytofix/Cytoperm (BD Biosciences) for 30 min on ice. Permeabilized cells are resuspended in BD Perm/Wash buffer (BD) and stained with APC-anti-IFN-γ or APC-anti-IL-2 antibody for 20 min. Samples are run on a BD LSRII flow cytometer and analyzed by BD FACSDiva software.

8.3.5 NIR Light Control of Calcium Signaling *In Vitro* and *In Vivo*

The lanthanide-doped upconversion nanoparticles (UCNPs) act as nanotransducers to convert tissue-penetrable near-infrared (NIR) light into visible light emission [11,13,38,39]. The UCNP has a sharp emission peak centered at 470 nm, which is suitable for Opto-CRAC photoactivation [19]. A Strep-tag was introduced into the second extracellular loop of ORAI1 channel without compromising its function to enrich streptavidin-conjugated UCNPs (UCNPs-Stv) to increase the specificity and efficiency of photoconversion. Next, HeLa cells are transfected with Opto-CRAC, mCherry-ORAI1$^{Strep\text{-}tag}$, and GCaMP6s-CAAX or GFP-NFAT using Lipofectamine 3000 and incubated with UCNPs-Stv (20 mg) prior to imaging. NIR light (980 nm, 15 mW/mm^2) stimulation was shown to induce both calcium influx (Figure 8.3e) and NFAT nuclear translocation (Figure 8.3f). The next step is to test the NIR light-triggered activation of the Opto-CRAC system in mouse. The mouse *in vivo* bioluminescence imaging is a quick and simple way to test the Opto-CRAC system. Engineered HeLa cells described earlier are transfected with NFAT-Luc, Opto-CRAC, and ORAI$^{Strep\text{-}tag}$. At 48 h post-transfection, cells are dissociated with trypsin, resuspended at a concentration of 2.5×10^6 per mL in DMEM medium, and subcutaneously injected into the upper thigh of mice after mixing with 1 μM PMA and 10 mg UCNPs-Stv. The implanted regions are subjected to NIR light stimulation with a 980 nm CW laser (50 mW/mm^2, 30 s every 1 min for a total of 25 min) during anesthesia using ketamine/xylazine (100, 10 mg/kg, i.v.). Five hours later, the implanted area is injected with D-luciferin (i.v., 100 μL, 15 mg/mL in PBS) and with the bioluminescence signals acquired with an IVIS-100 *in vivo* imaging system (2 min exposure; binning = 8) 20 min later [19].

8.4 CONCLUSIONS

The development of Opto-CRAC tools enables remote and convenient manipulation of intracellular calcium signals in non-excitable cells at an unprecedented spatiotemporal resolution. Most notably, in conjunction with upconversion nanotransducers [11,13,38,39], we are able to shift light-harvesting window to the NIR region where deep tissue penetration and cordless stimulation without the use of optical fibers are feasible. UCNPs can be readily customized and broadly applied to activate other existing optogenetic tools in chemical biology (e.g., ChR, light–oxygen–voltage domain 2 [LOV2], and cryptochrome 2 [CRY2]–based light switches) that are dependent on visible light–absorbing cofactors [14,40,41]. We anticipate that the broad adaptability of our engineering approach will open new opportunities to achieve noninvasive deep tissue optogenetic immunomodulation, as well as to be utilized in the interrogation of other light-controllable Ca^{2+}-dependent biological processes that are inaccessible by existing approaches (Figure 8.4).

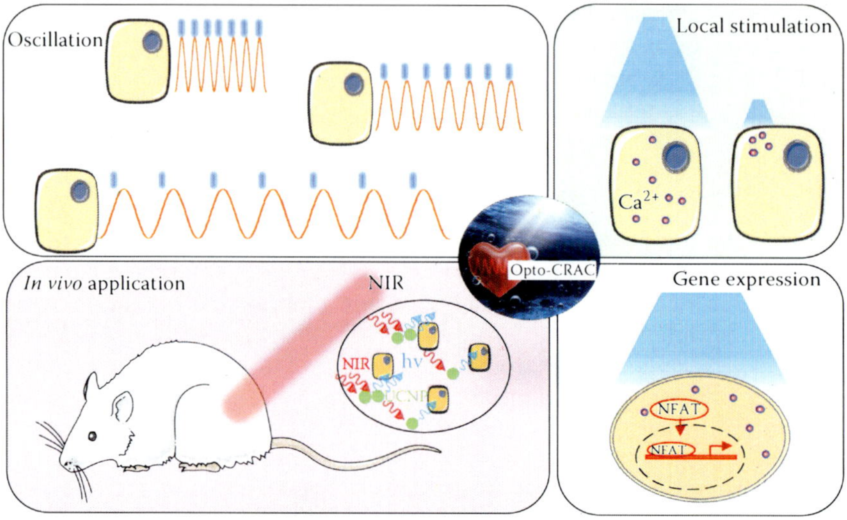

FIGURE 8.4 Applications of Opto-CRAC. The Opto-CRAC tools can be used to (1) manipulate the amplitude and frequency of Ca^{2+} oscillations, (2) elicit global or local Ca^{2+} signals through precise blue light illumination, and (3) photo-tune gene expression. When coupled with NIR-to-blue emitting upconversion nanoparticles, this NIR-inducible optogenetic system enables further applications *in vivo* to remotely control Ca^{2+}-modulated physiological responses.

REFERENCES

1. Berridge MJ, Lipp P, Bootman MD. The versatility and universality of calcium signalling. *Nat Rev Mol Cell Biol* 2000;1(1):11–21.
2. Clapham DE. Calcium signaling. *Cell* 2007;131(6):1047–1058.
3. Berridge MJ. The inositol trisphosphate/calcium signaling pathway in health and disease. *Physiol Rev* 2016;96(4):1261–1296.
4. Hogan PG, Lewis RS, Rao A. Molecular basis of calcium signaling in lymphocytes: STIM and ORAI. *Annu Rev Immunol* 2010;28:491–533.
5. Soboloff J, Rothberg BS, Madesh M, Gill DL. STIM proteins: Dynamic calcium signal transducers. *Nat Rev Mol Cell Biol* 2012;13(9):549–565.
6. Prakriya M, Lewis RS. Store-operated calcium channels. *Physiol Rev* 2015; 95(4):1383–1436.
7. Zhou Y, Srinivasan P, Razavi S, Seymour S, Meraner P, Gudlur A, Stathopulos PB, Ikura M, Rao A, Hogan PG. Initial activation of STIM1, the regulator of store-operated calcium entry. *Nat Struct Mol Biol* 2013;20(8):973–981.
8. Stathopulos PB, Zheng L, Li GY, Plevin MJ, Ikura M. Structural and mechanistic insights into STIM1-mediated initiation of store-operated calcium entry. *Cell* 2008; 135(1):110–122.
9. Fahrner M, Muik M, Schindl R, Butorac C, Stathopulos P, Zheng L, Jardin I, Ikura M, Romanin C. A coiled-coil clamp controls both conformation and clustering of stromal interaction molecule 1 (STIM1). *J Biol Chem* 2014;289(48):33231–33244.
10. Ma G, Wei M, He L, Liu C, Wu B, Zhang SL, Jing J et al. Inside-out Ca^{2+} signalling prompted by STIM1 conformational switch. *Nat Commun* 2015;6:7826.

11. Duan L, Che D, Zhang K, Ong Q, Guo S, Cui B. Optogenetic control of molecular motors and organelle distributions in cells. *Chem Biol* 2015;22(5):671–682.
12. Kyung T, Lee S, Kim JE, Cho T, Park H, Jeong YM, Kim D et al. Optogenetic control of endogenous Ca^{2+} channels *in vivo*. *Nat Biotechnol* 2015;33(10):1092–1096.
13. Zhang Y, Huang L, Li Z, Ma G, Zhou Y, Han G. Illuminating cell signaling with near-infrared light-responsive nanomaterials. *ACS Nano* 2016;10(4):3881–3885.
14. Fenno L, Yizhar O, Deisseroth K. The development and application of optogenetics. *Annu Rev Neurosci* 2011;34:389–412.
15. Mas P, Devlin PF, Panda S, Kay SA. Functional interaction of phytochrome B and cryptochrome 2. *Nature* 2000;408(6809):207–211.
16. Taslimi A, Vrana JD, Chen D, Borinskaya S, Mayer BJ, Kennedy MJ, Tucker CL. An optimized optogenetic clustering tool for probing protein interaction and function. *Nat Commun* 2014;5:4925.
17. Bugaj LJ, Choksi AT, Mesuda CK, Kane RS, Schaffer DV. Optogenetic protein clustering and signaling activation in mammalian cells. *Nat Methods* 2013;10(3):249–252.
18. Kennedy MJ, Hughes RM, Peteya LA, Schwartz JW, Ehlers MD, Tucker CL. Rapid blue-light-mediated induction of protein interactions in living cells. *Nat Methods* 2010;7(12):973–975.
19. He L, Zhang Y, Ma G, Tan P, Li Z, Zang S, Wu X et al. Near-infrared photoactivatable control of Ca^{2+} signaling and optogenetic immunomodulation. *Elife* 2015;4:e10024.
20. Chen TW, Wardill TJ, Sun Y, Pulver SR, Renninger SL, Baohan A, Schreiter ER et al. Ultrasensitive fluorescent proteins for imaging neuronal activity. *Nature* 2013; 499(7458):295–300.
21. Mank M, Griesbeck O. Genetically encoded calcium indicators. *Chem Rev* 2008;108(5):1550–1564.
22. Koldenkova VP, Nagai T. Genetically encoded Ca^{2+} indicators: Properties and evaluation. *Biochim Biophys Acta Mol Cell Res* 2013;1833(7):1787–1797.
23. Grynkiewicz G, Poenie M, Tsien RY. A new generation of Ca^{2+} indicators with greatly improved fluorescence properties. *J Biol Chem* 1985;260(6):3440–3450.
24. Wier WG, Cannell MB, Berlin JR, Marban E, Lederer WJ. Cellular and subcellular heterogeneity of $[Ca^{2+}]_i$ in single heart cells revealed by fura-2. *Science* 1987;235(4786):325–328.
25. Wu H, He CL, Fissore RA. Injection of a porcine sperm factor triggers calcium oscillations in mouse oocytes and bovine eggs. *Mol Reprod Dev* 1997;46(2):176–189.
26. Lipscombe D, Madison DV, Poenie M, Reuter H, Tsien RW, Tsien RY. Imaging of cytosolic Ca^{2+} transients arising from Ca^{2+} stores and Ca^{2+} channels in sympathetic neurons. *Neuron* 1988;1(5):355–365.
27. Hoth M, Penner R. Depletion of intracellular calcium stores activates a calcium current in mast cells. *Nature* 1992;355(6358):353–356.
28. Nakai J, Ohkura M, Imoto K. A high signal-to-noise Ca^{2+} probe composed of a single green fluorescent protein. *Nat Biotechnol* 2001;19(2):137–141.
29. Wu J, Liu L, Matsuda T, Zhao Y, Rebane A, Drobizhev M, Chang YF et al. Improved orange and red $Ca^{2\pm}$ indicators and photophysical considerations for optogenetic applications. *ACS Chem Neurosci* 2013;4(6):963–972.
30. Heim R, Tsien RY. Engineering green fluorescent protein for improved brightness, longer wavelengths and fluorescence resonance energy transfer. *Curr Biol* 1996;6(2):178–182.
31. Ormo M, Cubitt AB, Kallio K, Gross LA, Tsien RY, Remington SJ. Crystal structure of the *Aequorea victoria* green fluorescent protein. *Science* 1996;273(5280):1392–1395.
32. Akerboom J, Carreras Calderon N, Tian L, Wabnig S, Prigge M, Tolo J, Gordus A et al. Genetically encoded calcium indicators for multi-color neural activity imaging and combination with optogenetics. *Front Mol Neurosci* 2013;6:2.

33. Zhao Y, Araki S, Wu J, Teramoto T, Chang YF, Nakano M, Abdelfattah AS et al. An expanded palette of genetically encoded Ca^{2+} indicators. *Science* 2011;333(6051):1888–1891.
34. Rizzuto R, Brini M, Murgia M, Pozzan T. Microdomains with high Ca^{2+} close to IP3-sensitive channels that are sensed by neighboring mitochondria. *Science* 1993;262(5134):744–747.
35. Bautista DM, Lewis RS. Modulation of plasma membrane calcium-ATPase activity by local calcium microdomains near CRAC channels in human T cells. *J Physiol* 2004;556(Pt 3):805–817.
36. Dolmetsch RE, Xu K, Lewis RS. Calcium oscillations increase the efficiency and specificity of gene expression. *Nature* 1998;392(6679):933–936.
37. Muller MR, Rao A. NFAT, immunity and cancer: A transcription factor comes of age. *Nat Rev Immunol* 2010;10(9):645–656.
38. Shen J, Zhao L, Han G. Lanthanide-doped upconverting luminescent nanoparticle platforms for optical imaging-guided drug delivery and therapy. *Adv Drug Deliv Rev* 2013;65(5):744–755.
39. Chen G, Qiu H, Prasad PN, Chen X. Upconversion nanoparticles: Design, nanochemistry, and applications in theranostics. *Chem Rev* 2014;114(10):5161–5214.
40. Tischer D, Weiner OD. Illuminating cell signalling with optogenetic tools. *Nat Rev Mol Cell Biol* 2014;15(8):551–558.
41. Zhang K, Cui B. Optogenetic control of intracellular signaling pathways. *Trends Biotechnol* 2015;33(2):92–100.

9 Regulation of Orai/STIM Channels by pH

Albert S. Yu, Zhichao Yue, Jianlin Feng, and Lixia Yue

CONTENTS

9.1 INTRODUCTION

Ca^{2+} signaling is crucial in a variety of physiological/pathological processes associated with acidosis and alkalosis. In particular, capacitive Ca^{2+} entry [1] through the Ca^{2+} release–activated Ca^{2+} (CRAC) channels [2] plays an essential role in mediating intracellular and extracellular acidification and alkalinization–induced functional changes. Physiologically, intracellular alkalinization is associated with various physiological functions such as activity-dependent membrane depolarization [3], oocyte maturation [4], oocyte fertilization, sperm activation [5–7], mast cell degranulation [8], smooth muscle contraction [9], and growth factor–induced cell proliferation, differentiation, migration, and chemotaxis [10] (Figure 9.1). Pathologically, intracellular alkalinization is a hallmark of malignant cells associated with tumor progression [11,12], whereas acidic intracellular pH (pH_i) has been shown to promote apoptosis [13]. Additionally, extracellular acidosis is another hallmark of tumor progression [11,12], and also a major cause of immunodeficiency in clinical acidosis due to impaired lymphocyte proliferation and cytotoxicity [14]. Furthermore, extracellular low pH, which occurs under injury and ischemic conditions, inhibits a number of cellular responses, including cytosolic- and membrane-associated enzyme activities as well as ion transport and ion channel activities [14].

Like many other ion channels, native I_{CRAC} is inhibited by acidic but potentiated by basic extracellular or intracellular solutions in various cell types including macrophages [15], Jurkat T-lymphocytes [16], SH-SY5Y neuroblastoma cells [17], and smooth muscle cells [18]. Moreover, it has been shown that intracellular alkalinization–induced increase in intracellular Ca^{2+} is essential for platelet aggregation in response to thrombin [19]. Similarly, extracellular acidosis–induced inhibition, as well as alkalosis-induced stimulation of platelet aggregation, is mediated by changes in store-operated Ca^{2+} entry [20]. Furthermore, store-operated Ca^{2+} entry was suggested to mediate intracellular alkalinization in neutrophils [21], and a variety of growth factors have been demonstrated to induce cytosolic alkalinization together with Ca^{2+} entry [8]. Given the essential role of I_{CRAC} in acidosis- and alkalosis-associated physiological and pathological processes, there is a great interest in understanding the molecular basis of CRAC channel regulation by intracellular and extracellular pH.

The discovery of the molecular basis of I_{CRAC} and its gating mechanisms [22–29] provides a great opportunity to investigate the molecular basis of pH regulation of I_{CRAC}. CRAC channel activity can be influenced by alterations of either the coupling of Orai and STIM subunits or biophysical properties of the pore-forming Orai subunit. Thus, pH regulation of I_{CRAC} can be mediated by influencing this coupling process and/or by changing the biophysical characteristics of the pore-forming Orai subunits. Using Ca^{2+} imaging techniques, a previous study suggested that cytosolic alkalinization may lead to store depletion and therefore activate Orai1/STIM1 channels [30], whereas several other studies demonstrated that cytosolic alkalinization–induced Ca^{2+} release is not always related to Ca^{2+} entry [31–33]. Moreover, intracellular low pH caused by oxidative stress may result in uncoupling of Orai1 and STIM1, thereby inhibiting CRAC currents [34]. Recent studies have focused on using heterologously expressed Orai/STIM channels to fully understand how these channels are regulated by high and low internal and external pH, as well as the molecular basis of pH regulation of Orai/STIM [35–37].

This chapter summarizes recent advances in our understanding of pH regulation of I_{CRAC} by focusing on how Orai/STIM channels are regulated by pH and the potential molecular basis of pH regulation. We summarize the essential methods required to investigate the influence of pH on Orai/STIM channel function as well as molecular mechanisms of pH regulation and highlight future directions in this exciting research field.

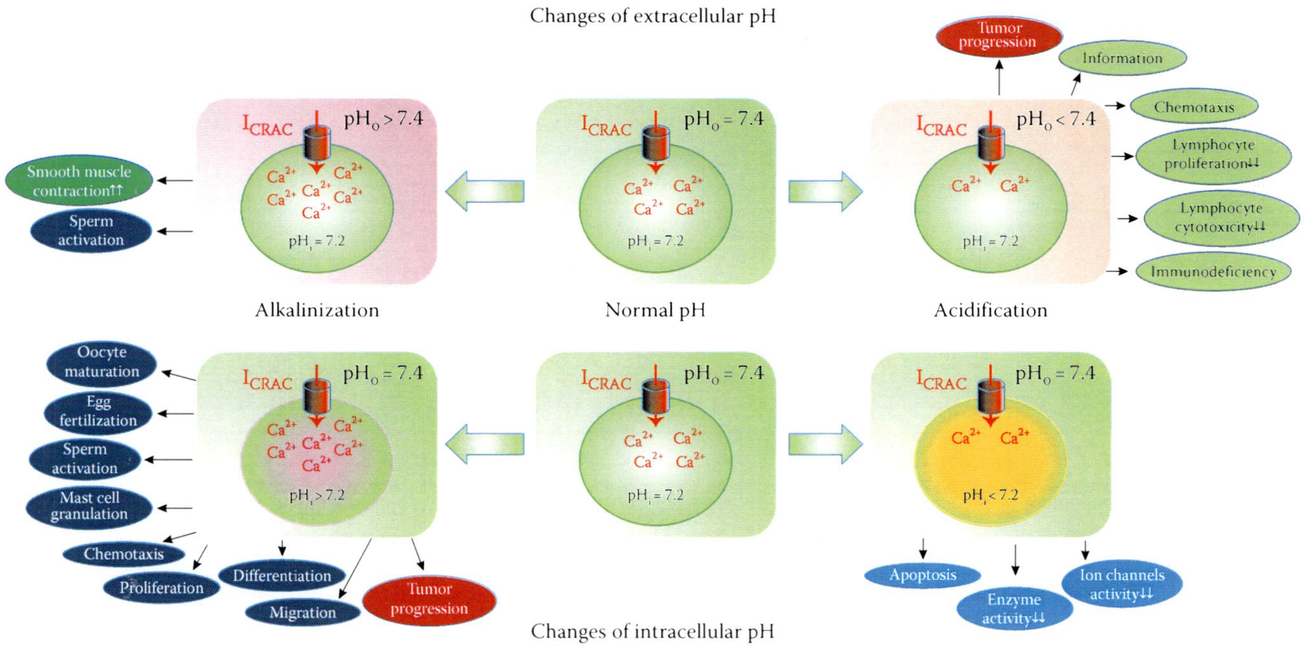

FIGURE 9.1 Diagram illustrating CRAC channel–mediated Ca^{2+} signaling and pathophysiological processes associated with intracellular and extracellular acidification and alkalinization. *Top*: Regulation of I_{CRAC} by acidic and basic extracellular pH and physiological/pathological changes. *Top left*: Basic extracellular pH potentiates I_{CRAC}, which is likely associated with sperm activation, smooth muscle activation, and other physiological/pathological functions. *Top right*: Acidic extracellular pH inhibits I_{CRAC} and is associated with tumor progression, inflammation, reduced chemotaxis, inhibited lymphocyte proliferation, cytotoxicity, and immunodeficiency. *Bottom*: Regulation of I_{CRAC} by acidic and basic intracellular pH and physiological/pathological function changes. *Bottom left*: Cytosolic alkalinization potentiates I_{CRAC}, which is likely responsible for physiological/pathological processes such as oocyte maturation, egg fertilization, sperm activation, mast cell degranulation, chemotaxis, proliferation and differentiation, migration, and tumor progression. *Bottom right*: A decrease in cytosolic pH leads to inhibition of I_{CRAC} and is associated with apoptosis, reduced enzyme activity, and inhibition of various ion channels and transporters.

9.2 BASIC METHODS

9.2.1 Cell Culture and Transfection

HEK-293 cells are cultured in DMEM/F12 medium supplemented with 10% FBS, penicillin, and streptomycin at 37°C in a humidity-controlled incubator with 5% CO_2. Cells are transiently transfected with wild-type (WT) Orai channels or its mutants and WT STIM1 using Lipofectamine 2000 (Invitrogen). 5 μL Lipofectamine 2000 is used for transfection of cells in a 35 mm culture dish. The GFP-containing pEGFP vector is transfected as a mock control. Successfully transfected cells can be identified by their green fluorescence when illuminated at 488 nm (eGFP excitation 488 nm, emission 510 nm). We cotransfected Orai1 (Addgene #12199), Orai2 (Addgene #16369), or Orai3 (Addgene #16370) together with STIM1 (Addgene #19754). After 24–36 h of transfection, Orai/STIM channel currents can be recorded using patch-clamp technique at room temperature (20°C–25°C).

Native I_{CRAC} can be readily recorded from various cell types including rat basophilic leukemia cells (RBL-2H3) [23]. RBL-2H3 cells are cultured using conditions similar to those of culturing HEK-293 cells.

9.2.2 Site-Directed Mutagenesis of Orai

Mutations at selected residues can be generated using site-directed mutagenesis. QuikChange™ Site-Directed Mutagenesis Kit (Agilent) is used for single, double, or triple mutations if the residues are close to each other [37–40]. If the amino acid residues to be mutagenized are distant from each other, one mutation can be generated first, and the mutant construct can serve as a template for the second and third mutations. All mutant constructs are sequenced to confirm that the mutagenized sequence has been generated and that there are no unexpected mutations introduced. We generated 6 mutations of Orai1 for investigating extracellular pH sensitivity, and 11 mutations for evaluating intracellular pH sensitivity [37].

9.2.3 Orai/STIM Current Recording by Patch-Clamp Electrophysiology

Orai/STIM currents can be recorded using patch-clamp methods (see Chapter 1). We use conventional whole-cell patch-clamp, an Axopatch 200B amplifier, Digidata 1320A, and pClamp9 software (Molecular Devices) for data acquisition and analysis. Data are usually digitized at 10 or 20 kHz and digitally filtered off-line at 1 kHz. Patch electrodes are pulled using P-97 (Sutter Instrument) puller from borosilicate glass capillaries and fire-polished to a resistance of 2–4 MΩ when filled with internal solutions. Series resistance (R_S) is compensated up to 90% to reduce R_S errors to <5 mV. Cells with R_S < 10 MΩ are used for analysis [41].

9.2.4 Ca^{2+} Imaging

Since Orai/STIM channels are highly Ca^{2+} selective, regulation of channel activity can also be evaluated by directly measuring Ca^{2+} influx. We use ratiometric calcium imaging to detect changes of intracellular Ca^{2+} concentration (see Chapters 1 and 16). Cells plated on 25 mm glass coverslips are loaded with 5 μM Fura-2/AM (Molecular Probes) for 45 min. Unincorporated dye is washed away using a HEPES-buffered saline solution (HBSS) containing (in mM) 1.2 $MgCl_2$, 1.3 $CaCl_2$, 1.2 KH_2PO_4, 4.7 KCl, 140 NaCl, 20 HEPES, and 10 glucose (pH 7.4). Changes in intracellular Ca^{2+} concentration can be monitored as changes of the ratio of fluorescence intensity at emission wavelength 510 nm and excitation wavelengths at 340 and 380 nm. Alternating the excitation wavelengths between 340 and 380 nm can be achieved using DG4 (Sutter), and fluorescence signals are acquired using CoolSNAP HQ2 (Photometrics). Data are analyzed off-line using NIS-Elements (Nikon). Changes in cytosolic Ca^{2+} concentration can be calculated using NIS-Elements. We also use 1 μM ionomycin as an internal control to monitor dye loading variability among different experimental groups. Changes of F340/380 induced by various experimental manipulations are normalized to F340/F380 elicited by 1 μM ionomycin as previously reported [42].

9.2.5 Measurement of Changes in Intracellular pH by Ratiometric pH_i Imaging

Several ion channels are not only modulated by protons but also permeable to these ions at acidic external pH, which may cause changes in intracellular pH [43,44]. Intracellular pH changes can be monitored by ratiometric pH imaging [45,46] using membrane-permeable pH indicators, such as 2′,7′-bis-(2-carboxyethyl)-5-(and-6)-carboxyfluorescein (BCECF/AM), or the genetically encoded ratiometric pH indicator SypHer, as demonstrated in recent studies [47,48]. For BCECF/AM, the ratio of fluorescence intensity detected at 535 nm when BCECF is excited at 490 versus 440 nm is used for calculation of intracellular pH. For calibration, cells are loaded with 0.5–0.8 μM BCECF/AM (Molecular Probes) at 37°C for 30 min. Excitation spectra of free and protonated BCECF can be determined in HEK-293 cells equilibrated in a high K^+ solution containing 128 mM KCl, 10 mM NaCl, 1 mM $MgCl_2$, 1 mM $CaCl_2$, 10 mM glucose, 0.2% (w/v) bovine serum albumin, and 10 mm HEPES, supplemented with nigericin and monensin (10 μM each) and pH adjusted to 9.0 or 4.0, respectively [45]. The relative concentrations of free and protonated BCECF in the probe are calculated by multivariate linear regression analysis [49] to analyze the spectral properties of experimental data with stored spectra of the free and protonated indicator. Experimental data can then be converted to pH_i from the equation $pH_i = 6.97 - \log([BCECF\text{-}H]/[BCECF^-])$ [45].

9.3 REGULATION OF ORAI/STIM CHANNEL BY INTERNAL AND EXTERNAL pH

Many ion channels and transporters are modulated by changes of internal and external pH. Patch-clamp electrophysiology is a commonly used method to evaluate the pH regulation of ion channel activities, whereas mutation of the presumptive amino acid residues responsible for pH sensitivity is an effective approach to understand the molecular mechanisms of pH regulation.

9.3.1 Influence of Changes of Internal and External pH on Orai/STIM Channel Activation

A unique feature of Orai/STIM channels is that the activation of these channels involves coupling of Orai and STIM proteins. To determine if activation of Orai/STIM channels is influenced by pH changes, perforated patch recording can be performed in order to maintain the integrity of intracellular contents. We used pipette solutions containing 145 mM Cs-methanesulfonate ($CsSO_3CH_3$), 8 NaCl, 3 $MgCl_2$, 10 EGTA, and 10 HEPES (pH adjusted to 7.4 with CsOH) and the antibiotic nystatin (180 μg/mL) for this recording configuration. Increases of cytoplasmic pH can be achieved by superfusing the cells with 30 mM NH_4Cl added to Tyrode's buffer [37]. We found that changes of either intracellular or extracellular pH can regulate but not activate Orai/STIM channels [37].

9.3.2 Orai/STIM Channel Regulation by External Protons

The native I_{CRAC} is inhibited by extracellular low pH in macrophages [15]. To test pH effects on Orai/STIM channels heterologously expressed in HEK-293 cells, we passively depleted the calcium stores to activate Orai/STIM using pipette solutions with free Ca^{2+} concentration chelated to subnanomolar levels (calculated with MaxChelator software http://web.stanford.edu/~cpatton/webmaxcS.htm). We used divalent-free (DVF) solution at various pH values as the extracellular solution. The internal pipette solution for whole-cell current recordings contained 145 mM Cs-methanesulfonate ($CsSO_3CH_3$), 8 mM NaCl, 10 EGTA, and 20 HEPES, with pH 7.2 adjusted with CsOH. $MgCl_2$ (3 mM) was included in the pipette solution to block endogenous TRPM7 currents present in the HEK-293 cells or other cell types [50]. The standard extracellular Tyrode's buffer for whole-cell recordings contained (mM) 145 NaCl, 5 KCl, 2 $CaCl_2$, 1 $MgCl_2$, 10 HEPES, and 10 glucose; pH was adjusted to 7.4 with NaOH. For altering the pH, HEPES, MES (2-(N-morpholino)ethanesulfonic acid) and citric acid, and Tris can be used as the pH buffer for solutions at pH 6–8, pH < 6.0, and pH > 8.0, respectively. We used HCl and 5 mM MES for acidic extracellular solutions and N-methyl-D-glucamine (NMDG) for basic extracellular solutions. Divalent-free external solution contained (mM) 145 Na-SO_3CH_3, 20 HEPES, 5 EGTA, 2 EDTA, and 10 glucose, with free $[Ca^{2+}] \leq 1$ nM and free $[Mg^{2+}] \approx 10$ nM estimated at pH 7.4. When higher Ca^{2+} concentrations are required in the external solution, Na^+ concentration is reduced accordingly to keep the same osmolality. For extracellular acidic solutions, pH was adjusted with citric acid and MES, and for alkaline solutions, NMDG was used in order to maintain the same monovalent cation concentrations [51]. NMDG used for adjusting pH in solutions did not significantly influence Orai1/STIM1 current amplitudes by itself [37].

Orai/STIM channels can be activated by active or passive store depletion (Chapter 1). We activated the channels by passive store deletion after the pipette solution diffused into the cells. Moreover, we used DVF external solution since Na^+-carried CRAC current is larger than the Ca^{2+} current in normal Tyrode's [23]. I_{CRAC} is elicited by a command voltage ramp ranging from −120 to +100 mV of 200 ms duration. Inward current amplitude was used to evaluate the effects of various external pH solutions on Orai/STIM channel activity. NMDG can be used to monitor leak current when switching between various external solutions. Since the effect of pH on these channels is reversible, I_{CRAC} in one cell can be used to test the effects of various external pH solutions to avoid current amplitude variations among cells [37]. A complete recovery after each pH stimulation can be observed by superfusing the cells with normal Tyrode's solution at pH_o 7.4 (Figure 9.3).

For both native I_{CRAC} and overexpressed Orai/STIM, acidic pH inhibited whereas basic pH potentiated the current amplitude [35–37]. The effects of external protons on Orai1/STIM1, Orai2/STIM1, and Orai3/STIM1 are similar [37] and also similar to their effects on I_{CRAC} in RBL-2H3 cells. Moreover, I_{CRAC} arising from overexpression of different Orai isoforms with STIM2 also showed a similar response to external protons [36].

9.3.3 Regulation of Orai/STIM Channel Activity by Internal Protons

The effects of intracellular pH (pH_i) can be tested by fixing pipette solutions at various pH. Acidic pH_i can be adjusted with citrate and MES, and basic pH_i can be adjusted with CsOH [37]. Alternatively, internal pH changes can be achieved by superfusing the cells with ammonium [50,52]. For current recording under various pH_i conditions, normal Tyrode's or DVF solution can be used as the external solution. We used passive store depletion by including high concentrations of Ca^{2+} chelator in the pipette solution to activate I_{CRAC} and used DVF as the external solution to obtain larger I_{CRAC} currents [1]. At acidic pH_i, I_{CRAC} amplitude recorded from overexpression of Orai1/STIM1 is significantly smaller than that at neutral pH_i 7.2; whereas at basic pH_i, I_{CRAC} current amplitude of Orai1/STIM1 is much larger than at neutral pH_i 7.2. The increased channel activity at basic pH_i supports the notion that CRAC channel–mediated Ca^{2+} signaling may play a crucial role in various physiological functions (Figure 9.1) including alkalinization associated with oocyte maturation and fertilization [6,53–55] as well as mast cell degranulation [8].

The effects of internal pH on Orai/STIM channel activity can also be assessed by the Ca^{2+} imaging approach, as conventional patch-clamp can only test one pH_i value in a cell. Store depletion can be induced by thapsigargin, a SERCA pump inhibitor [1,23]. After store depletion and activation of Orai/STIM channels, increases of intracellular pH can be achieved by adding different concentrations of NH_4Cl [50,52], whereas acidic pH_i can be produced by adding $NaHCO_3$ [56]. The concentrations of NH_4Cl or $NaHCO_3$ required for a specific pH_i can be defined from ratiometric pH imaging [45,46].

9.3.4 Evaluation of Proton Permeation

For Ca^{2+}-permeable channels, external protons not only modulate ion channel activity or channel gating but also become permeant cations for some ion channels, such as TRPM7 [44,57,58] and TRPV1 [45]. Proton permeation can be determined by electrophysiology or measurement of intracellular pH. To determine if protons are permeant cations by patch-clamp, Na^+ and K^+ cations in the external solution can be replaced with NMDG and the concentration of protons systematically altered.

For TRPM7, the channel activity is potentiated by external protons [57–59], and the channel becomes permeable to protons at acidic pH_o. A noticeable proton current (~200 pA) through TRPM7 channels can be readily recorded at pH_o 4.0 [57,59]. Proton permeation can also be determined by measuring intracellular pH changes. Proton permeation through TRPV1 leads to cellular acidification at pH_o 5.5 [45]. TRPM2 was also suggested to be permeable to protons based on the altered I–V shape [60]. Orai/STIM channel permeability to protons is yet to be determined.

9.4 MOLECULAR MECHANISMS OF pH SENSITIVITY

Protons can regulate ion channel activities by changing channel permeation properties or gating properties. For Orai/STIM channels, channel gating is governed by coupling of Orai and STIM proteins [23]. Since changes of pH cannot activate Orai/STIM without store depletion, we reasoned that it is unlikely that protons regulate coupling of Orai and STIM proteins [35–37]. Therefore, we focused on how intracellular and extracellular protons regulate Orai/STIM channel activity through altering the function of the pore-forming Orai subunits.

9.4.1 Key Amino Acid Residues Responsible for Extracellular pH Sensitivity

To understand the mechanism by which external protons regulate Orai/STIM channel activities, mutations of candidate amino acid residues were generated by site-directed mutagenesis. We chose the negatively charged residues in the outer mouth of the pore (D110, D112, and D114) as well as the residues along the channel pore (E106 and E190) (Figure 9.2) and neutralized those residues to evaluate their sensitivity to extracellular protons. Among the residues proposed for external pH sensitivity, E106, D110, and E190 are conserved in Orai1, Orai2, and Orai3, albeit D110 in Orai1 is replaced by E110 in Orai2 and Orai3 (Figure 9.2). The residues D112 and D114 of Orai1, however, are replaced with neutral residues Q112 and Q114 in Orai2 but are conserved as negatively charged residues in Orai3 with D112 and E114 (Figure 9.2). In addition to neutralizing the negatively charged residues, we also generated mutants E106D and E190D that preserved charge, given that the length of the side chain of the residues may also contribute to altered channel activity. Mutants of Orai1 were then coexpressed with STIM1 for evaluation of channel activity by patch-clamp. The mutants and the oligo primers used for mutagenesis are listed in Table 9.1.

To compare the pH sensitivity of WT Orai1/STIM1 to mutant channels, dose–response curves can be constructed by measuring the Orai/STIM current amplitude for every pH_o. The dose–response curve can be fitted with the Boltzmann equation to obtain the proton concentration required for inducing 50% of maximal response, EC_{50} (or IC_{50}) or pK_a (Figure 9.3). A shift of dose–response curves and changes of pK_a indicate altered pH sensitivity [37]. Using dose–response curves to evaluate changes of pH sensitivity allows evaluation of both the potency and efficacy of protons in Orai/STIM channels.

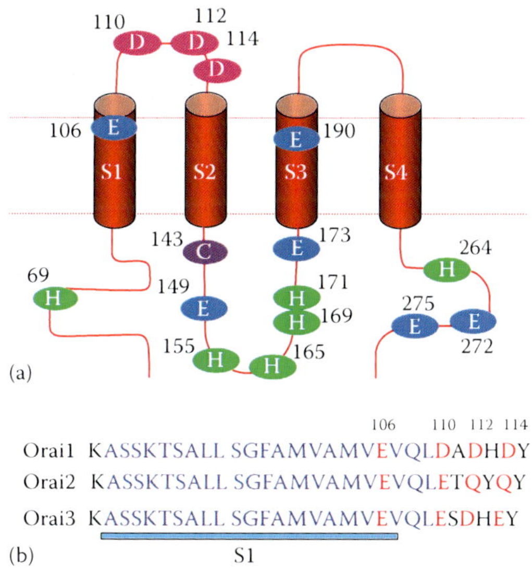

FIGURE 9.2 Mutations of intracellular and extracellular residues and pH sensitivity. (a) Schematic diagram of the pore-forming subunit Orai1. (b) Alignment of the first transmembrane domain (S1) and the extracellular TM1–TM2 loop of Orai1, Orai2, and Orai3. The residues highlighted in red are the mutagenized residues. Note that the negatively charged D112 and D114 of Orai1 are replaced by neutral glutamines in Orai2 but conserved in Orai3. (Modified from Tsujikawa, H. et al., *Sci. Rep.*, 5, 16747, 2015.)

Using monovalent cations as charge carriers is a common approach to study pH effects on different ion channels, including Ca^{2+}-selective ion channels [44,61–67]. Interestingly, for Orai/STIM channels, the pH_o sensitivity is determined by E190 when Na^+ is the charge carrier but by E106 when Ca^{2+} is the charge carrier [37]. Mutants D110N and D112N/D114N produced minor changes in potentiation of current amplitude at very high pH_o but did not change pK_a or shift proton dose–response curves.

9.4.2 Molecular Basis of Intracellular pH Sensitivity of Orai/STIM Channels

To identify the amino acid residues responsible for intracellular pH sensitivity of Orai1/STIM1 channels, we made a series of mutations at the N- and C-termini and at the loop between TM2 and TM3 of Orai1 (Figure 9.2). We chose titratable residues such as histidine and glutamic acid, as well as cysteine residues, which have been shown to be involved in internal pH sensing in other channels [68–70], and generated 10 single or double mutants by neutralizing positive or negative charges.

Since the effects of intracellular pH can only be tested one value at a time, an initial screening of the mutants responsible for altered pH_i sensitivity can be performed by using a high and a low pH stimulation. For example, pH_i 9.0 and 5.5 can induce

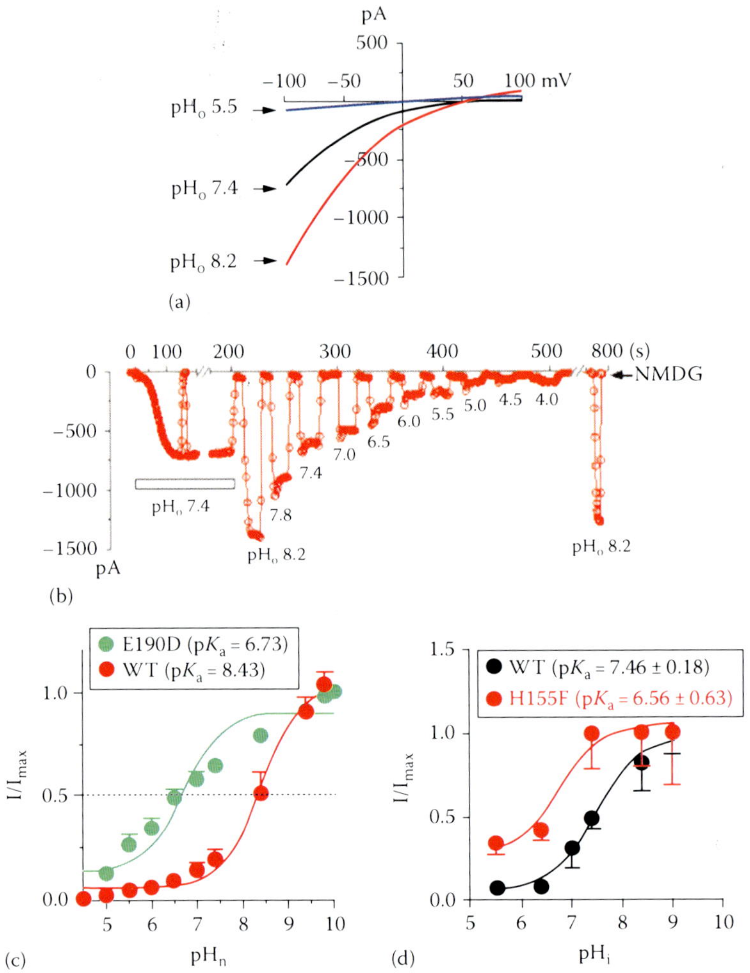

FIGURE 9.3 Regulation of Orai/STIM channel activity by internal and external pH. (a) Representative recordings of Orai/STIM current at pH_o 5.5, 7.4, and 8.2. (b) Concentration-dependent effects of external pH on WT Orai/STIM channels. Note that the effect of external pH is reversible. (c) Dose–response curves of WT and E190D mutant. The noticeable shift of dose–response curve and significant difference of pK_a between WT and E190D suggest that E190 is responsible for external pH sensitivity. (d) Dose–response curves for WT and H155F mutant. H155F significantly shifted the dose–response curve to the left and pK_a from 8.43 to 6.73, indicating that H155 is the intracellular pH sensor of Orai/STIM channels. (Modified from Tsujikawa, H. et al., *Sci. Rep.*, 5, 16747, 2015.)

maximal potentiation and inhibition of Orai/STIM channel activities, respectively. The effects of pH_i 9.0 and pH_i 5.5 on different mutants were compared to those of WT Orai/STIM channels, and the mutants that showed altered pH response were further analyzed by testing concentration-dependent effects using additional pH values. pK_a can be obtained by analyzing concentration-dependent effects of protons on the

TABLE 9.1
Primers Used for Site-Directed Mutagenesis of Putative Residues Responsible for Internal and External pH Sensitivity

Mutants	Forward Primers	Reverse Primers
E106D	GGTGGCAATGGTGGACGTGCAGCTGGACG	CGTCCAGCTGCACGTCCACCATTGCCACC
E106Q	GGTGGCAATGGTGCAGGTGCAGCTGGACG	CGTCCAGCTGCACCTGCACCATTGCCACC
D110N	GGAGGTGCAGCTGAACGCTGACCACGACTACC	GGTAGTCGTGGTCAGCGTTCAGCTGCACCTCC
D112N/D114N	GCAGCTGGACGCTGACCACGACTACCCACCGG	CCGGTGGGTAGTCGTGGTCAGCGTCCAGCTGC
E190D	CTCTTCCTAGCTGACGTGGTGCTGCTCTG	CAGAGCAGCACCACGTCAGCTAGGAAGAG
E190Q	CTCTTCCTAGCTCAGGTGGTGCTGCTCTG	CAGAGCAGCACCACCTGAGCTAGGAAGAG
H69N	GATGAGCCTCAACGAGAACTCCATGCAGGCGC	GCGCCTGCATGGAGTTCTCGTTGAGGCTCATC
C143S	GCTCATGATCAGCACCAGCATCCTGCCCAAC	GTTGGGCAGGATGCTGGTGCTGATCATGAGC
E149Q	CCTGCCCAACATCCAGGCGGTGAGCAACG	CGTTGCTCACCGCCTGGATGTTGGGCAGG
H155F	GGCGGTGAGCAACGTGTTCAATCTCAACTCGG	CCGAGTTGAGATTGAACACGTTGCTCACCGCC
H155N	GCGGTGAGCAACGTGAACAATCTCAACTCG	CGAGTTGAGATTGTTCACGTTGCTCACCGC
H165F	CAAGGAGTCCCCCTTTGAGCGCATGCACCG	CGGTGCATGCGCTCAAAGGGGGACTCCTTG
H165N	GGAGTCCCCCAATGAGCGCATGCACCG	CGGTGCATGCGCTCATTGGGGGACTCC
H169N	CCCATGAGCGCATGAACCGCCACATCGAG	CTCGATGTGGCGGTTCATGCGCTCATGGG
H169F/171F	CATGAGCGCATGTTCCGCTTCATCGAGCTGGCC	GGCCAGCTCGATGAAGCGGAACATGCGCTCATG
E173Q	CCGCCACATCCAGCTGGCCTGGGCC	GGCCCAGGCCAGCTGGATGTGGCGG
H256F	CGTCTTCGCCGTCTTCTTCTACCGCTCAC	GTGAGCGGTAGAAGAAGACGGCGAAGACG
H264F	GCTCACTGGTTAGCTTTAAGACTGACCGACAGTTC	GAACTGTCGGTCAGTCTTAAAGCTAACCAGTGAGC
E275Q/E275Q	CGACAGTTCCAGGAGCTCAACGAGCTGGCGGAG	CTCCGCCAGCTCGTTGAGCTCCTGGAACTGTCG

mutant channels. Among all the mutants tested, H115 appears to be the residue governing internal pH regulation of Orai1/STIM1 channel activities.

9.5 FUNCTIONAL ASSESSMENT OF pH REGULATION OF ORAI/STIM CHANNELS

I_{CRAC} is involved in various physiological/pathological functions. An increase in I_{CRAC} is associated with various cellular functions and can easily be detected. For example, the increase of I_{CRAC} by alkalinization can lead to mast cell release of histamine, which is readily measureable [8]. Mast cells can be isolated from peritoneal cavities of rats [71] or mice and purified using isotonic Percoll centrifugation at 400×*g* for ~10 min. Purified mast cells are then maintained in a cell culture incubator. NH_4Cl is used for inducing cellular alkalinization. After 10 min incubation with NH_4Cl, supernatants of cell culture are collected for the histamine release assay. Since histamine may not be completely released from cells into the supernatant, mast cells can also be collected for residual histamine measurement. Histamine concentrations can be determined using commercially available ELISA-based histamine kits or the conventional histamine assay method [72].

9.6 SUMMARY AND FUTURE RESEARCH DIRECTIONS

As a major Ca^{2+} channel in many non-excitable cell types, the CRAC channel plays a pivotal role in various physiological functions [23]. In acidosis and alkalosis, I_{CRAC} is likely a major mediator of Ca^{2+} signaling in several pathophysiological processes [11–13]. For example, the increased I_{CRAC} by cytosolic alkalinization may contribute to tumor progression [11,12], whereas reduced I_{CRAC} in acidosis may cause impaired lymphocyte functions thereby leading to immunodeficiency associated with clinical acidosis [14]. Therefore, understanding the molecular basis of pH regulation on Orai/STIM channels may provide important information for identifying pharmacological tools to modulate Orai/STIM channel activity under acidic and basic pH conditions. Although we find that internal and external protons modulate Orai/STIM channel activities after the channels are activated, a thorough investigation of pH influence on the coupling of Orai and STIM subunits and channel gating is necessary to fully understand how the activation process of Orai1/STIM is influenced by pH changes. Furthermore, the results of pH regulation of Orai/STIM channels point toward the importance of exploring the therapeutic potential of manipulating I_{CRAC} in reverting or preventing pathological consequences of acidosis and alkalosis, such as immunodeficiency and tumor progression.

ACKNOWLEDGMENTS

We would like to thank the current and former lab members for helpful discussions. This work was supported by the National Institutes of Health, National Heart, Lung and Blood Institute (NHLBI, 2R01HL078960), and American Heart Association (AHA, 16GRNT26430113) to L.Y.

REFERENCES

1. Parekh, A.B. and Putney, J.W., Jr. 2005. Store-operated calcium channels. *Physiol Rev* 85:757–810.
2. Hoth, M. and Penner, R. 1992. Depletion of intracellular calcium stores activates a calcium current in mast cells. *Nature* 355:353–356.
3. Lyall, V. and Biber, T.U. 1994. Potential-induced changes in intracellular pH. *Am J Physiol* 266:F685–F696.
4. Mahnensmith, R.L. and Aronson, P.S. 1985. The plasma membrane sodium–hydrogen exchanger and its role in physiological and pathophysiological processes. *Circ Res* 56:773–788.
5. Lishko, P.V. and Kirichok, Y. 2010. The role of Hv1 and CatSper channels in sperm activation. *J Physiol* 588:4667–4672.
6. Webb, D.J. and Nuccitelli, R. 1981. Direct measurement of intracellular pH changes in Xenopus eggs at fertilization and cleavage. *J Cell Biol* 91:562–567.
7. Nuccitelli, R., Webb, D.J., Lagier, S.T., and Matson, G.B. 1981. 31P NMR reveals increased intracellular pH after fertilization in Xenopus eggs. *Proc Natl Acad Sci USA* 78:4421–4425.
8. Alfonso, A., Cabado, A.G., Vieytes, M.R., and Botana, L.M. 2000. Calcium-pH crosstalks in rat mast cells: Cytosolic alkalinization, but not intracellular calcium release, is a sufficient signal for degranulation. *Br J Pharmacol* 130:1809–1816.
9. Heaton, R.C., Taggart, M.J., and Wray, S. 1992. The effects of intracellular and extracellular alkalinization on contractions of the isolated rat uterus. *Pflügers Arch* 422:24–30.
10. Srivastava, J., Barber, D.L., and Jacobson, M.P. 2007. Intracellular pH sensors: Design principles and functional significance. *Physiology (Bethesda)* 22:30–39.
11. Reshkin, S.J., Bellizzi, A., Caldeira, S., Albarani, V., Malanchi, I., Poignee, M., Alunni-Fabbroni, M., Casavola, V., and Tommasino, M. 2000. Na^+/H^+ exchanger-dependent intracellular alkalinization is an early event in malignant transformation and plays an essential role in the development of subsequent transformation-associated phenotypes. *FASEB J* 14:2185–2197.
12. Lindner, D. and Raghavan, D. 2009. Intra-tumoural extra-cellular pH: A useful parameter of response to chemotherapy in syngeneic tumour lines. *Br J Cancer* 100:1287–1291.
13. Lagadic-Gossmann, D., Huc, L., and Lecureur, V. 2004. Alterations of intracellular pH homeostasis in apoptosis: Origins and roles. *Cell Death Differ* 11:953–961.
14. Lardner, A. 2001. The effects of extracellular pH on immune function. *J Leukoc Biol* 69:522–530.
15. Malayev, A. and Nelson, D.J. 1995. Extracellular pH modulates the Ca^{2+} current activated by depletion of intracellular Ca^{2+} stores in human macrophages. *J Membr Biol* 146:101–111.
16. Guse, A.H., Roth, E., and Emmrich, F. 1994. Ca^{2+} release and Ca^{2+} entry induced by rapid cytosolic alkalinization in Jurkat T-lymphocytes. *Biochem J* 301(Pt 1):83–88.
17. Laskay, G., Kalman, K., Van Kerkhove, E., Steels, P., and Ameloot, M. 2005. Store-operated Ca^{2+}-channels are sensitive to changes in extracellular pH. *Biochem Biophys Res Commun* 337:571–579.
18. Iwasawa, K., Nakajima, T., Hazama, H., Goto, A., Shin, W.S., Toyo-oka, T., and Omata, M. 1997. Effects of extracellular pH on receptor-mediated Ca^{2+} influx in A7r5 rat smooth muscle cells: Involvement of two different types of channel. *J Physiol* 503(Pt 2):237–251.
19. Siffert, W. and Akkerman, J.W. 1987. Activation of sodium-proton exchange is a prerequisite for Ca^{2+} mobilization in human platelets. *Nature* 325:456–458.
20. Marumo, M., Suehiro, A., Kakishita, E., Groschner, K., and Wakabayashi, I. 2001. Extracellular pH affects platelet aggregation associated with modulation of store-operated Ca^{2+} entry. *Thromb Res* 104:353–360.

21. Sandoval, A.J., Riquelme, J.P., Carretta, M.D., Hancke, J.L., Hidalgo, M.A., and Burgos, R.A. 2007. Store-operated calcium entry mediates intracellular alkalinization, ERK1/2, and Akt/PKB phosphorylation in bovine neutrophils. *J Leukoc Biol* 82:1266–1277.
22. Roos, J., DiGregorio, P.J., Yeromin, A.V., Ohlsen, K., Lioudyno, M., Zhang, S., Safrina, O. et al. 2005. STIM1, an essential and conserved component of store-operated Ca^{2+} channel function. *J Cell Biol* 169:435–445.
23. Prakriya, M. and Lewis, R.S. 2015. Store-operated calcium channels. *Physiol Rev* 95:1383–1436.
24. Yeromin, A.V., Zhang, S.L., Jiang, W., Yu, Y., Safrina, O., and Cahalan, M.D. 2006. Molecular identification of the CRAC channel by altered ion selectivity in a mutant of Orai. *Nature* 443:226–229.
25. Prakriya, M., Feske, S., Gwack, Y., Srikanth, S., Rao, A., and Hogan, P.G. 2006. Orai1 is an essential pore subunit of the CRAC channel. *Nature* 443:230–233.
26. Vig, M., Peinelt, C., Beck, A., Koomoa, D.L., Rabah, D., Koblan-Huberson, M., Kraft, S. et al. 2006. CRACM1 is a plasma membrane protein essential for store-operated Ca^{2+} entry. *Science* 312:1220–1223.
27. Park, C.Y., Hoover, P.J., Mullins, F.M., Bachhawat, P., Covington, E.D., Raunser, S., Walz, T., Garcia, K.C., Dolmetsch, R.E., and Lewis, R.S. 2009. STIM1 clusters and activates CRAC channels via direct binding of a cytosolic domain to Orai1. *Cell* 136:876–890.
28. Yuan, J.P., Zeng, W., Dorwart, M.R., Choi, Y.J., Worley, P.F., and Muallem, S. 2009. SOAR and the polybasic STIM1 domains gate and regulate Orai channels. *Nat Cell Biol* 11:337–343.
29. Wang, X., Wang, Y., Zhou, Y., Hendron, E., Mancarella, S., Andrake, M.D., Rothberg, B.S., Soboloff, J., and Gill, D.L. 2014. Distinct Orai-coupling domains in STIM1 and STIM2 define the Orai-activating site. *Nat Commun* 5:3183.
30. Li, S., Hao, B., Lu, Y., Yu, P., Lee, H.C., and Yue, J. 2012. Intracellular alkalinization induces cytosolic Ca^{2+} increases by inhibiting sarco/endoplasmic reticulum Ca^{2+}-ATPase (SERCA). *PLoS One* 7:e31905.
31. Yodozawa, S., Speake, T., and Elliott, A. 1997. Intracellular alkalinization mobilizes calcium from agonist-sensitive pools in rat lacrimal acinar cells. *J Physiol* 499(Pt 3): 601–611.
32. Nitschke, R., Riedel, A., Ricken, S., Leipziger, J., Benning, N., Fischer, K.G., and Greger, R. 1996. The effect of intracellular pH on cytosolic Ca^{2+} in HT29 cells. *Pflügers Arch* 433:98–108.
33. Dettbarn, C. and Palade, P. 1991. Effects of alkaline pH on sarcoplasmic reticulum Ca^{2+} release and Ca^{2+} uptake. *J Biol Chem* 266:8993–9001.
34. Mancarella, S., Wang, Y., Deng, X., Landesberg, G., Scalia, R., Panettieri, R.A., Mallilankaraman, K., Tang, X.D., Madesh, M., and Gill, D.L. 2011. Hypoxia-induced acidosis uncouples the STIM-Orai calcium signaling complex. *J Biol Chem* 286:44788–44798.
35. Scrimgeour, N.R., Wilson, D.P., and Rychkov, G.Y. 2012. Glu^{106} in the Orai1 pore contributes to fast Ca^{2+}-dependent inactivation and pH dependence of Ca^{2+} release-activated Ca^{2+} (CRAC) current. *Biochem J* 441:743–753.
36. Beck, A., Fleig, A., Penner, R., and Peinelt, C. 2014. Regulation of endogenous and heterologous Ca^{2+} release-activated Ca^{2+} currents by pH. *Cell Calcium* 56:235–243.
37. Tsujikawa, H., Yu, A.S., Xie, J., Yue, Z., Yang, W., He, Y., and Yue, L. 2015. Identification of key amino acid residues responsible for internal and external pH sensitivity of Orai1/STIM1 channels. *Sci Rep* 5:16747.
38. Li, M., Du, J., Jiang, J., Ratzan, W., Su, L.T., Runnels, L.W., and Yue, L. 2007. Molecular determinants of Mg^{2+} and Ca^{2+} permeability and pH sensitivity in TRPM6 and TRPM7. *J Biol Chem* 282:25817–25830.

39. Xie, J., Sun, B., Du, J., Yang, W., Chen, H.C., Overton, J.D., Runnels, L.W., and Yue, L. 2011. Phosphatidylinositol 4,5-bisphosphate (PIP(2)) controls magnesium gatekeeper TRPM6 activity. *Sci Rep* 1:146.
40. Du, J., Xie, J., and Yue, L. 2009. Modulation of TRPM2 by acidic pH and the underlying mechanisms for pH sensitivity. *J Gen Physiol* 134:471–488.
41. Yue, L., Navarro, B., Ren, D., Ramos, A., and Clapham, D. 2002. The cation selectivity filter of the bacterial sodium channel, NaChBac. *J Gen Physiol* 160:845–853.
42. Du, J., Xie, J., Zhang, Z., Tsujikawa, H., Fusco, D., Silverman, D., Liang, B., and Yue, L. 2010. TRPM7-mediated Ca^{2+} signals confer fibrogenesis in human atrial fibrillation. *Circ Res* 106:992–1003.
43. DeCoursey, T.E. 2013. Voltage-gated proton channels: Molecular biology, physiology, and pathophysiology of the H(V) family. *Physiol Rev* 93:599–652.
44. Sabovcik, R., Li, J., Kucera, P., and Prod'hom, B. 1995. Extracellular protons modulate the Ca^{2+} block of a Ca^{2+}-blockable monovalent cation channel in chick embryo. *Pflügers Arch* 430:599–601.
45. Hellwig, N., Plant, T.D., Janson, W., Schäfer, M., Schultz, G., and Schaefer, M. 2004. TRPV1 acts as proton channel to induce acidification in nociceptive neurons. *J Biol Chem* 279:34553–34561.
46. Han, J. and Burgess, K. 2010. Fluorescent indicators for intracellular pH. *Chem Rev* 110:2709–2728.
47. Matlashov, M.E., Bogdanova, Y.A., Ermakova, G.V., Mishina, N.M., Ermakova, Y.G., Nikitin, E.S., Balaban, P.M. et al. 2015. Fluorescent ratiometric pH indicator SypHer2: Applications in neuroscience and regenerative biology. *Biochim Biophys Acta* 1850:2318–2328.
48. Shirmanova, M.V., Druzhkova, I.N., Lukina, M.M., Matlashov, M.E., Belousov, V.V., Snopova, L.B., Prodanetz, N.N., Dudenkova, V.V., Lukyanov, S.A., and Zagaynova, E.V. 2015. Intracellular pH imaging in cancer cells in vitro and tumors in vivo using the new genetically encoded sensor SypHer2. *Biochim Biophys Acta* 1850:1905–1911.
49. Lenz, J.C., Reusch, H.P., Albrecht, N., Schultz, G., and Schaefer, M. 2002. Ca^{2+}-controlled competitive diacylglycerol binding of protein kinase C isoenzymes in living cells. *J Cell Biol* 159:291–302.
50. Kozak, J.A., Matsushita, M., Nairn, A.C., and Cahalan, M.D. 2005. Charge screening by internal pH and polyvalent cations as a mechanism for activation, inhibition, and rundown of TRPM7/MIC Channels. *J Gen Physiol* 126:499–514.
51. Yamashita, M. and Prakriya, M. 2014. Divergence of Ca^{2+} selectivity and equilibrium Ca^{2+} blockade in a Ca^{2+} release-activated Ca^{2+} channel. *J Gen Physiol* 143:325–343.
52. Galler, S. and Moser, H. 1986. The ionic mechanism of intracellular pH regulation in crayfish muscle fibres. *J Physiol* 374:137–151.
53. Gomez-Fernandez, C., Pozo-Guisado, E., Ganan-Parra, M., Perianes, M.J., Alvarez, I.S., and Martin-Romero, F.J. 2009. Relocalization of STIM1 in mouse oocytes at fertilization: Early involvement of store-operated calcium entry. *Reproduction* 138:211–221.
54. Gomez-Fernandez, C., Lopez-Guerrero, A.M., Pozo-Guisado, E., Alvarez, I.S., and Martin-Romero, F.J. 2012. Calcium signaling in mouse oocyte maturation: The roles of STIM1, ORAI1 and SOCE. *Mol Hum Reprod* 18:194–203.
55. Martin-Romero, F.J., Lopez-Guerrero, A.M., Alvarez, I.S., and Pozo-Guisado, E. 2012. Role of store-operated calcium entry during meiotic progression and fertilization of mammalian oocytes. *Int Rev Cell Mol Biol* 295:291–328.
56. Lagadic-Gossmann, D., Buckler, K.J., and Vaughan-Jones, R.D. 1992. Role of bicarbonate in pH recovery from intracellular acidosis in the guinea-pig ventricular myocyte. *J Physiol* 458:361–384.
57. Jiang, J., Li, M., and Yue, L. 2005. Potentiation of TRPM7 inward currents by protons. *J Gen Physiol* 126:137–150.

58. Numata, T. and Okada, Y. 2008. Proton conductivity through the human TRPM7 channel and its molecular determinants. *J Biol Chem* 283:15097–15103.
59. Numata, T. and Okada, Y. 2008. Molecular determinants of sensitivity and conductivity of human TRPM7 to Mg^{2+} and Ca^{2+}. *Channels (Austin)* 2:283–286.
60. Starkus, J.G., Fleig, A., and Penner, R. 2010. The calcium-permeable non-selective cation channel TRPM2 is modulated by cellular acidification. *J Physiol* 588:1227–1240.
61. Yeh, B.I., Kim, Y.K., Jabbar, W., and Huang, C.L. 2005. Conformational changes of pore helix coupled to gating of TRPV5 by protons. *EMBO J* 24:3224–3234.
62. Yeh, B.-I., Sun, T.-J., Lee, J.Z., Chen, H.-H., and Huang, C.-L. 2003. Mechanism and molecular determinant for regulation of rabbit Transient Receptor Potential Type 5 (TRPV5) channel by extracellular pH. *J Biol Chem* 278:51044–51052.
63. Chen, X.H., Bezprozvanny, I., and Tsien, R.W. 1996. Molecular basis of proton block of L-type Ca^{2+} channels. *J Gen Physiol* 108:363–374.
64. Chen, X.-H. and Tsien, R.W. 1997. Aspartate substitutions establish the concerted action of P-region glutamates in repeats I and III in forming the protonation site of L-type Ca^{2+} Channels. *J Biol Chem* 272:30002–30008.
65. Jordt, S.E., Tominaga, M., and Julius, D. 2000. Acid potentiation of the capsaicin receptor determined by a key extracellular site. *Proc Natl Acad Sci USA* 97:8134–8139.
66. Ryu, S., Liu, B., and Qin, F. 2003. Low pH potentiates both capsaicin binding and channel gating of VR1 receptors. *J Gen Physiol* 122:45–61.
67. Ryu, S., Liu, B., Yao, J., Fu, Q., and Qin, F. 2007. Uncoupling proton activation of vanilloid receptor TRPV1. *J Neurosci* 27:12797–12807.
68. Dhaka, A., Uzzell, V., Dubin, A.E., Mathur, J., Petrus, M., Bandell, M., and Patapoutian, A. 2009. TRPV1 is activated by both acidic and basic pH. *J Neurosci* 29:153–158.
69. Zong, X., Stieber, J., Ludwig, A., Hofmann, F., and Biel, M. 2001. A single histidine residue determines the pH sensitivity of the pacemaker channel HCN2. *J Biol Chem* 276:6313–6319.
70. Fujita, F., Uchida, K., Moriyama, T., Shima, A., Shibasaki, K., Inada, H., Sokabe, T., and Tominaga, M. 2008. Intracellular alkalization causes pain sensation through activation of TRPA1 in mice. *J Clin Invest* 118:4049–4057.
71. Botana, L.M., Espinosa, J., and Eleno, N. 1987. Adrenergic activity on rat pleural and peritoneal mast cells. Loss of beta-receptors during the purification procedure. *Gen Pharmacol* 18:141–148.
72. Shore, P.A. 1971. The chemical determination of histamine. *Methods Biochem Anal* 1971(Suppl):89–97.

10 Non-Orai Partners of STIM Proteins

Role in ER–PM Communication and Ca^{2+} Signaling

Klaus Groschner, Niroj Shrestha, and Nicola Fameli

CONTENTS

10.1 INTRODUCTION

The discovery of stromal interaction molecules (STIM), STIM1 and STIM2, as signaling elements that govern Ca^{2+} entry by reversibly bridging the intermembrane gap between the cortical endoplasmic reticulum (ER) and plasma membrane (PM) has resulted in a considerable gain in knowledge of the molecular mechanisms underlying ER–PM junctional communication. Progress in understanding the structural basis of the Ca^{2+} entry pathway mediated by coupling of STIM1 to the PM Ca^{2+} channel Orai is paralleled by rapidly accumulating information on STIM-associated signaling events and on alternative (non-Orai) or auxiliary interaction partners of STIM (see Chapters 3 and 10). The highly divergent target molecules of STIMs reside within as well as outside of junctional ER structures and include cation channel effectors as well as junctional regulators and scaffolds. The signaling function of STIM embraces not only ER luminal Ca^{2+} sensing and the regulation of Ca^{2+} handling proteins but also dynamic architectural ER remodeling. Here we aim to provide an outline of the emerging picture of STIM proteins as multifunctional and highly versatile ER-resident molecular switches involved in cellular Ca^{2+} homeostasis and beyond.

10.2 ER–PM Ca^{2+} SIGNALING BY STIM

When two decades ago a gene associated with childhood malignancies was cloned [1] and independently identified to encode a stromal surface molecule, which recognizes B cell precursors and affects their survival/proliferation [2], a more general role of this gene in immune responses was already anticipated, while its involvement in Ca^{2+} handling was initially not suspected. In fact, this gene product, designated as stromal interaction molecule1 (STIM1), as well as its homologue STIM2 [3] represents a single pass (type 1) transmembrane protein that has more recently been recognized as a highly dynamic ER-resident molecular switch involved in a prominent process of immune cell Ca^{2+} signaling termed "capacitative" or "store-operated Ca^{2+} entry" (SOCE) (see Chapter 3) [4]. The history and a state-of-the-art overview of this aspect are provided in Chapters 2 and 16. The groundbreaking discovery of the role of STIM and Orai molecules in immune cell Ca^{2+} handling and pathophysiology [4–6] was undoubtedly a key step toward understanding the SOCE phenomenon as well as its electrophysiological representation, originally termed Ca^{2+} release-activated Ca^{2+} (CRAC) current (see Chapter 1) [7]. This fundamental insight into a ubiquitously expressed Ca^{2+} signaling process included the recognition of STIM1 and its homologue STIM2 as ER Ca^{2+} sensors, which are able to communicate the ER Ca^{2+} level to PM Ca^{2+} channels [8]. Moreover, this concept was further refined by the demonstration of STIM1 being able to bridge the ER and PM by a mechanism involving physical contact with the channel protein Orai1 [9,10]. These findings have considerably advanced our understanding of junctional ER communication with the PM and other organelles. The direct molecular recognition of PM-resident molecules by cytosolic C-terminal structures in active STIM1 oligomers has been shown by classical biochemical methods and FRET microscopy (see Chapter 7).

The sequence of events in STIM-mediated information transfer to the PM starts with the dissociation of luminal Ca^{2+} from the N-terminal EF hand-SAM domain where Ca^{2+} is bound to a canonical EF hand motif with an affinity of roughly 200–600 μM [11,12]. A drop of ER Ca^{2+} levels that leads to the dissociation of Ca^{2+} from this EF hand triggers structural destabilization and rearrangement in the luminal N-terminus to promote the rapid oligomerization and activation of STIM into a configuration in which the cytoplasmic C-terminus adopts an extended conformation exposing structures (i.e., STIM–Orai activating region—SOAR; Orai activating small fragment—OASF) for binding to cytoplasmic Orai domains. STIM oligomers that display an extended C-terminal conformation and enter nanojunctional ER–PM contact sites by lateral diffusion become trapped by their physical interaction with junctional Orai complexes [13].

In addition to this coupling to and gating of Orai complexes, which initiates highly selective Ca^{2+} entry, STIM was found to recognize an array of non-Orai interaction partners. Some of these molecules, like members of the 6-transmembrane domain family of transient receptor potential canonical (TRPC) channels, appear as signaling effectors in alternative mechanisms of junctional Ca^{2+} signaling [14,15], and others like the ER protein SOCE-associated regulatory factor (SARAF) participate in and modulate the STIM–Orai gating machinery (see Chapter 4) [16].

Importantly, the membrane-bridging function of STIM involves not only protein but also membrane lipid recognition and binding. Moreover, cytoskeletal interactions of STIM, in terms of a microtubule plus-end tracking protein (TIP), which prevail in its resting Ca^{2+}-bound form, were found to be crucial for STIM-dependent ER structural dynamics [17]. Although, STIM coupling to microtubules appears nonessential for its primary function in Orai gating, the additional role of STIM in microtubule-growth-dependent dynamics of the cortical ER architecture is likely to have an impact in a more general manner on junctional nanospace Ca^{2+} signaling. It is unclear so far if only a single type of STIM signaling-competent junction exists or if STIM serves multiple signaling functions within a more heterogeneous set of ER–PM nanojunctions.

In this chapter we aim to provide an overview of non-Orai–STIM interaction partners, covering SOCE modulators, alternative effectors, and scaffolds with an outline of the experimental evidence and details of their physical integration into STIM signal complexes, as well as a discussion of the (patho)physiological significance of these interactions.

10.3 MOLECULAR RECOGNITION OF NON-ORAI SIGNALING MOLECULES BY STIM

Structural motifs that confer the principal functions of STIM1 and STIM2 in SOCE have been clearly delineated, and, more recently, high-resolution structural information has been obtained for parts of STIM–Orai coupling domains [18,19]. The structure–function relations for luminal Ca^{2+} binding, autoregulatory oligomerization, and cytoplasmic binding to and gating of Orai are discussed in detail elsewhere in this book (Chapters 2 and 3). A graphical representation of interactions and molecules that represent potentially Orai-independent effectors of STIMs is provided in Figure 10.1. Nonetheless, these interaction sites inevitably overlap with the classical

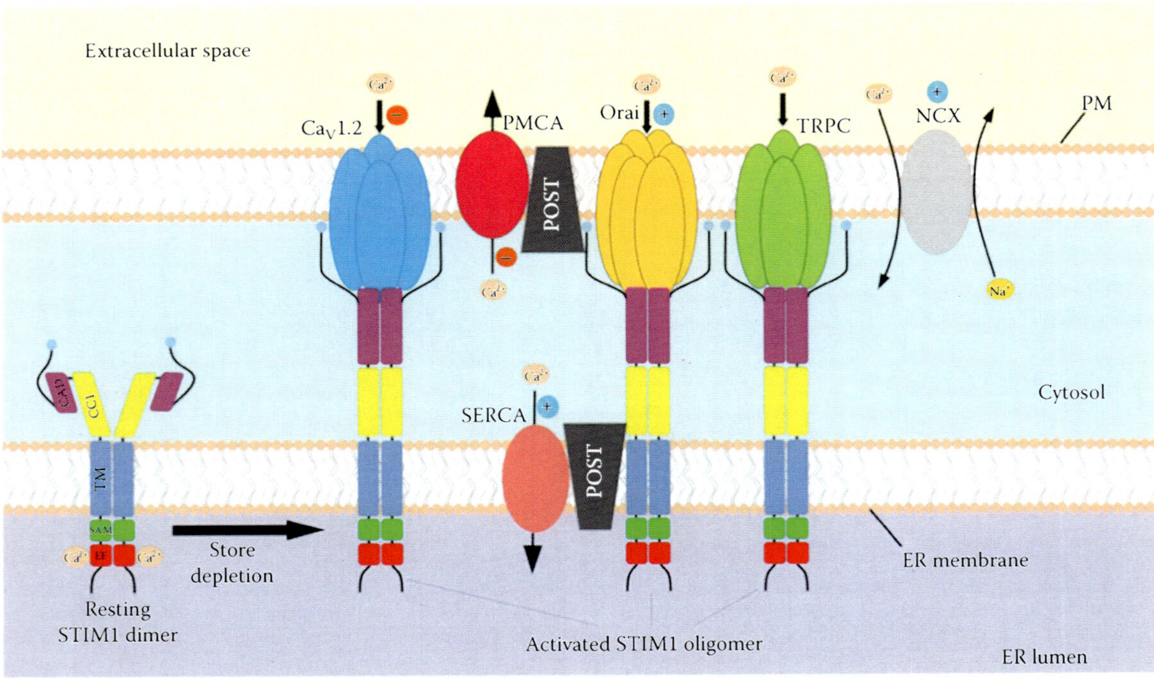

FIGURE 10.1 Orai and non-Orai effectors of STIM1 within ER–PM junctions. At rest, STIM1 exists as a dimer with its CAD (SOAR) domain bound to CC1. Upon ER Ca^{2+} store depletion, Ca^{2+} dissociates from the EF hand of STIM1, leading to STIM1 oligomerization and translocation into ER–PM junctions. STIM1 interacts with PM lipids through its lysine-rich domain and is trapped within ER–PM junctions forming higher oligomers (not shown). STIM1 activates Orai1 or alternatively inhibits $Ca_V1.2$ channels by interactions involving the CAD domain. STIM1 recruits POST to the junction, which inhibits PMCA-mediated Ca^{2+} efflux from the cell and SERCA-mediated ER Ca^{2+} loading. STIM1 also interacts with TRPC, resulting in a highly tissue- and stoichiometry-dependent Ca^{2+} signaling. STIM1 promotes Ca^{2+} entry via reverse mode NCX through unclear mechanisms.

STIM–Orai coupling machinery and, thus, a strict separation of signaling pathways is barely feasible. A typical example is the lysine-rich (polybasic) C-terminal tail of STIM (K-domain), which is critical for STIM–Orai communication and has early on been recognized as a PM targeting motif interacting with PM anionic phospholipids [20]. Lipid recognition and ER–PM bridging via the STIM K-domain represents a potential mechanism of nanojunctional scaffolding. Moreover, this motif interacts with negatively charged residues in TRPC proteins and may be essential for Orai-independent signaling events. Hence, this region of STIM is a paradigm for multifunctionality and likely to serve interdependent signaling mechanisms. In this section, we outline current knowledge on molecular recognition by STIM along with available structural information on these interactions. Potential signaling partners of STIM for which functional but not direct physical coupling has been demonstrated are also discussed.

10.3.1 Non-Orai PM Ca^{2+} Transport Molecules as STIM Effectors

10.3.1.1 TRPC Channels

When STIM1 and Orai were discovered as the essential molecular components of SOCE [21], TRPC proteins had already been extensively characterized as Ca^{2+} entry channels, and a linkage to ER store depletion as well as a role in SOCE had repeatedly been proposed for specific tissues [22,23]. While a general store-operated mechanism of gating may be excluded for TRPC channels [24], a close relationship between some of the members of the canonical TRPC family and SOCE was convincingly demonstrated [25–27] and a rather complex ER–TRPC coupling machinery has been uncovered [28]. TRPC channels, in contrast to Orai, were found subject to store-dependent regulation/activation only in specific cellular settings, presumably defined by certain stoichiometries within their signaling complexes. The linkage of TRPC channels to the ER filling state has been attributed to multiple essential steps, including the Ca^{2+}-mediated promotion of surface presentation [29], trafficking and targeting into ER–PM junctional domains [30] and co-clustering as well as gating competent interaction with STIM1. Therefore, TRPC proteins have been clearly identified as signaling partners of STIM–Orai complexes within ER–PM junctions and as downstream effector targets of STIM proteins. The best understood TRPC molecule in terms of linkage to STIM signaling is TRPC1. As a PM protein erstwhile recognized as a potential store-operated channel [25], TRPC1 was more recently shown to associate via a PM proximal ezrin–radixin–moesin (ERM) domain including the SOAR structure within the STIM C-terminus [31], while gating transition in the TRPC channel is triggered by additional electrostatic interactions between lysine residues in the distal STIM C-terminus and a conserved set of aspartate residues in the TRPC1 C-terminus [32] (Figure 10.1). The interaction between STIM and TRPC1 was demonstrated by classical biochemical methods, functional tests, and the delineation of co-localization by fluorescence microscopy. Of note, interaction of STIM with TRPC1 enables heteromerization with other TRPC proteins, specifically TRPC3 [14], and thereby confers STIM regulation to a larger set of TRPC proteins. Physical coupling to STIM has more recently been localized to the C-terminal coiled coil domain (CCD) in TRPC molecules, and for TRPC3, TRPC1-mediated activation by STIM was found to be associated with rearrangements and destabilization in N-C-terminal CCD interactions [15]. Together, the current understanding of TRPC channel complexes as junctional effectors of STIM and thereby linked to ER Ca^{2+} depletion suggests a highly tissue- and stoichiometry-dependent role in Ca^{2+} signaling and (patho)physiology. Importantly, communication between Orai and TRPC channels is expected to occur due to overlapping interactions with STIM and by Ca^{2+}-mediated cross talk in junctional nanodomains. Thus, it appears possible that specific TRPC complexes function as either Orai-independent or Orai-dependent effectors of STIM signaling [28].

10.3.1.2 Voltage-Gated Ca^{2+} Channels (Ca_V)

SOCE and the STIM–Orai pathway has been considered to be of a particular significance for non-excitable tissues, which typically lack prominent voltage-gated Ca^{2+} entry pathways (see Chapter 1). Increasing evidence and information on the function and cellular role of SOCE in electrically excitable tissues have focused attention on the cellular coordination of these fundamental Ca^{2+} signaling mechanisms. In experiments using the smooth muscle cell line A7r5, STIM was demonstrated to act as an inhibitory regulator of $Ca_V1.2$ (L-type) Ca^{2+} channels, which were identified as PM effectors of STIM, and therefore as part of Ca^{2+} handling within STIM competent ER–PM contact sites [33]. A population of muscle $Ca_V1.2$ channels were found to co-localize with Orai in ER–PM junctional regions. These voltage-gated Ca^{2+} channels were subject to inhibitory regulation by ER depletion due to interaction with STIM, which did not require functional Orai channels or cytoplasmic Ca^{2+} signals. This is an important aspect since the coordination of receptor/store-operated Ca^{2+} entry and voltage-gated Ca^{2+} entry is highly complex and, specifically in smooth muscle, likely to involve also the local Ca^{2+} modulation of regulatory phosphorylation. In parallel, the inhibition and physical coupling between STIM1 and $Ca_V1.2$ was convincingly demonstrated in neuronal cells as well as in immune cells. Physical coupling and inhibitory regulation was found to involve the association of the OASF (CAD) domain with a motif (aa 1809–1908) in the $Ca_V1.2$ C-terminus [34] (Figure 10.1). Similar to the STIM–TRPC coupling, experimental evidence included classical biochemical approaches along with functional tests and co-localization analysis by high-resolution fluorescence microscopy (Table 10.1). Importantly, these studies suggest STIM as the "master regulator" of a set of Ca^{2+} transport systems within ER–PM nanojunctions to generate spatially restricted and temporally distinct Ca^{2+} signaling events.

10.3.1.3 Plasma Membrane Ca^{2+} ATPase (PMCA) and Na^+/Ca^{2+} Exchanger (NCX)

In addition to the control of Ca^{2+}-permeable cation channels in the PM, STIM molecules govern the junctional Ca^{2+} signal by effects on transporters that extrude Ca^{2+} from the junctional space. PMCA was found to accumulate and co-localize with STIM–Orai complexes in activated immune cells [35]. This phenomenon is associated with reduced activity of Ca^{2+} pumps and was found to result from the interaction of PMCA with a STIM complex including a 10 transmembrane segment protein

TABLE 10.1
Non-Orai Partners of STIM1, the Molecular Basis of Interactions and Functional Consequences

Partner	Location	Interaction	Evidence/Method	Functional Consequences	References
TRPC	PM	Physical binding via ERM of STIM1 and the electrostatic interaction between lysine residues of STIM1 and aspartate residues in TRPC	Co-IP, GST-pulldown assay, whole cell patch clamp, and TIRFM	Ca^{2+} and Na^{+} entry; highly tissue- and stoichiometry-dependent role in Ca^{2+} signaling	[14,15,31,32]
Voltage-gated Ca^{2+} channels ($Ca_V1.2$)	PM	CAD of STIM1 and C-terminal region (between amino acids 1809 and 1908) of $Ca_V1.2$	Co-IP, Ca^{2+} imaging, whole cell patch clamp, and confocal microscopy	Inhibition of Ca^{2+} entry via $Ca_V1.2$ after store depletion	[33,34]
PMCA	PM	Indirect; coupling via POST	Co-IP and Ca^{2+} imaging	Inhibition of Ca^{2+} extrusion from the junctional space	[35,36]
NCX 1.3	PM	Unclear	Whole cell patch clamp	Promotion of Ca^{2+} entry	[37]
SARAF	ER resident, TM protein	OASF region of STIM1 and cytoplasmic N-terminus of SARAF	TIRFM, Ca^{2+} imaging, Co-IP, and whole cell patch clamp	Cytosolic Ca^{2+}-dependent inactivation of Orai	[38]
STIMATE	ER resident, TM protein	CC1 domain (amino acids 233–343) of STIM1 and C-termini of STIMATE	Co-IP, GST-pulldown assay, Ca^{2+} imaging, and FRET analysis	Promotion of SOCE and NFAT signaling	[41]
POST	ER resident, TM protein (predominant), PM (minor)	Coupling with STIM1 upon store depletion	Co-IP, Ca^{2+} imaging, whole cell patch clamp, and TIRFM	No effect on SOCE Promotion of STIM1binding to SERCA2, PMCA, Na^{+}/K^{+}-ATPase and nuclear transporters Importin-β1 and Exportin-1 and attenuation of PMCA activity	[36]

(Continued)

TABLE 10.1 (*Continued*)
Non-Orai Partners of STIM1, the Molecular Basis of Interactions and Functional Consequences

Partner	Location	Interaction	Evidence/Method	Functional Consequences	References
Junctate	ER resident, TM protein	N-terminus of STIM1 and ER luminal region (residues 71–236) of junctate	Co-IP, GST-pulldown assay, Ca^{2+} imaging, TIRFM	Structural platform to recruit STIM1 into PM-ER junctions	[42]
ORMDL3	ER resident, TM protein	Unclear	Co-localization via confocal microscopy	Negative modulator of SOCE	[45]
ERp57	ER resident, TM protein	Two conserved cysteine residues, C49 and C56 of STIM1	Surface plasmon resonance screen and FRET analysis	Negative modulator of SOCE	[46]
CRACR2A	Cytoplasmic protein	Ternary complex with Orai1 and STIM1at low junctional Ca^{2+} levels	Co-IP, GST-pulldown assay, and TIRFM	Promotion of SOCE and dissociation of the STIM1–Orai1 complex after Ca^{2+} influx	[47]
Golli	Cytoplasmic protein	C-terminus of STIM1	GST-pulldown assay, affinity chromatography, and mass spectroscopy	Interaction with STIM1–Orai1 complex to attenuate SOCE upon Ca^{2+} influx after store depletion	[48,49]

Notes: Co-IP, co-immunoprecipitation; ER, endoplasmic reticulum; FRET, fluorescence resonance energy transfer; GST, glutathione S-transferase; PM, plasma membrane; TIRFM, total internal reflection fluorescence microscopy.

partner of STIM (POST, or TMEM20). POST (see also following text) represents a scaffold that interacts with STIM in a store-operated manner and links STIM to a series of target proteins within but also outside ER–PM junctional regions [36]. Hence, an array of cellular transport molecules are linked to and governed by the ER luminal Ca^{2+} level via STIM. The store-operated inhibition of PMCA is likely to prolong and shape junctional Ca^{2+} signals, and thereby impact the downstream signaling of SOCE. Notably, a key Na^+ handling system in the PM, the junctional Na^+/K^+ATPase, is also subject to STIM–POST modulation. This mechanism may affect junctional Na^+ levels and thereby, indirectly another essential Ca^{2+} transport system, the Na^+/Ca^{2+} exchanger (NCX). Experiments in airway smooth muscle suggest a STIM-dependent reverse mode Ca^{2+} entry via NCX1.3 [37]. It is so far unclear if this Ca^{2+} entry is initiated by complex formation between STIM and the exchanger or by a STIM-mediated accumulation of cytoplasmic Na^+. STIM-mediated changes in junctional Na^+ may be explained by either the inhibition of Na^+/K^+ATPase or the opening of junctional Na^+ entry channels of the TRPC family. Collectively, the available information suggests that STIM is able to control Ca^{2+} handling in ER–PM junctions by the modulation of multiple Ca^{2+} entry mechanisms as well as Ca^{2+} extrusion from the junctional space.

10.3.2 ER Proteins as Regulators and Effectors

A divergent array of proteins, primarily resident in the ER membrane, that associate with and modulate the function of STIM has been found. Among these interaction partners, some have been clearly identified as important regulatory factors of STIM–Orai signaling, whereas others serve scaffold functions to link STIM to specific downstream signaling components of SOCE within the ER–PM junction. Hence, ER protein partners of STIM may be considered as regulators of the STIM–Orai pathway and/or alternative effectors.

10.3.2.1 SARAF

Although STIM and Orai are considered sufficient to provide the molecular basis of the SOCE phenomenon within an ER–PM junctional setting, cell-specific modulatory and ancillary components exist, some of which reside within the ER. A single transmembrane-spanning ER protein was recently identified as a STIM–Orai complex partner and regulator of the SOCE machinery [38]. The 339 amino acid protein was termed SARAF for SOCE-associated regulatory factor and was recognized to promote Ca^{2+}-dependent inhibitory regulation (Ca^{2+}-dependent inactivation—CDI) of Orai [39]. The cytoplasmic N-terminus of SARAF was subsequently found to interact with and occlude the OASF region in STIM to induce an inactivated state. This STIM–SARAF association is governed by a short C-terminal inhibitory domain (CTID) in STIM and requires STIM–Orai interaction within a distinct junctional micro/nano environment involving membrane tethering of the STIM C-terminus to anionic lipids via its polybasic tail. Moreover, it was demonstrated that association is favored in PI(4,5)P2-rich domains and the STIM–Orai activation/inactivation cycle involves transitions of the complex between membrane domains of different PI(4,5)P2 levels [16]. Consistently, SARAF binding to STIM was found to depend on the presence of the lipid-binding proteins Septin4 and extended Synaptotagmin 1 (E-Syt1), a key determinant of junctional stability and architecture (see also succeeding text).

10.3.2.2 STIMATE

The STIM-activating enhancer STIMATE (TMEM110), which emerged initially in a genome-wide RNAi screen as a determinant of nuclear factor of activated T cells (NFAT) signaling [40], was identified as an ER-resident multipass transmembrane protein that promotes SOCE by interacting with STIM [41]. STIMATE was found capable of forming oligomers within the ER membrane presumably via a GXXXG association motif in the transmembrane region, and fluorescence protease protection experiments revealed the cytoplasmic localization of both N- and C-termini. Localization and interaction analysis by fluorescence microscopy and expression of GFP fusion proteins clearly demonstrated co-localization and physical interaction of STIMATE and STIM within the ER of resting cells [41]. Interestingly, store-depletion-induced translocation of STIM was not quantitatively joined by STIMATE, while total internal reflection fluorescence (TIRF) microscopy uncovered that a fraction of STIMATE was indeed co-localized with STIM at ER–PM junctions after store depletion. The overexpression of STIMATE, particularly in the presence of exogenous STIM, resulted in the constitutive clustering of STIM and activation of Ca^{2+} entry in the absence of any store-depleting stimulus. STIMATE was found to promote ER–PM junction formation. Consistently, STIM cluster formation in response to store depletion was promoted with significantly larger punctae being formed. Moreover, the physical interaction between STIM and STIMATE was demonstrated by both biochemical and fluorescence microscopic methods (Table 10.1). This interaction involves the association of STIMATE C-terminal structures with the STIM CC1 domain (aa 233–343), thereby promoting a fully activated/extended state and exposure of SOAR for interaction with junctional Orai channels. Hence, STIMATE represents an efficient modulator of both ER–PM nanojunctional stability and the efficiency of the STIM–Orai Ca^{2+} signaling machinery.

10.3.2.3 POST

POST (TMEM20) is predominantly expressed in the ER membrane with a minor fraction (5%–10%) present in the PM [36]. Unlike SARAF or STIMATE, POST forms a complex with STIM upon store depletion without profoundly affecting SOCE. POST is a scaffold molecule that associates with Orai in a manner independent of the ER Ca^{2+} content and binds STIM to translocate into ER–PM junctions upon cell stimulation. A series of target molecules of STIM–POST complexes have been identified. Interestingly, these POST targets reside not only in the PM (Na^+/K^+ATPase) and in the ER membrane (SERCA2) but also outside junctional ER regions such as the nuclear envelope (importin-beta1, exportin-1). POST has been implicated in the regulation of these targets (as also outlined earlier) and is likely to convey ER store-dependent regulation to targets beyond the ER–PM junctional region.

10.3.2.4 Junctate

Junctate is an ER membrane protein that is prevalent in ER–PM junctions in a Ca^{2+}-bound form and stabilizes the ER–PM junctional interaction [42]. Results from affinity purification-mass spectrometry and immunoprecipitation experiments demonstrated that junctate is an interacting partner of the Orai1–STIM1 complex. Junctate has a Ca^{2+}-binding EF hand domain in the ER lumen that detects the ER Ca^{2+} levels. Mutation in this domain leads to the clustering of STIM1 and elevation of cytoplasmic Ca^{2+} concentration in T cells independently of store depletion. Upon store depletion, Orai1 and STIM1 clustered into the ER–PM junctions, whereas the localization of junctate was barely altered, suggesting a role of junctate as a structural platform to recruit STIM1 into ER–PM junctions. Junctate can recruit STIM1 into the ER–PM junction independent of phosphoinositide and Orai1 binding [42,43]. Importantly, junctate is a molecule reported to interact also with IP3 receptors and TRPC channels, and is therefore potentially involved in the cross talk of STIMs with non-Orai alternative Ca^{2+} signaling mechanisms [44].

10.3.2.5 ORMDL3

Orosomucoid-like 3 (ORMLD3) is an ER membrane-resident protein with both N- and C-termini facing the cytosol. ORMDL3 overexpression attenuated the CRAC current (I_{CRAC}) and SOCE in Jurkat and HEK293 cells, whereas ORMDL3 gene silencing increased I_{CRAC} and SOCE. Using confocal microscopy, ORMDL3 was found to co-localize with STIM1 under basal conditions and move with STIM1 puncta toward ER–PM junctions upon store depletion although co-immunoprecipitation between ORMDL3 and STIM1 was not detected. Thus, ORMDL3 acts as a negative modulator of SOCE [45].

10.3.2.6 ERp57

ERp57 is an ER luminal oxidoreductase involved in ER luminal regulation of STIM1. Using a surface plasmon resonance screen and FRET analysis, ERp57 was found to interact with the ER luminal domain of STIM1 via two conserved cysteine residues, C49 and C56, under resting conditions and store depletion. ERp57-deficient mouse embryonic fibroblasts (MEFs) showed enhanced SOCE, suggesting a suppressive role of ERp57 toward SOCE [46].

10.3.2.7 Sarco/Endoplasmic Reticulum Ca^{2+} ATPase (SERCA)

SERCA represents a key downstream target of SOCE in terms of receiving the junctional Ca^{2+} entered through STIM–Orai channels for privileged refilling of the ER. This process requires certain spatial relations between SOCE and SERCA as well as a degree of functional coordination. As outlined earlier, this appears in part to be achieved by a physical linkage between STIM and SERCA2a via POST.

10.3.3 Cytoplasmic Junctional Regulators

10.3.3.1 CRACR2A

Calcium release-activated Ca^{2+} channel regulator 2A (CRACR2A) is a cytoplasmic protein with two predicted EF hand motifs that is highly expressed in T cells. Using immunoprecipitation-glutathione S-transferase (GST)-pulldown analysis and TIRF microscopy, CRACR2A was found to interact directly with Orai1 and STIM1 to form a ternary complex at low junctional Ca^{2+} levels. As Ca^{2+} levels increase at ER–PM junctions, CRACR2A binds to Ca^{2+} and promotes STIM1–Orai1 complex dissociation. The knockdown of CRACR2A inhibited STIM1 clustering upon store depletion, whereas the overexpression of an EF hand mutant of CRACR2A enhanced STIM1 clustering and elevated cytoplasmic Ca^{2+}. Thus, CRACR2A is a key regulator of STIM–Orai mediated SOCE [47] (see Chapter 4).

10.3.3.2 Golli

Golli proteins are alternatively spliced isoforms of myelin basic protein (MBP) family, located in the cytoplasm. Golli-deficient T cells were found to display enhanced Ca^{2+} influx upon store depletion, whereas its overexpression inhibited SOCE. The myristoylation of golli at its N-terminus and association with PM was required for this inhibitory effect on SOCE [48]. Using GST pulldown and bimolecular fluorescence complementation assays combined with affinity chromatography and mass spectroscopy, golli was found to interact with the C-terminus of STIM1 but not with Orai1, modulated by cytosolic Ca^{2+} concentration. Upon store depletion in HeLa cells, golli co-localized with STIM1–Orai1 complexes. The overexpression of golli reduced SOCE in HeLa cells, but STIM1 coexpression overcame this inhibition [49]. Thus, Ca^{2+} influx upon store depletion promotes the interaction of golli proteins with STIM1–Orai1 complexes to attenuate SOCE.

10.3.4 Microtubules

As discussed earlier, STIM1 plays a role in ER structural dynamics and may by this mechanism influence pan-junctional Ca^{2+} signaling. STIM1 binds to the microtubule plus-end tracking protein EB1 and forms comet-like accumulations at sites where polymerizing microtubule ends come in contact with the ER network. STIM1 overexpression stimulates ER extension together with the EB1-positive end of a growing microtubule whereas the depletion of STIM1 and EB1 decreases ER protrusion [17]. Of note, a recent study demonstrated the differential regulation of STIM function in cancer cells via microtubule-associated histone deacetylase 6 (HDAC6) in an EB1-independent manner [43].

10.3.5 Membrane Lipid Domains

A specific and intimate molecular interaction partner of STIM is the PM lipid bilayer. The key role of PM phospholipids in SOCE was recognized early on [50,51] and negatively charged phospholipids, specifically $PI(4,5)P_2$ and $PI(3,4,5)P_3$, were found to be essential for PM targeting and puncta formation by STIMs [20,52,53].

The structure responsible for lipid recognition in STIM was clearly identified as the polybasic, lysine-rich cluster (K-domain) in its very C-terminus [20,52]. This electrostatic mechanism of lipid recognition, which is common to many membrane-targeted proteins (e.g., [54]), is an essential step in the STIM–Orai coupling cycle and may contribute in part to the diffusion trap mechanism of STIM puncta formation within ER–PM junctions [13]. Membrane lipids and their lateral segregation have been found to be essential for junctional Ca^{2+} signaling. The lipid composition within junctional domains governs not only membrane bridging and junctional trapping of STIM but also recruitment of other scaffolds and structural determinants of ER–PM contact sites. Septins represent a family of GTPases able to form higher molecular assemblies that bind negatively charged PM lipids [55], thereby creating lateral diffusion barriers to control membrane-delimited signaling events. Specifically, septin 4 and 5 were identified in an RNA_i screen as essential components of SOCE–NFAT signaling and subsequently shown to enable the effective regulation of Orai within junctional contacts [40]. The interaction of septin filaments with lipids in turn has an impact on the membrane distribution of phosphoinositides. Thereby, septins not only promote the junctional targeting of Orai and stabilize STIM–Orai interactions but also interfere with the recruitment and function of regulators such as SARAF [16]. Consequently, membrane phospholipids and lipid-associated scaffolds are critically involved in Ca^{2+}-dependent inactivation of Orai (see also previous text). In line with the pivotal role of PM phosphoinositides as determinants of STIM–Orai signaling, the translocation of Orai between microdomains of different phospholipid ($PI(4,5)P_2$) levels has been implicated in the STIM–Orai regulatory cycle. This is reminiscent of the suggested dynamic reorganization of non-Orai STIM target channels within PM lipid rafts [56]. It is important to note the tight linkage between SOCE and PM lipid metabolism as well as junctional nanoarchitecture. SOCE is typically associated with alterations in PM lipid composition and recruitment of extended synaptotagmins (E-Syt1) [35], which is potentially involved in intermembrane phospholipid transport and in tethering and stabilizing ER–PM contacts with as yet ill-defined consequences. Collectively, phospholipids and phospholipid-associated scaffolds, along with phospholipid-binding and structural elements of ER–PM junctions such as extended synaptotagmins (E-Syt1), provide the platform for efficient cellular control of STIM–Orai complexes and potentially for non-Orai–STIM signaling pathways.

10.4 ER–PM JUNCTIONAL Ca^{2+} SIGNALING HUB—MOLECULAR RECOGNITION MEETS NANOARCHITECTURE

As we learned in the previous sections, STIM is pivotal not only in kick-starting the SOCE process by functioning as a luminal Ca^{2+} detector but also in ensuring efficient communication with the actual CRAC channel (Orai). This is accomplished by way of a multitude of interactions with other molecular partners both for activation of Ca^{2+} entry upon ER depletion and deactivation thereof upon physiological Ca^{2+} level restoration.

Along with the strict biochemical link between STIM and CRAC channels in SOCE, what transpires clearly since the earliest attempts at understanding both extracellular Ca^{2+} entry in general and within the SOCE mechanism in particular is the necessity for the transport of extracellular Ca^{2+} into the ER lumen to occur through PM patches, which are closely apposed with the ER membrane, that is, ER–PM junctions [57–59]. This conjecture was compatible with earlier ultrastructural observations of junctions between the S/ER and the PM in a variety of cell systems [60,61].

In the context of SOCE, two key studies by Lewis and colleagues in Jurkat T cells [13,62] showed that the STIM puncta, formed around the cell periphery following ER Ca^{2+} release, were localized at ER–PM junctions. Importantly, TIRF (for its ability to probe only the subplasmalemmal region) and electron microscopy (for its high resolution) were employed to obtain these observations, which then allowed a quantitative characterization of these junctions. They found that STIM puncta accumulated at junctions where the ER membrane ran roughly parallel to the PM at a mean distance of about 17 nm. The study also highlighted the dynamic character of these junctions in response to Ca^{2+} store depletion, reporting a mean junctional lateral extension of about 150 nm in resting cells and 300 nm in Ca^{2+}-depleted ones. Lastly, these results indicated that STIM puncta accumulation could occur at preexisting as well as newly formed ER–PM junctions and that these same junctions are the places at which the STIM puncta activate CRAC channels.

A more recent and valuable study by the same group based on single molecule tracking techniques and particle diffusion analysis in HEK293 cells measured the dramatically diminished mobility of both STIM1 and Orai1 down to similar diffusivity values following store depletion [13]. They also determined that in stimulated cells both proteins appear to become trapped in subplasmalemmal regions whose size is compatible with ER–PM junctions, thereby corroborating earlier observations that such junctions are pivotal in supporting SOCE and also other forms of junctional communication.

From the ultrastructural standpoint, visually and quantitatively stunning confirmation that STIM clusters at ER–PM junctions (in proximity to Orai) after store depletion was provided by Franzini-Armstrong's group in HEK293 cells [63]. Therein they employed transmission electron microscopy and freeze-fracture to characterize the architecture of ER–PM junctions and the protein organization at the nanoscale level before and after ER Ca^{2+} depletion. The study suggests that ER–PM gaps are slightly narrower than those measured by Wu et al. (see above) at 12 nm, although still compatible with them. Current understanding suggests ER–PM junctions are highly dynamic nanostructures that harbor a wide range of proteins that either control STIM–Orai functions or serve as alternative effectors. Importantly, STIM interactions are expected to participate in the architectural dynamics of ER–PM junctions. An example of such dynamic remodeling is the Ca^{2+}-dependent recruitment of structural determinants such as E-Syt1 and Nir2 [64], which are part of a regulatory cycle that involves changes in both membrane phospholipid composition and nanoarchitecture. Changes in lipid domain structure, formation of lateral diffusion limits, and alterations in the width of the intermembrane cleft are likely the basis of the coordination of signaling processes within the junctional space. Of note, dynamic changes and the enlargement of ER–PM contact sites in response to SOCE have recently been visualized by live cell imaging [65].

As is evident from some of the articles mentioned in this section, the understanding of experimental outcome was aided by *ad hoc* modeling. It is likely that full comprehension of the ER–PM junctional dynamics behind SOCE, including STIM trapping, communication with Orai channels and alternative targets, as well as interference with lipid domains and structural determinants will require quantitative models capable of condensing the wealth of experimental information into an organic and dynamic view of the interactions arising in the ER–PM junctions and among the rich signaling machinery revolving around STIM proteins.

10.5 CONCLUSION AND PERSPECTIVES

A highly multifunctional role of STIM in cellular Ca^{2+} handling has been uncovered over the past decade. Signaling downstream of STIM involves not only the control of Orai channels, but also other Ca^{2+} transport molecules expressed in a more tissue-specific manner as well as a yet incompletely understood impact of the ER on architectural dynamics. By undergoing a multitude of dynamic protein–protein and protein–lipid interactions as outlined in this chapter, STIM controls important signaling processes within and outside cytoplasmic nanospaces formed by the ER–PM contact sites. Non-Orai interaction partners of STIM involved in these processes appear as crucial and even indispensable components of coordinated Ca^{2+} signaling within cortical ER nanospaces and are likely to emerge as future therapeutic targets.

REFERENCES

1. Parker NJ, Begley CG, Smith PJ, Fox RM. Molecular cloning of a novel human gene (D11S4896E) at chromosomal region 11p15.5. *Genomics*. 1996;37(2):253–256.
2. Oritani K, Kincade PW. Identification of stromal cell products that interact with pre-B cells. *The Journal of Cell Biology*. 1996;134(3):771–782.
3. Williams RT, Manji SS, Parker NJ, Hancock MS, Van Stekelenburg L, Eid JP et al. Identification and characterization of the STIM (stromal interaction molecule) gene family: Coding for a novel class of transmembrane proteins. *The Biochemical Journal*. 2001;357(Pt 3):673–685.
4. Liou J, Kim ML, Heo WD, Jones JT, Myers JW, Ferrell JE, Jr. et al. STIM is a Ca^{2+} sensor essential for Ca^{2+}-store-depletion-triggered Ca^{2+} influx. *Current Biology*. 2005;15(13):1235–1241.
5. Feske S, Gwack Y, Prakriya M, Srikanth S, Puppel SH, Tanasa B et al. A mutation in Orai1 causes immune deficiency by abrogating CRAC channel function. *Nature*. 2006;441(7090):179–185.
6. Roos J, DiGregorio PJ, Yeromin AV, Ohlsen K, Lioudyno M, Zhang S et al. STIM1, an essential and conserved component of store-operated Ca^{2+} channel function. *The Journal of Cell Biology*. 2005;169(3):435–445.
7. Hoth M, Penner R. Depletion of intracellular calcium stores activates a calcium current in mast cells. *Nature*. 1992;355(6358):353–356.
8. Soboloff J, Rothberg BS, Madesh M, Gill DL. STIM proteins: Dynamic calcium signal transducers. *Nature Reviews Molecular Cell Biology*. 2012;13(9):549–565.
9. Muik M, Frischauf I, Derler I, Fahrner M, Bergsmann J, Eder P et al. Dynamic coupling of the putative coiled-coil domain of ORAI1 with STIM1 mediates ORAI1 channel activation. *The Journal of Biological Chemistry*. 2008;283(12):8014–8022.

10. Park CY, Hoover PJ, Mullins FM, Bachhawat P, Covington ED, Raunser S et al. STIM1 clusters and activates CRAC channels via direct binding of a cytosolic domain to Orai1. *Cell*. 2009;136(5):876–890.
11. Zheng L, Stathopulos PB, Li GY, Ikura M. Biophysical characterization of the EF-hand and SAM domain containing Ca^{2+} sensory region of STIM1 and STIM2. *Biochemical and Biophysical Research Communications*. 2008;369(1):240–246.
12. Stathopulos PB, Li GY, Plevin MJ, Ames JB, Ikura M. Stored Ca^{2+} depletion-induced oligomerization of stromal interaction molecule 1 (STIM1) via the EF-SAM region: An initiation mechanism for capacitive Ca^{2+} entry. *The Journal of Biological Chemistry*. 2006;281(47):35855–35862.
13. Wu MM, Covington ED, Lewis RS. Single-molecule analysis of diffusion and trapping of STIM1 and Orai1 at endoplasmic reticulum-plasma membrane junctions. *Molecular Biology of the Cell*. 2014;25(22):3672–3685.
14. Yuan JP, Zeng W, Huang GN, Worley PF, Muallem S. STIM1 heteromultimerizes TRPC channels to determine their function as store-operated channels. *Nature Cell Biology*. 2007;9(6):636–645.
15. Lee KP, Choi S, Hong JH, Ahuja M, Graham S, Ma R et al. Molecular determinants mediating gating of Transient Receptor Potential Canonical (TRPC) channels by stromal interaction molecule 1 (STIM1). *The Journal of Biological Chemistry*. 2014;289(10):6372–6382.
16. Maleth J, Choi S, Muallem S, Ahuja M. Translocation between PI(4,5)P2-poor and PI(4,5)P2-rich microdomains during store depletion determines STIM1 conformation and Orai1 gating. *Nature Communications*. 2014;5:5843.
17. Grigoriev I, Gouveia SM, van der Vaart B, Demmers J, Smyth JT, Honnappa S et al. STIM1 is a MT-plus-end-tracking protein involved in remodeling of the ER. *Current Biology*. 2008;18(3):177–182.
18. Yang X, Jin H, Cai X, Li S, Shen Y. Structural and mechanistic insights into the activation of Stromal interaction molecule 1 (STIM1). *Proceedings of the National Academy of Sciences of the United States of America*. 2012;109(15):5657–5662.
19. Stathopulos PB, Schindl R, Fahrner M, Zheng L, Gasmi-Seabrook GM, Muik M et al. STIM1/Orai1 coiled-coil interplay in the regulation of store-operated calcium entry. *Nature Communications*. 2013;4:2963.
20. Liou J, Fivaz M, Inoue T, Meyer T. Live-cell imaging reveals sequential oligomerization and local plasma membrane targeting of stromal interaction molecule 1 after Ca^{2+} store depletion. *Proceedings of the National Academy of Sciences of the United States of America*. 2007;104(22):9301–9306.
21. Hogan PG, Lewis RS, Rao A. Molecular basis of calcium signaling in lymphocytes: STIM and ORAI. *Annual Review of Immunology*. 2010;28:491–533.
22. Abramowitz J, Yildirim E, Birnbaumer L. The TRPC family of ion channels: Relation to the TRP superfamily and role in receptor- and store-operated calcium entry. In: Liedtke WB, Heller S, eds. *TRP Ion Channel Function in Sensory Transduction and Cellular Signaling Cascades*. Frontiers in Neuroscience. Boca Raton, FL: CRC Press & Taylor & Francis Group, 2007.
23. Cheng KT, Ong HL, Liu X, Ambudkar IS. Contribution and regulation of TRPC channels in store-operated Ca^{2+} entry. *Current Topics in Membranes*. 2013;71:149–179.
24. DeHaven WI, Jones BF, Petranka JG, Smyth JT, Tomita T, Bird GS et al. TRPC channels function independently of STIM1 and Orai1. *The Journal of Physiology*. 2009;587 (Pt 10):2275–2298.
25. Liu X, Bandyopadhyay BC, Singh BB, Groschner K, Ambudkar IS. Molecular analysis of a store-operated and 2-acetyl-sn-glycerol-sensitive non-selective cation channel. Heteromeric assembly of TRPC1-TRPC3. *The Journal of Biological Chemistry*. 2005;280(22):21600–21606.

26. Groschner K, Hingel S, Lintschinger B, Balzer M, Romanin C, Zhu X et al. Trp proteins form store-operated cation channels in human vascular endothelial cells. *FEBS Letters*. 1998;437(1–2):101–106.
27. Freichel M, Suh SH, Pfeifer A, Schweig U, Trost C, Weissgerber P et al. Lack of an endothelial store-operated Ca^{2+} current impairs agonist-dependent vasorelaxation in TRP4–/– mice. *Nature Cell Biology*. 2001;3(2):121–127.
28. Ong HL, Ambudkar IS. The dynamic complexity of the TRPC1 channelosome. *Channels*. 2011;5(5):424–431.
29. Cheng KT, Liu X, Ong HL, Swaim W, Ambudkar IS. Local Ca^{2+} entry via Orai1 regulates plasma membrane recruitment of TRPC1 and controls cytosolic Ca^{2+} signals required for specific cell functions. *PLoS Biology*. 2011;9(3):e1001025.
30. de Souza LB, Ong HL, Liu X, Ambudkar IS. Fast endocytic recycling determines TRPC1-STIM1 clustering in ER-PM junctions and plasma membrane function of the channel. *Biochimica et Biophysica Acta*. 2015;1853(10 Pt A):2709–2721.
31. Huang GN, Zeng W, Kim JY, Yuan JP, Han L, Muallem S et al. STIM1 carboxyl-terminus activates native SOC, I(crac) and TRPC1 channels. *Nature Cell Biology*. 2006;8(9):1003–1010.
32. Zeng W, Yuan JP, Kim MS, Choi YJ, Huang GN, Worley PF et al. STIM1 gates TRPC channels, but not Orai1, by electrostatic interaction. *Molecular Cell*. 2008;32(3):439–448.
33. Wang Y, Deng X, Mancarella S, Hendron E, Eguchi S, Soboloff J et al. The calcium store sensor, STIM1, reciprocally controls Orai and $Ca_V1.2$ channels. *Science*. 2010; 330(6000):105–109.
34. Park CY, Shcheglovitov A, Dolmetsch R. The CRAC channel activator STIM1 binds and inhibits L-type voltage-gated calcium channels. *Science*. 2010;330(6000):101–105.
35. Ritchie MF, Samakai E, Soboloff J. STIM1 is required for attenuation of PMCA-mediated Ca^{2+} clearance during T-cell activation. *The EMBO Journal*. 2012;31(5):1123–1133.
36. Krapivinsky G, Krapivinsky L, Stotz SC, Manasian Y, Clapham DE. POST, partner of stromal interaction molecule 1 (STIM1), targets STIM1 to multiple transporters. *Proceedings of the National Academy of Sciences of the United States of America*. 2011;108(48):19234–19239.
37. Liu B, Peel SE, Fox J, Hall IP. Reverse mode Na^+/Ca^{2+} exchange mediated by STIM1 contributes to Ca^{2+} influx in airway smooth muscle following agonist stimulation. *Respiratory Research*. 2010;11:168.
38. Palty R, Raveh A, Kaminsky I, Meller R, Reuveny E. SARAF inactivates the store operated calcium entry machinery to prevent excess calcium refilling. *Cell*. 2012;149(2):425–438.
39. Jha A, Ahuja M, Maleth J, Moreno CM, Yuan JP, Kim MS et al. The STIM1 CTID domain determines access of SARAF to SOAR to regulate Orai1 channel function. *The Journal of Cell Biology*. 2013;202(1):71–79.
40. Sharma S, Quintana A, Findlay GM, Mettlen M, Baust B, Jain M et al. An siRNA screen for NFAT activation identifies septins as coordinators of store-operated Ca^{2+} entry. *Nature*. 2013;499(7457):238–242.
41. Jing J, He L, Sun A, Quintana A, Ding Y, Ma G et al. Proteomic mapping of ER-PM junctions identifies STIMATE as a regulator of Ca^{2+} influx. *Nature Cell Biology*. 2015;17(10):1339–1347.
42. Srikanth S, Jew M, Kim KD, Yee MK, Abramson J, Gwack Y. Junctate is a Ca^{2+}-sensing structural component of Orai1 and stromal interaction molecule 1 (STIM1). *Proceedings of the National Academy of Sciences of the United States of America*. 2012;109(22):8682–8687.
43. Chen YT, Chen YF, Chiu WT, Liu KY, Liu YL, Chang JY et al. Microtubule-associated histone deacetylase 6 supports the calcium store sensor STIM1 in mediating malignant cell behaviors. *Cancer Research*. 2013;73(14):4500–4509.

44. Stamboulian S, Moutin MJ, Treves S, Pochon N, Grunwald D, Zorzato F et al. Junctate, an inositol 1,4,5-triphosphate receptor associated protein, is present in rodent sperm and binds TRPC2 and TRPC5 but not TRPC1 channels. *Developmental Biology*. 2005;286(1):326–337.
45. Carreras-Sureda A, Cantero-Recasens G, Rubio-Moscardo F, Kiefer K, Peinelt C, Niemeyer BA et al. ORMDL3 modulates store-operated calcium entry and lymphocyte activation. *Human Molecular Genetics*. 2013;22(3):519–530.
46. Prins D, Groenendyk J, Touret N, Michalak M. Modulation of STIM1 and capacitative Ca^{2+} entry by the endoplasmic reticulum luminal oxidoreductase ERp57. *EMBO Reports*. 2011;12(11):1182–1188.
47. Srikanth S, Jung HJ, Kim KD, Souda P, Whitelegge J, Gwack Y. A novel EF-hand protein, CRACR2A, is a cytosolic Ca^{2+} sensor that stabilizes CRAC channels in T cells. *Nature Cell Biology*. 2010;12(5):436–446.
48. Feng JM, Hu YK, Xie LH, Colwell CS, Shao XM, Sun XP et al. Golli protein negatively regulates store depletion-induced calcium influx in T cells. *Immunity*. 2006;24(6):717–727.
49. Walsh CM, Doherty MK, Tepikin AV, Burgoyne RD. Evidence for an interaction between Golli and STIM1 in store-operated calcium entry. *The Biochemical Journal*. 2010;430(3):453–460.
50. Broad LM, Braun FJ, Lievremont JP, Bird GS, Kurosaki T, Putney JW, Jr. Role of the phospholipase C-inositol 1,4,5-trisphosphate pathway in calcium release-activated calcium current and capacitative calcium entry. *The Journal of Biological Chemistry*. 2001;276(19):15945–15952.
51. Rosado JA, Sage SO. Phosphoinositides are required for store-mediated calcium entry in human platelets. *The Journal of Biological Chemistry*. 2000;275(13):9110–9113.
52. Ercan E, Momburg F, Engel U, Temmerman K, Nickel W, Seedorf M. A conserved, lipid-mediated sorting mechanism of yeast Ist2 and mammalian STIM proteins to the peripheral ER. *Traffic*. 2009;10(12):1802–1818.
53. Walsh CM, Chvanov M, Haynes LP, Petersen OH, Tepikin AV, Burgoyne RD. Role of phosphoinositides in STIM1 dynamics and store-operated calcium entry. *The Biochemical Journal*. 2010;425(1):159–168.
54. Heo WD, Inoue T, Park WS, Kim ML, Park BO, Wandless TJ et al. PI(3,4,5)P3 and PI(4,5)P2 lipids target proteins with polybasic clusters to the plasma membrane. *Science*. 2006;314(5804):1458–1461.
55. Mostowy S, Cossart P. Septins: The fourth component of the cytoskeleton. *Nature Reviews Molecular Cell Biology*. 2012;13(3):183–194.
56. Ong HL, Ambudkar IS. Molecular determinants of TRPC1 regulation within ER-PM junctions. *Cell Calcium*. 2015;58(4):376–386.
57. Brading AF, Burnett M, Sneddon P. The effect of sodium removal on the contractile response of the guinea-pig taenia coli to carbachol. *The Journal of Physiology*. 1980;306:411–429.
58. Van Breemen C. Calcium requirement for activation of intact aortic smooth muscle. *The Journal of Physiology*. 1977;272(2):317–329.
59. Putney JW, Jr. A model for receptor-regulated calcium entry. *Cell Calcium*. 1986;7(1):1–12.
60. Gabella G. Caveolae intracellulares and sarcoplasmic reticulum in smooth muscle. *Journal of Cell Science*. 1971;8(3):601–609.
61. Devine CE, Somlyo AV, Somlyo AP. Sarcoplasmic reticulum and excitation-contraction coupling in mammalian smooth muscles. *The Journal of Cell Biology*. 1972;52(3):690–718.

62. Luik RM, Wu MM, Buchanan J, Lewis RS. The elementary unit of store-operated Ca^{2+} entry: Local activation of CRAC channels by STIM1 at ER-plasma membrane junctions. *The Journal of Cell Biology*. 2006;174(6):815–825.
63. Perni S, Dynes JL, Yeromin AV, Cahalan MD, Franzini-Armstrong C. Nanoscale patterning of STIM1 and Orai1 during store-operated Ca^{2+} entry. *Proceedings of the National Academy of Sciences of the United States of America*. 2015;112(40):E5533–E5542.
64. Chang CL, Hsieh TS, Yang TT, Rothberg KG, Azizoglu DB, Volk E et al. Feedback regulation of receptor-induced Ca^{2+} signaling mediated by E-Syt1 and Nir2 at endoplasmic reticulum-plasma membrane junctions. *Cell Reports*. 2013;5(3):813–825.
65. Poteser M, Leitinger G, Pritz E, Platzer D, Frischauf I, Romanin C et al. Live-cell imaging of ER-PM contact architecture by a novel TIRFM approach reveals extension of junctions in response to store-operated Ca^{2+}-entry. *Scientific Reports*. 2016;6:35656.

11 Store-Independent Orai Channels Regulated by STIM

Xuexin Zhang, Maxime Gueguinou, and Mohamed Trebak

CONTENTS

11.1 INTRODUCTION

The identification of Orai and STIM proteins has opened up new avenues of research in the field of receptor-regulated calcium signaling. The ligation of phospholipase C (PLC)-coupled receptors can activate either the common *s*tore-*o*perated *c*alcium *e*ntry (SOCE) pathway or *s*tore-*i*ndependent *c*alcium *e*ntry (SICE) pathway. The representative conductance of the SOCE pathway is the *c*alcium *r*elease-*a*ctivated *c*alcium (CRAC) channel encoded by Orai (CRACM) proteins. The SICE pathway biophysical manifestation is currents activated by arachidonic acid (AA) or the AA metabolite leukotriene C_4 (LTC_4) and termed *a*rachidonate-*r*egulated or *L*TC_4-*r*egulated *c*alcium (ARC/LRC) current encoded by channels composed of both Orai1 and Orai3 proteins.

About three decades ago, Putney first proposed the capacitative Ca^{2+} entry model (subsequently known as SOCE) [1]. Orai1 protein, the pore forming subunit of the CRAC channel, was discovered almost simultaneously by three groups in 2006

[2–4]. The Orai family of channels contains three different proteins (Orai1, 2, and 3) encoded by independent genes [5]. A large number of agonists can act on G protein-coupled receptors (GPCRs) to activate PLC. PLC hydrolyzes phosphatidylinositol-4,5-bisphopshate (PIP_2) into diacylglycerol (DAG) and inositol-1,4,5-trisphosphate (IP_3) [6]. The latter binds to IP_3 receptors (IP_3R) on the membrane of the endoplasmic reticulum (ER), resulting in Ca^{2+} store emptying. The action of ER Ca^{2+} store emptying causes stromal interaction molecule 1 (STIM1), a calcium sensor to lose Ca^{2+} from its N-terminal low affinity EF hand located in the lumen of the ER [7,8]. This causes STIM1 to aggregate and to move to highly specialized areas where the ER comes close to the plasma membrane to physically trap and interact with Orai1 channels and activate Ca^{2+} entry [8]. STIM1 has one homologue, STIM2, which mediates Orai1 channel activation under resting conditions in the absence of agonist stimulation [9]. In most cells studied so far, SOCE is mediated by STIM1 and Orai1 proteins [10]. However, we reported Orai3-mediated SOCE in a subset of estrogen receptor positive breast cancer cells [11–13].

The ARC channel and its role in calcium signaling have been first reported and intensely studied by Shuttleworth and colleagues [14–17]. This group identified and characterized a conductance in HEK293 cells activated by relatively low exogenous concentrations of arachidonic acid or by low concentrations of a muscarinic agonist [18]. Polyunsaturated fatty acids were described as poor activators of these channels, and mono unsaturated or saturated fatty acids were ineffective [19]. Unlike a number of channels of the *T*ransient *R*eceptor *P*otential *C*anonical (TRPC3/6/7) family [20,21], ARC channels are not activated by high concentrations of DAG (100 μM) [19]. Mignen and colleagues showed that despite many similarities with CRAC currents present in the same cell type studied, ARC channels possess distinct pharmacological characteristics and biophysical properties [18]. For instance, unlike CRAC channels, ARC channels do not show the typical fast *Ca*$^{2+}$-*d*ependent *i*nactivation (CDI), are not inhibited by a reduction in extracellular pH from 7.2 to 6.7, and are insensitive to 2-aminoethoxydiphenyl borate (2-APB) [18,22]. As is the case with CRAC channels [34,35], the absence of divalent cations in the extracellular recording medium induces the permeability of ARC channels to monovalent cations, such as Na^+ [23]. However, this monovalent macroscopic current has different characteristics from those observed for CRAC channels especially from the perspective of their depotentiation and permeability. By blocking monovalent currents by increasing extracellular calcium concentrations as a relative measure of selectivity of calcium channels, Mignen and colleagues proposed that ARC channels have high Ca^{2+} selectivity and are 50 times more Ca^{2+}-selective than CRAC channels [18,22]. These authors argued that ARC channels are the predominant calcium channels activated when cells are stimulated with low concentrations of agonists that induce repetitive calcium oscillations [24]. Using an M3 muscarinic receptor-expressing HEK293 cells and murine parotid and pancreatic acinar cells, they reported the activation of ARC channels mediating intracellular calcium oscillations by low concentrations (0.2–1 μM) of carbachol [25]. In the same cells they described the activation of the AA-producing enzyme, phospholipase A2 type IV, upon stimulation with low concentrations of carbachol.

Earlier work by the Shuttleworth group suggested that the pharmacological inhibition of PLA2 with isotetrandrine blocks the activation of ARC channels, while the pharmacological inhibition of the lipoxygenase and cyclooxygenase pathways had no effect on ARC activation [14,26], indicating that AA is produced by receptor-mediated activation of PLA2 and that AA processing into downstream metabolites is not required for ARC channel activation. After identification of STIM and Orai proteins, Shuttleworth and colleagues showed that both Orai1 and Orai3 are required for ARC channel activation [27], in addition to the minor pool of STIM1 located in the plasma membrane [28]. More recent work from our laboratory identified a SICE channel in primary aortic vascular smooth muscle cells (VSMC). We found that this conductance is activated by AA, but AA metabolism into LTC_4 by the enzymatic activity of LTC_4 synthase (LTC_4S) provided a more robust activation of these channels; LTC_4 acts intracellularly when applied through the patch pipette but not extracellularly when added to the bath solution. We named this channel *L*TC_4-*r*egulated *c*alcium (LRC) channel [23,29–31]. Collectively, our data in VSMC showed that receptor activation causes production of AA through sequential activation of PLC and DAG lipase and that AA metabolism by 5-lipooxygenase and LTC_4S into LTC_4 is required for LRC channel activation [29,31]. A molecular knockdown on LTC_4 synthase (LTC_4S) abrogated receptor-mediated LRC channel activation (using the PAR1 agonist thrombin), while direct application of LTC_4 through the patch pipette robustly activated LRC currents. The biophysical properties of LRC channels were identical to those of ARC channels, prompting us to undertake a side by side comparison in VSMC and HEK293 cells to determine whether these two conductances are mediated by the same or by different cellular pools of STIM and Orai proteins [23]. Briefly, using protein knockdown, pharmacological inhibitors, and a nonmetabolizable form of AA, we found that regardless of the cell type considered (HEK293 cells or VSMC), ARC and LRC currents are the manifestation of the same channel that can be activated by AA but is more robustly activated by LTC_4 [23]. We also found that in both cell types, ARC/LRC currents depended on Orai1, Orai3, and STIM1 [23,29], but unlike findings from the Shuttleworth group, we were able to rescue ARC/LRC activity in HEK293 cells and VSMC with expressed STIM1 constructs that do not traffic to the plasma membrane when using Fura-2 calcium imaging and perforated patch recording in intact cells but not in whole-cell recordings. These results suggest a facilitatory role for PM-STIM1 in ARC/LRC channel activation [23]. Orai1 exists in two variants generated through alternative translation-initiation of the Orai1 mRNA: a longer Orai1α form contains an additional N-terminal (NT) 63 amino acids upstream of the conserved start site of a shorter Orai1β [32]. A study from our group showed that while Orai1α and Orai1β are interchangeable for forming CRAC channels, only Orai1α can support ARC/LRC channels by forming a unique heteromeric channel with Orai3. Studies by the Shuttleworth group were performed before Orai1α variant was discovered; it is therefore unclear which Orai1 subtype was used [33]. We also showed that a specific interaction of STIM1 second C-terminal (CT) coiled-coil (CC2) with Orai3 CT region is required for LRC channel activation by LTC_4 [31].

In summary, the SICE pathway appears to be mediated by one channel entity. In succeeding text, we will refer to this channel as either ARC or LRC, depending on whether we are referring to experiments that used either AA or LTC_4 to activate this conductance. ARC/LRC channels are encoded by Orai1 and Orai3 and regulated by STIM1. There are two major points of contention between our findings and those of the Shuttleworth group: (1) the requirement for AA metabolic conversion into LTC_4 and (2) the cellular pool of STIM1 required for ARC/LRC activation, that is, ER-resident versus PM-resident STIM1.

11.2 BIOPHYSICAL PROPERTIES AND MOLECULAR COMPOSITION OF SICE CHANNELS

In 1996, it was first demonstrated that AA-activated noncapacitative Ca^{2+} entry through an unknown channel in avian nasal gland cells [14]. Four years later, using the whole-cell patch clamp technique, Ca^{2+} currents activated by exogenous application of AA were recorded under conditions where cytosolic Ca^{2+} was buffered to ~100 nM in the pipette solution [18]. This novel non-store-operated channel was named the arachidonate-regulated calcium (ARC) channel [18]. A recording of ARC channel current (I_{ARC}) in M3 muscarinic receptor-expressing HEK293 cells has revealed that, like CRAC, ARC channels possess a small, highly calcium-selective conductance [18] (see Chapter 1). When external solution contained 10 mM Ca^{2+} and the pipette solution was buffered to 100 nM Ca^{2+}, application of 8 μM AA to the bath solution activated I_{ARC} that averaged 0.56 ± 0.05 pA/pF at −80 mV. I_{ARC} displays marked inward rectification, reversal potentials greater than +30 mV, and inhibition by La^{3+}. Unlike CRAC channels, however, ARC channels are insensitive to 2-APB, are unaffected by reduction of extracellular pH, and do not show Ca^{2+}-dependent fast inactivation [18]. Moreover, like most voltage-gated Ca^{2+} channels and CRAC channels [34,35], in external divalent cation-free (DVF) solutions, ARC channels start conducting large monovalent currents [22,23]. However, unlike CRAC channels [36], ARC channels do not show the typical inactivation during a short pulse of DVF bath solution, called depotentiation [23] (see Figure 11.1). However, there is an example where overexpressed STIM1 and Orai1 generate CRAC currents that did not depotentiate in DVF solutions (e.g., see [37]). This could be due to different STIM1/Orai1 stoichiometries and/or limiting endogenous regulatory proteins during

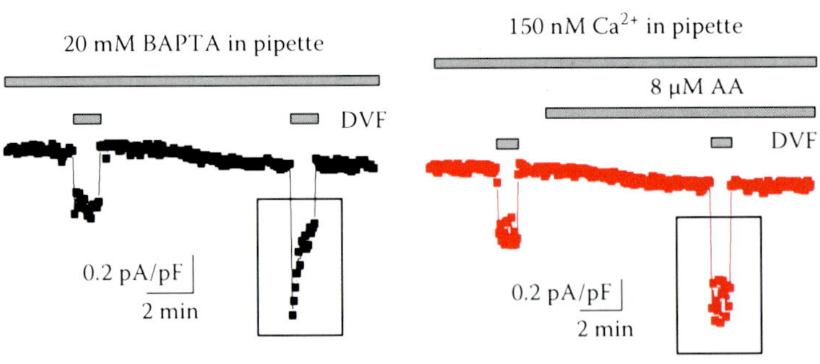

FIGURE 11.1 Whole-cell patch clamp electrophysiological recording from HEK293 cells shows that CRAC current activated by 20 mM BAPTA in the patch pipette exhibits depotentiation during a short pulse of divalent cation-free (DVF) bath solution, whereas ARC currents activated by exogenous 8 μM AA do not.

overexpression. The Shuttleworth group used arachidonyl coenzyme-A (ACoA), a membrane-impermeant analog of AA to show that activation of the ARC channels reflects an action of the fatty acid specifically at the inner surface of the plasma membrane [19]. A more recent study proposed that the N-terminus of Orai3 is required for AA action on ARC channels [38], but how AA interacts with ARC channels to gate them is entirely unknown.

Recordings performed in our lab revealed similar LRC channel biophysical properties, such as small conductance, inward rectification, >+30 mV reversal potentials, no depotentiation during the pulse of divalent-free bath (DVF) solution, and inhibition by Gd^{3+}. Comparisons of the properties of ARC, LRC, and CRAC are presented in Table 11.1. We also showed that N-methyl LTC_4 ($NMLTC_4$), a nonmetabolizable form of LTC_4 delivered through the patch pipette was fully capable of activating LRC currents, suggesting the metabolism of LTC_4 into downstream metabolites is not required. However, $NMLTC_4$ did not activate LRC channels when it was applied extracellularly, indicating that LTC_4 acts through the inner side of the cell plasma membrane [23]. Exactly how LTC_4 interacts and gates the channel remains unclear.

The crystal structure of the founding member of Orai channels, *Drosophila* Orai (*dOrai*), was recently resolved at a resolution of 3.35 Å [39]. This *dOrai* structure shows a drastically different molecular organization from other ion channels and

TABLE 11.1
Biophysical Properties of ARC, LRC, and CRAC Channels

	ARC Channels	LRC Channels	CRAC Channels
Permeability	Ca^{2+}-selective channel	Ca^{2+}-selective channel	Ca^{2+}-selective channel
Activation	Arachidonic acid	LTC_4	Store depletion
Inhibition	Inhibited by La^{3+} and Gd^{3+}, insensitive to 2-APB	Inhibited by Gd^{3+}, insensitive to 2-APB	Inhibited by Gd^{3+} and high concentration (30–50 μM) of 2-APB
Endogenous current size	0.4–0.6 pA/pF at −80 mV	0.1–0.2 pA/pF at −100 mV	0.1–0.2 pA/pF at −100 mV
Direction of current	Inward current	Inward current	Inward current
Component of channel	Orai1, Orai3	Orai1, Orai3	Orai1
Stoichiometry of channel	2 Orai3 and 3 Orai1 (31113 or 31311)	n/d	6 Orai1
pH sensitive	Insensitive to low pH = 6.7	n/d	Sensitive to low pH = 6.7
Fast CDI	Absent	n/d	Present
PKA	Regulated by PKA	n/d	Non-regulated by PKA
Interaction	PM-STIM1	ER-STIM1	ER-STIM1
Site of action	Inner surface of PM	Inner surface of PM	n/a

Notes: CDI, Ca^{2+}-dependent inactivation; PM, plasma membrane; ER, endoplasmic reticulum; n/d, not determined; n/a, not applicable.

reveals *dOrai* as a homohexameric channel arranged around a central pore with a ring of six extracellular glutamate residues (E106 in human Orai1) representing the selectivity filter (see Chapters 2, 3, 14). While this structure strongly suggests that human Orai most likely also form hexameric channels, how many subunits of each of Orai1 and Orai3 are required to form functional ARC/LRC channels that fully recapitulate the properties of native channels remains a contentious issue. The Shuttleworth group suggested that CRAC channels are homotetramers of four Orai1 [40]. Subsequent studies from the same group proposed that native ARC channels are heteropentamers of Orai1 and Orai3 (either organized as 31113 or 31311) [41] (discussed in Chapter 2). They also published a report questioning the hexameric assembly of *dOrai* as a possible artifact of crystallization [42]. It is worth noting that studies challenging the crystal structure were performed with concatemers ectopically expressed in HEK293 cells expressing endogenous Orai isoforms. In this case, native Orai channels could potentially assemble with concatenated tetramers/pentamers to form hexamers.

11.3 METHODS FOR MEASURING SICE CHANNEL FUNCTION

11.3.1 Whole-Cell Patch Clamp Recording

The patch clamp technique is the best approach to study ion channel function. Due to the tiny unitary conductance of Orai channels, for instance, single CRAC channel conductance was estimated using stationary noise analysis around 24 fS [43] (Chapter 1). Therefore, for CRAC as well as ARC channel current recordings, the whole-cell configuration of the patch clamp technique is the best choice.

11.3.1.1 Equipment Setup for Patch Clamp Recording

11.3.1.1.1 Amplifier, Low-Noise Digitizer, and Software

The Axopatch 200B amplifier (Molecular Devices) works together with either the Digidata 1550B or 1440A low-noise digitizer (Molecular Devices). For connecting the amplifier and digitizer, the manuals provide step-by-step instructions. For acquiring and analyzing data, the software of Clampex 10 and Clampfit 10 are used, respectively. Similar equipment used by other investigators is also available from HEKA Elektronik, Germany. The HEKA EPC 10 USB Single is a good choice for a patch clamp amplifier. A fully equipped setup includes the EPC 10 amplifier combined with a computer and PATCHMASTER software, a digital storage oscilloscope, a variable analog filter, and a sophisticated pulse generator.

11.3.1.1.2 Microscope

A Nikon ECLIPSE Ti quantitative phase contrast microscope equipped with 20× fluorescence objective is a good choice for the patch clamp setup. Other suitable microscopes are available from various vendors, including Olympus, Leica, and Zeiss.

11.3.1.1.3 Micromanipulator

The MP-225 micromanipulator (Sutter Instrument) is designed primarily for positioning patch and intracellular recording pipettes. Its speed and resolution of movement are easily selected with a multiple position thumbwheel, allowing fast/coarse movement and slow/ultrafine movement in 10 increments. Two commonly used robotic movements have been incorporated for user convenience. A single button press can initiate a move to a home position for pipette exchange or to a user-defined work position for the quick positioning of the pipette near the recording location. Other manufacturers provide similar micromanipulators.

11.3.1.1.4 Vibration Isolation Table

Both Kinetic and TMC brand products are suitable. The minimum size for patch clamp setup is 30 × 36 in. to give enough space for the microscope, perfusion system, and micromanipulator controller with a gas cylinder connected to the air table near the setup.

11.3.1.1.5 Faraday Cage

Proper grounding is essential for obtaining low-noise patch clamp recording from small conductance ARC/LRC channels. Therefore, careful grounding of all instruments including microscope, perfusion system, amplifier, digitizer, micromanipulator, and computer through low-resistance ground cables and use of Faraday cage will minimize noise. The Faraday cage (Kinetic systems) mounted on the top of the vibration isolation table should also be grounded.

11.3.1.1.6 Micropipette Puller

A well-designed micropipette puller can help you deliver a successful whole-cell patch clamp experiment. For the micropipette puller, the Sutter micropipette pullers MP-97 or MP-1000 (Sutter instrument) are a good choice to use with borosilicate glass capillaries (World Precision Instruments) with 1.5 mm OD and 0.86 mm ID to obtain patch pipettes with 0.5–1 μm diameter tips.

11.3.1.1.7 Microforge

For obtaining superior gigaohm (GΩ) seals between the patch pipette and the plasma membrane of cells, tips of patch pipettes should be polished using a microforge controller, for example, DMF1000 (World Precision Instruments) under a microscope, like Revelation III (LW Scientific). After polishing, the seal resistance obtained can be improved by up to 5–10-fold, compared to unpolished pipettes. Therefore, this step is necessary for low noise recordings of small currents such as ARC/LRC.

11.3.1.1.8 Computer

The computer system requirements of patch clamp setup are similar to the calcium imaging system described in the following text. For example, an Intel Pentium-4 processor or faster, Microsoft Windows XP or later, CD-ROM drive, 1 GB or more system memory (RAM), 128 GB or more disk space, and 24-bit graphics display can essentially meet the required needs.

11.3.1.2 Solutions for Electrophysiological Recordings

DVF solution composition: 155 mM Na-methanesulfonate, 10 mM HEDTA, 1 mM EDTA, and 10 mM HEPES (pH 7.4, adjusted with NaOH). We and other researchers have used DVF external solutions to amplify the ARC and CRAC channel currents [18,22,23,29,31,44].

11.3.1.2.1 For Activation of Currents Using Exogenous AA Delivered in the Bath

Ca^{2+}-containing bath solution: 115 mM Na-methanesulfonate, 10 mM CsCl, 1.2 mM $MgSO_4$, 10 mM HEPES, 20 mM $CaCl_2$, and 10 mM glucose (pH adjusted to 7.4 with NaOH).

Ca^{2+}-containing pipette solution: 115 mM Cs-methanesulfonate, 10 mM Cs-BAPTA, 5 mM $CaCl_2$, 8 mM $MgCl_2$, and 10 mM HEPES (pH adjusted to 7.2 with CsOH). Calculated free Ca^{2+} was 150 nM using Maxchelator software (http://maxchelator.stanford.edu/).

11.3.1.2.2 For Activation of Currents Using Intracellular LTC_4 Delivered through the Patch Pipette

Ca^{2+}-containing bath solution: 115 mM Na-methanesulfonate, 10 mM CsCl, 1.2 mM $MgSO_4$, 10 mM HEPES, 20 mM $CaCl_2$, and 10 mM glucose (pH was adjusted to 7.4 with NaOH).

Ca^{2+}-containing pipette solution: 115 mM Cs-methanesulfonate, 10 mM Cs-BAPTA, 5 mM $CaCl_2$, 8 mM $MgCl_2$, and 10 mM HEPES (pH adjusted to 7.2 with CsOH). 50–100 nM LTC_4 is added.

11.3.1.2.3 For Activation of Store Depletion-Activated CRAC Currents

Bath solution (same as the previous): 115 mM Na-methanesulfonate, 10 mM CsCl, 1.2 mM $MgSO_4$, 10 mM HEPES, 20 mM $CaCl_2$, and 10 mM glucose (pH was adjusted to 7.4 with NaOH).

Ca^{2+}-free pipette solution: 115 mM Cs-methanesulfonate, 20 mM Cs-BAPTA, 8 mM $MgCl_2$, and 10 mM HEPES (pH adjusted to 7.2 with CsOH).

Note: CRAC current recordings are used as controls to highlight the biophysical, pharmacological, and molecular distinction between I_{CRAC} and I_{ARC}.

11.3.1.3 Experimental Procedures

11.3.1.3.1 Seeding Cells

Twelve to twenty-four hours before patch clamp experiments, cells are seeded onto 30 mm round glass coverslips (Thermo Scientific) in 6-well tissue culture plates (VWR) at a low density to allow easy identification of single cells for recordings.

11.3.1.3.2 Preparing Patch Pipettes

Pipettes should be pulled on the day of recordings and every pipette should be inspected under the microforge microscope for imperfections before fire-polishing. Polished patch pipettes are stored in a vacuum container for the rest of the day.

11.3.1.3.3 Performing Patch Clamp Electrophysiology Experiments

Before starting the recording, coverslips with attached cells are mounted in recording chambers, each containing a 1 mL bath solution, and transferred to the microscope

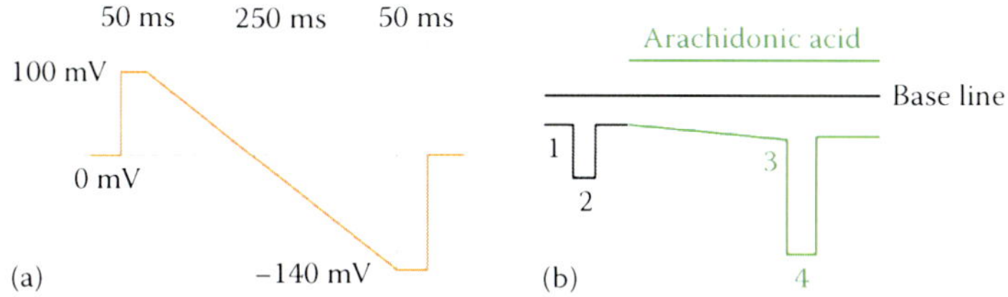

FIGURE 11.2 Example of ramp protocol (a) and typical current development at −100 mV (b) in whole-cell mode. Voltage ramps ranging from +100 to −140 mV lasting 250 ms are applied every 2 s. First, a DVF application (before AA addition) is performed to gauge the leak or basal current and after AA addition (green) and maximal current development, a second DVF application is performed to obtain the maximum AA-activated Na^+ current. A typical time course of AA-activated current is plotted from current data taken at −100 mV.

stage. After identifying a single cell under the microscope, the patch pipette is filled with filtered pipette solution and mounted into the pipette holder. Resistances of filled glass pipettes are 1–3 MΩ. The liquid-junction potential offsets are corrected before each recording. Since ARC/LRC current amplitudes are usually small (several picoamperes) [31], only cells forming tight seals (>16 GΩ) are selected for whole-cell configuration. Immediately after establishing the whole-cell patch clamp configuration, the recording is initiated by applying voltage ramps (typically from 100 to −140 mV) lasting 250 ms at 0.5 Hz (Figure 11.2a). An initial DVF application is performed before the current has been activated (by addition of AA) or before the current has developed on inclusion of LTC_4 in the patch pipette. The first DVF pulse allows the determination of basal currents or "leaks" that should be subsequently subtracted from total currents obtained after full activation. Specifically, the initial I–voltage (V) relations obtained in Ca^{2+}-containing bath solutions (position 1 in Figure 11.2b) and DVF bath solutions (position 2) represent background currents that are subtracted from AA- or LTC_4-activated Ca^{2+} currents (position 3) and Na^+ currents (obtained in DVF bath solutions; position 4), respectively. After currents are fully activated by AA or LTC_4, I–V curves are obtained for Ca^{2+} currents (in Ca^{2+}-containing bath solutions) and Na^+ currents (in DVF bath solutions). Using Origin software (OriginLab), I–V curves corresponding to background currents obtained in Ca^{2+} and Na^+ are subtracted from the I–V curves obtained in Ca^{2+} and Na^+ after AA/LTC_4 stimulation and maximal current activation. Namely, for Ca^{2+} currents (curve 3–curve 1) and Na^+ currents (curve 4–curve 2), respectively. For recording ARC/LRC, cells are maintained at a 0 or +30 mV holding potential [28]. Reverse ramps from positive to negative voltages are recommended in certain cell types in order to inhibit voltage-gated Na^+ channels expressed in these cells. Inclusion of 8 mM $MgCl_2$ in the pipette solution is designed to inhibit TRPM7 currents that are expressed in most cell lines, including HEK293 cells [35]. Experiments are typically performed at room temperature.

11.3.2 Calcium Imaging

ARC channels are highly Ca^{2+}-selective channels, and fluorescence imaging microscopy is also a useful tool for studying Ca^{2+} influx through these channels in living cells. Typical imaging workstations, solutions, dyes, and protocols are described elsewhere [45–48] (see also Chapters 1 and 16).

11.3.2.1 Equipment Setup for Fluorescence Calcium Measurement

Figure 11.3 shows a schematic of the typical fluorescence calcium imaging setup used for Ca^{2+} measurements in cells.

11.3.2.1.1 Light Source

A xenon lamp as a light source is a cost-effective choice for a calcium imaging system. The Lambda LS (Sutter) xenon lamp has a built-in motor-driven six-positioned filter wheel; switching between neighboring filters (e.g., between 340 and 380 nm, as required for Fura-2) is relatively rapid, completed within 55 ms. A shutter control facilitates graded power output; an external controller enables manual switching. Filtered light from Lambda is transmitted to the fluorescence microscope via a 2 m long liquid light guide to avoid transmission of heat and vibration to the microscope. The Lambda LS xenon lamp produces light between 330 and 650 nm wavelength, which is suitable for a wide variety of dyes, including most fluorescent proteins (e.g., GFP, YFP, and RFP). It is necessary to allow the xenon lamp to warm up for at least 30 min before taking measurements. The light bulb of the Lambda LS lamp can last between 400 and 2000 h, depending on maintenance and the number of on/off switches.

11.3.2.1.2 Fluorescence Microscope

For basic Fura-2-based calcium imaging experiments, it is not necessary to purchase a high-end fluorescence microscope. For example, the Nikon ECLIPSE TS-100 (Nikon) fluorescence microscope is a good choice. The Nikon ECLIPSE TS-100 should be equipped with a 20× fluorescence objective and a Fura-2 filter set (Chroma 74500—with a dichroic mirror and an emission filter, the 340 and 380 nm excitation filters placed in the Lambda filter wheel).

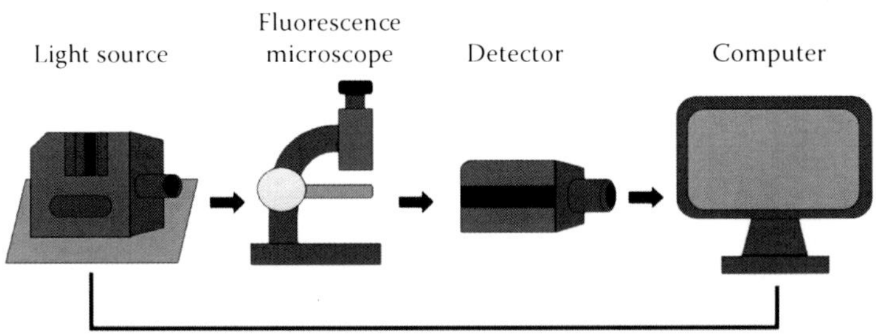

FIGURE 11.3 Fluorescence calcium imaging setup typically used to measure intracellular Ca^{2+}. Typically, a light source providing light with 340 and 380 nm Fura-2 excitation filters. A light guide connects the light source to cells viewed under a microscope equipped emission and dichroic filters. Emitted fluorescence is collected by a CCD camera on a pixel by pixel basis and data acquisition and analysis are handled by imaging software.

11.3.2.1.3 The Detector

Detection is achieved by a CCD camera (BASLER scA640, Basler AG). The BASLER scA640 camera resolution is 658 × 492 pixels, and it is equipped with an ICX414 sensor, which has a frame rate of 79 fps. The free pylon software can be downloaded at http://www.baslerweb.com/de/produkte/software. Between the CCD camera and the inverted microscope, there is a HR055-CMT 0.55× High Resolution C-Mount Adapter (Diagnostic Instruments).

11.3.2.1.4 Computer-Controlled Filter Changer

The Lambda 10-B Optical Filter Changer (Sutter) is ideal for imaging applications requiring a single filter wheel. Lambda 10-B uses advanced motor technology to achieve 40 ms switching times between adjacent filters. It features USB and serial port interfaces, as well as keypad control.

11.3.2.1.5 Computer and Software

The computer system requirements of fluorescence calcium image system: Intel Pentium-4 processor or later, Microsoft Windows XP or later, CD-ROM drive, 1 GB or more system memory (RAM), 128 GB or more free disk space, and 24-bit graphics display. Data acquisition and analysis is achieved by specialized commercially available software.

11.3.2.2 Ca^{2+} Indicators

Fura-2 and Indo-1 are widely used UV-excitable fluorescence Ca^{2+} indicators (see Table 11.2). The synthesis and properties of Indo-1 and Fura-2 were presented by Tsien and colleagues in 1985 [49,50]. Fura-2 is a ratiometric Ca^{2+} indicator, which is one of the most popular Ca^{2+} indicators and is widely used for quantitative intracellular Ca^{2+} measurements. Its peak absorbance shifts from 335 to 363 nm in the Ca^{2+}-bound and Ca^{2+}-free state, respectively. Fluorescence emission occurs at a peak wavelength of 512 nm for excitation at either UV wavelength. The use of the ratio automatically cancels out confounding variables, such as variable dye concentration and cell thickness, making Fura-2 one of the most appreciated tools to quantify Ca^{2+}

TABLE 11.2
Properties of Ca^{2+} Indicators for Fluorescence Calcium Measurement

Indicator	Excitation Peak (nm)	Emission Peak (nm)	Kd for Ca^{2+} (nM)	Membrane Permeability	Notes
Indo-1 AM	338	405/485	230	Yes	Single excitation/dual emission
Fura-2 AM	335/363	512	145	Yes	Dual excitation/single emission

Note: See [62] for more details.

levels. Fura-2 has a Ca^{2+} affinity (Kd ~ 145 nM) that is comparable to endogenous resting Ca^{2+} levels [51,52].

Indo-1 is also widely used ratiometric Ca^{2+} indicator. It differs from Fura-2 in that it is single excitation and dual emission. When Ca^{2+} binding occurs, its emission exhibits a large change, which shifts from 485 nm without Ca^{2+} to 405 nm with Ca^{2+} when excited at about 338 nm. The use of the 405/485 nm emission ratio for indo-1 allows accurate measurements of the intracellular Ca^{2+} concentration. Both Fura-2 and Indo-1 have been used to measure calcium entry induced by arachidonic acid [19,53].

11.3.2.3 Solutions

For measuring calcium entry through ARC channels, the following bathing solutions are used:

Solution 1: Ca^{2+} free Hanks' balanced salt solution (HBSS) solution (in mM): 140 NaCl, 1.13 $MgCl_2$, 4.7 KCl, 10 D-glucose, and 10 HEPES, with pH adjusted to 7.4 with NaOH (20 mL)

Solution 2: 2 mM Ca^{2+} HBSS solution (in mM): 140 NaCl, 1.13 $MgCl_2$, 2 mM $CaCl_2$, 4.7 KCl, 10 D-glucose, and 10 HEPES, with pH adjusted to 7.4 with NaOH (50 mL)

Solution 3: Ca^{2+}-free HBSS solution + 8 μM AA (20 mL)

Solution 4: 2 mM Ca^{2+} HBSS solution + 8 μM AA (20 mL)

Solution 5: 2 mM Ca^{2+} HBSS solution + 50 μM 2-APB (20 mL)

Solution 6: 2 mM Ca^{2+} HBSS solution + 10 μM ionomycin (20 mL)

11.3.2.4 Experimental Procedures

11.3.2.4.1 Seeding Cells

Twelve to twenty-four hours before performing calcium imaging experiments, cells are seeded onto 35 mm glass bottom dishes (MatTek Corporation) or 30 mm round glass coverslips in 6-well tissue culture plates (VWR).

11.3.2.4.2 Loading Cells

- One milliliter of media from the culture dish is transferred to a 15 mL centrifuge tube (Corning), then 1 μL, 1000×, 2 mM Fura-2AM (dissolved in DMSO, Life Technologies) is added to a centrifuge tube and mixed to achieve a 2 μM final concentration.
- A coverslip with attached cells is transferred to an imaging chamber and incubated at 37°C for 40 min to 1 h (incubation times depend on cell types) in culture media containing 2 μM Fura-2AM.
- The coverslip is washed with solution 2 for 3–4 times, then 1 mL of solution 2 is added, and the cells are left in the dark at room temperature for 10 min to allow cellular esterases to cleave the acetoxymethyl ester groups in Fura-2AM. Fura-2 acid capable of binding calcium is produced and trapped in the cytosol.

11.3.2.4.3 Performing Calcium Imaging Experiments

Before starting the recording, cells of interest are chosen. Excitation filter is switched between F340 and F380, and fluorescence intensities are measured. The measurement of Ca^{2+} influx through ARC channels is performed as described as follows:

At 1 min, gently remove Solution 2 by suction, and add 1 mL Solution 4.
At 4 min, switch to Solution 5.
At 7 min, switch to Solution 6, and wait another 1 min, then end the experiment.

11.4 SICE CHANNEL FUNCTION IN HEALTH AND DISEASE

Compared to SOCE channels, little is known about the role of STIM/Orai-mediated SICE pathways in cell functions and their contribution to disease. Pla and colleagues first reported that low concentrations of arachidonic acid are able to evoke a store-independent Ca^{2+} influx, exerting a mitogenic role in bovine aortic endothelial cells [54]. Endogenous ARC currents in primary murine parotid and pancreatic acinar cells were reported, and it was shown that they play a critical role in modulating calcium entry responses to physiological agonists [25]. Another study has proposed that ARC channels may play a role in the regulation of insulin secretion in rat pancreatic β cells [55]. A recent study reported that AA-activated ARC currents from airway smooth muscle (ASM) cells isolated from asthmatic individuals are significantly greater than in ASM cells of normal controls, suggesting that ARC channels could potentially contribute to dysregulated calcium signaling in diseases such as asthma [56]. Using immunofluorescence and biotinylation, it was demonstrated that Orai3 expression in the plasma membrane is triggered by vascular endothelial growth factor (VEGF) stimulation of endothelial cells. VEGF-mediated Orai3 membrane accumulation involves activation of phospholipase Cγ1, cytosolic group IV phospholipase A2α leading to AA and AA metabolism into LTC_4 [57]. Saliba and colleagues showed that adult cardiomyocytes express a calcium-permeable conductance activated by AA, mediated by Orai3, and regulated by STIM1. This LRC/ARC-like conductance is increased during cardiac hypertrophy and was proposed to mediate the effects of STIM1 in driving pathological remodeling in heart during cardiac hypertrophy [58]. Studies from our group showed upregulation of Orai3 and LRC/ARC currents during vascular smooth muscle remodeling *in vivo*, namely, in vessels of rats subjected to balloon angioplasty [29]. The knockdown of Orai3 in balloon-injured carotid arteries using lentivirus-encoding shRNA prevented Orai3 upregulation, inhibited LRC/ARC currents, and decreased neointima formation, supporting the idea that remodeling of Orai1/Orai3 LRC/ARC channels contributes to neointima formation after vascular injury [29]. In a subsequent study, we showed that the knockdown of either LTC_4S or Orai3 inhibits VSMC migration with no effect on proliferation and that *in vivo* knockdown of LTC_4S inhibits neointima formation [30]. A similar remodeling of Orai1/Orai3 expression was reported in prostate cancer [59]. This remodeling was proposed to involve increased expression of Orai3, favoring the formation of heteromultimers of Orai1/Orai3 to increase an ARC-like conductance and promote decrease in apoptosis and increase in proliferation of prostate cancer. Another study in prostate cancer proposed that Orai1/Orai3 heteromultimers are store-operated. They proposed that SOCE in human prostate epithelial cells and prostate cancer cells is mediated by Orai1/Orai3 heteromers and that there is a correlation between the Orai1/Orai3 ratio and the redox sensitivity of SOCE and therefore, cell viability. An increased Orai1/Orai3 ratio in cells derived from prostate cancer tumors may contribute to the higher sensitivity of these cells to reactive oxygen species (ROS) [60]. Indeed, earlier work by Bogeski and colleagues showed that Orai1 is more resistant to ROS-mediated inhibition of channel function than Orai3 [61].

In summary, much work is needed to fully understand the physiological functions of different Orai channel isoforms, their differential regulation and multimerization patterns, and their contribution to pathological conditions. Particularly, the regulation, exact subunit composition of Orai1/Orai3 SICE channels, and their dysregulation during disease are far from being completely understood. The existence of three Orai isoforms encoded by three independent genes and of translational variants (such as Orai1α and β) and likely yet to be identified splice variants suggest that various associations between these different isoforms likely contribute to enhancing the diversity as well as the subcellular localization of Orai channels for the purpose of selective calcium signaling. While Orai1 has been clearly implicated in the SOCE pathway in virtually all cell types, the exact functions of Orai2 channels remain mostly obscure, and Orai3 has been uniquely implicated in store-independent calcium entry, along with Orai1. The association of two exclusively mammalian proteins, Orai1α and Orai3, to form SICE channels likely constitutes a highly specialized conductance that mediates cellular responses to subtle environmental or humoral cues. The upregulation of Orai1 and Orai3 observed in various disease states suggests the potential use of channels formed by these two proteins as targets for therapy for those diseases. Future studies into the exact oligomeric state of native Orai1α/Orai3 channels, their cellular distribution, and mechanisms of regulation are likely to bring us closer to using these channels as targets in human ARC/LRC-related disease therapy.

REFERENCES

1. Putney, J.W., Jr. 1986. A model for receptor-regulated calcium entry. *Cell Calcium* 7:1–12.
2. Feske, S., Gwack, Y., Prakriya, M., Srikanth, S., Puppel, S.H., Tanasa, B., Hogan, P.G., Lewis, R.S., Daly, M., and Rao, A. 2006. A mutation in Orai1 causes immune deficiency by abrogating CRAC channel function. *Nature* 441:179–185.
3. Vig, M., Peinelt, C., Beck, A., Koomoa, D.L., Rabah, D., Koblan-Huberson, M., Kraft, S. et al. 2006. CRACM1 is a plasma membrane protein essential for store-operated Ca^{2+} entry. *Science* 312:1220–1223.
4. Zhang, S.L., Yeromin, A.V., Zhang, X.H., Yu, Y., Safrina, O., Penna, A., Roos, J., Stauderman, K.A., and Cahalan, M.D. 2006. Genome-wide RNAi screen of Ca^{2+} influx identifies genes that regulate Ca^{2+} release-activated Ca^{2+} channel activity. *Proc Natl Acad Sci USA* 103:9357–9362.
5. Prakriya, M. and Lewis, R.S. 2015. Store-operated calcium channels. *Physiol Rev* 95:1383–1436.
6. Berridge, M.J. 1993. Inositol trisphosphate and calcium signalling. *Nature* 361:315–325.

7. Roos, J., DiGregorio, P.J., Yeromin, A.V., Ohlsen, K., Lioudyno, M., Zhang, S., Safrina, O. et al. 2005. STIM1, an essential and conserved component of store-operated Ca^{2+} channel function. *J Cell Biol* 169:435–445.
8. Liou, J., Kim, M.L., Heo, W.D., Jones, J.T., Myers, J.W., Ferrell, J.E., Jr., and Meyer, T. 2005. STIM is a Ca^{2+} sensor essential for Ca^{2+}-store-depletion-triggered Ca^{2+} influx. *Curr Biol* 15:1235–1241.
9. Brandman, O., Liou, J., Park, W.S., and Meyer, T. 2007. STIM2 is a feedback regulator that stabilizes basal cytosolic and endoplasmic reticulum Ca^{2+} levels. *Cell* 131:1327–1339.
10. Trebak, M. 2012. STIM/Orai signalling complexes in vascular smooth muscle. *J Physiol* 590:4201–4208.
11. Motiani, R.K., Abdullaev, I.F., and Trebak, M. 2010. A novel native store-operated calcium channel encoded by Orai3: Selective requirement of Orai3 versus Orai1 in estrogen receptor-positive versus estrogen receptor-negative breast cancer cells. *J Biol Chem* 285:19173–19183.
12. Motiani, R.K., Zhang, X., Harmon, K.E., Keller, R.S., Matrougui, K., Bennett, J.A., and Trebak, M. 2013. Orai3 is an estrogen receptor alpha-regulated Ca^{2+} channel that promotes tumorigenesis. *FASEB J* 27:63–75.
13. Motiani, R.K., Stolwijk, J.A., Newton, R.L., Zhang, X., and Trebak, M. 2013. Emerging roles of Orai3 in pathophysiology. *Channels (Austin)* 7:392–401.
14. Shuttleworth, T.J. 1996. Arachidonic acid activates the noncapacitative entry of Ca^{2+} during $[Ca^{2+}]_i$ oscillations. *J Biol Chem* 271:21720–21725.
15. Shuttleworth, T.J. 2012. Orai3—The 'exceptional' Orai? *J Physiol* 590:241–257.
16. Shuttleworth, T.J. 2012. STIM and Orai proteins and the non-capacitative ARC channels. *Front Biosci* 17:847–860.
17. Thompson, J.L., Mignen, O., and Shuttleworth, T.J. 2013. The ARC channel—An endogenous store-independent Orai channel. *Curr Top Membr* 71:125–148.
18. Mignen, O. and Shuttleworth, T.J. 2000. I(ARC), a novel arachidonate-regulated, noncapacitative Ca^{2+} entry channel. *J Biol Chem* 275:9114–9119.
19. Mignen, O., Thompson, J.L., and Shuttleworth, T.J. 2003. Ca^{2+} selectivity and fatty acid specificity of the noncapacitative, arachidonate-regulated Ca^{2+} (ARC) channels. *J Biol Chem* 278:10174–10181.
20. Gonzalez-Cobos, J.C. and Trebak, M. 2010. TRPC channels in smooth muscle cells. *Front Biosci (Landmark Ed)* 15:1023–1039.
21. Trebak, M., Vazquez, G., Bird, G.S., and Putney, J.W., Jr. 2003. The TRPC3/6/7 subfamily of cation channels. *Cell Calcium* 33:451–461.
22. Mignen, O. and Shuttleworth, T.J. 2001. Permeation of monovalent cations through the non-capacitative arachidonate-regulated Ca^{2+} channels in HEK293 cells. Comparison with endogenous store-operated channels. *J Biol Chem* 276:21365–21374.
23. Zhang, X., Zhang, W., Gonzalez-Cobos, J.C., Jardin, I., Romanin, C., Matrougui, K., and Trebak, M. 2014. Complex role of STIM1 in the activation of store-independent Orai1/3 channels. *J Gen Physiol* 143:345–359.
24. Shuttleworth, T.J. and Mignen, O. 2003. Calcium entry and the control of calcium oscillations. *Biochem Soc Trans* 31:916–919.
25. Mignen, O., Thompson, J.L., Yule, D.I., and Shuttleworth, T.J. 2005. Agonist activation of arachidonate-regulated Ca^{2+}-selective (ARC) channels in murine parotid and pancreatic acinar cells. *J Physiol* 564:791–801.
26. Mignen, O., Thompson, J.L., and Shuttleworth, T.J. 2001. Reciprocal regulation of capacitative and arachidonate-regulated noncapacitative Ca^{2+} entry pathways. *J Biol Chem* 276:35676–35683.
27. Mignen, O., Thompson, J.L., and Shuttleworth, T.J. 2008. Both Orai1 and Orai3 are essential components of the arachidonate-regulated Ca^{2+}-selective (ARC) channels. *J Physiol* 586:185–195.

28. Mignen, O., Thompson, J.L., and Shuttleworth, T.J. 2007. STIM1 regulates Ca^{2+} entry via arachidonate-regulated Ca^{2+}-selective (ARC) channels without store depletion or translocation to the plasma membrane. *J Physiol* 579:703–715.
29. Gonzalez-Cobos, J.C., Zhang, X., Zhang, W., Ruhle, B., Motiani, R.K., Schindl, R., Muik, M. et al. 2013. Store-independent Orai1/3 channels activated by intracrine leukotriene C4: Role in neointimal hyperplasia. *Circ Res* 112:1013–1025.
30. Zhang, W., Zhang, X., Gonzalez-Cobos, J.C., Stolwijk, J.A., Matrougui, K., and Trebak, M. 2015. Leukotriene-C4 synthase, a critical enzyme in the activation of store-independent Orai1/Orai3 channels, is required for neointimal hyperplasia. *J Biol Chem* 290:5015–5027.
31. Zhang, X., Gonzalez-Cobos, J.C., Schindl, R., Muik, M., Ruhle, B., Motiani, R.K., Bisaillon, J.M. et al. 2013. Mechanisms of STIM1 activation of store-independent leukotriene C4-regulated Ca^{2+} channels. *Mol Cell Biol* 33:3715–3723.
32. Fukushima, M., Tomita, T., Janoshazi, A., and Putney, J.W. 2012. Alternative translation initiation gives rise to two isoforms of Orai1 with distinct plasma membrane mobilities. *J Cell Sci* 125:4354–4361.
33. Desai, P.N., Zhang, X., Wu, S., Janoshazi, A., Bolimuntha, S., Putney, J.W., and Trebak, M. 2015. Multiple types of calcium channels arising from alternative translation initiation of the Orai1 message. *Sci Signal* 8:ra74.
34. Hoth, M. and Penner, R. 1993. Calcium release-activated calcium current in rat mast cells. *J Physiol* 465:359–386.
35. Kerschbaum, H.H. and Cahalan, M.D. 1998. Monovalent permeability, rectification, and ionic block of store-operated calcium channels in Jurkat T lymphocytes. *J Gen Physiol* 111:521–537.
36. Prakriya, M. and Lewis, R.S. 2003. CRAC channels: Activation, permeation, and the search for a molecular identity. *Cell Calcium* 33:311–321.
37. Zhang, S.L., Kozak, J.A., Jiang, W., Yeromin, A.V., Chen, J., Yu, Y., Penna, A., Shen, W., Chi, V., and Cahalan, M.D. 2008. Store-dependent and -independent modes regulating Ca^{2+} release-activated Ca^{2+} channel activity of human Orai1 and Orai3. *J Biol Chem* 283:17662–17671.
38. Thompson, J., Mignen, O., and Shuttleworth, T.J. 2010. The N-terminal domain of Orai3 determines selectivity for activation of the store-independent ARC channel by arachidonic acid. *Channels (Austin)* 4:398–410.
39. Hou, X., Pedi, L., Diver, M.M., and Long, S.B. 2012. Crystal structure of the calcium release-activated calcium channel Orai. *Science* 338:1308–1313.
40. Mignen, O., Thompson, J.L., and Shuttleworth, T.J. 2008. Orai1 subunit stoichiometry of the mammalian CRAC channel pore. *J Physiol* 586:419–425.
41. Mignen, O., Thompson, J.L., and Shuttleworth, T.J. 2009. The molecular architecture of the arachidonate-regulated Ca^{2+}-selective ARC channel is a pentameric assembly of Orai1 and Orai3 subunits. *J Physiol* 587:4181–4197.
42. Thompson, J.L. and Shuttleworth, T.J. 2013. How many Orai's does it take to make a CRAC channel? *Sci Rep* 3:1961.
43. Zweifach, A. and Lewis, R.S. 1993. Mitogen-regulated Ca^{2+} current of T lymphocytes is activated by depletion of intracellular Ca^{2+} stores. *Proc Natl Acad Sci USA* 90:6295–6299.
44. Dehaven, W., Jones, B., Petranka, J., Smyth, J., Tomita, T., Bird, G., and Putney, J. 2009. TRPC channels function independently of STIM1 and Orai1. *J Physiol* 587:2275–2298.
45. Trebak, M., Bird, G.S., McKay, R.R., and Putney, J.W., Jr. 2002. Comparison of human TRPC3 channels in receptor-activated and store-operated modes. Differential sensitivity to channel blockers suggests fundamental differences in channel composition. *J Biol Chem* 277:21617–21623.

46. Trebak, M., St, J.B.G., McKay, R.R., Birnbaumer, L., and Putney, J.W., Jr. 2003. Signaling mechanism for receptor-activated canonical transient receptor potential 3 (TRPC3) channels. *J Biol Chem* 278:16244–16252.
47. Trebak, M., Lemonnier, L., DeHaven, W.I., Wedel, B.J., Bird, G.S., and Putney, J.W., Jr. 2009. Complex functions of phosphatidylinositol 4,5-bisphosphate in regulation of TRPC5 cation channels. *Pflugers Arch* 457:757–769.
48. Lemonnier, L., Trebak, M., Lievremont, J.P., Bird, G.S., and Putney, J.W., Jr. 2006. Protection of TRPC7 cation channels from calcium inhibition by closely associated SERCA pumps. *FASEB J* 20:503–505.
49. Tsien, R.Y., Rink, T.J., and Poenie, M. 1985. Measurement of cytosolic free Ca^{2+} in individual small cells using fluorescence microscopy with dual excitation wavelengths. *Cell Calcium* 6:145–157.
50. Grynkiewicz, G., Poenie, M., and Tsien, R.Y. 1985. A new generation of Ca^{2+} indicators with greatly improved fluorescence properties. *J Biol Chem* 260:3440–3450.
51. Hurley, T.W., Ryan, M.P., and Brinck, R.W. 1992. Changes of cytosolic Ca^{2+} interfere with measurements of cytosolic Mg^{2+} using mag-fura-2. *Am J Physiol* 263:C300–C307.
52. Pesco, J., Salmon, J.M., Vigo, J., and Viallet, P. 2001. Mag-indo1 affinity for Ca^{2+}, compartmentalization and binding to proteins: The challenge of measuring Mg^{2+} concentrations in living cells. *Anal Biochem* 290:221–231.
53. Mignen, O., Thompson, J.L., and Shuttleworth, T.J. 2003. Calcineurin directs the reciprocal regulation of calcium entry pathways in nonexcitable cells. *J Biol Chem* 278:40088–40096.
54. Fiorio Pla, A. and Munaron, L. 2001. Calcium influx, arachidonic acid, and control of endothelial cell proliferation. *Cell Calcium* 30:235–244.
55. Yeung-Yam-Wah, V., Lee, A.K., Tse, F.W., and Tse, A. 2010. Arachidonic acid stimulates extracellular Ca^{2+} entry in rat pancreatic beta cells via activation of the noncapacitative arachidonate-regulated Ca^{2+} (ARC) channels. *Cell Calcium* 47:77–83.
56. Thompson, M.A., Prakash, Y.S., and Pabelick, C.M. 2014. Arachidonate-regulated Ca influx in human airway smooth muscle. *Am J Respir Cell Mol Biol* 51:68–76.
57. Li, J., Bruns, A.F., Hou, B., Rode, B., Webster, P.J., Bailey, M.A., Appleby, H.L. et al. 2015. Orai3 surface accumulation and calcium entry evoked by vascular endothelial growth factor. *Arterioscler Thromb Vasc Biol* 35:1987–1994.
58. Saliba, Y., Keck, M., Marchand, A., Atassi, F., Ouille, A., Cazorla, O., Trebak, M. et al. 2015. Emergence of Orai3 activity during cardiac hypertrophy. *Cardiovasc Res* 105:248–259.
59. Dubois, C., Vanden Abeele, F., Lehen'kyi, V., Gkika, D., Guarmit, B., Lepage, G., Slomianny, C. et al. 2014. Remodeling of channel-forming ORAI proteins determines an oncogenic switch in prostate cancer. *Cancer Cell* 26:19–32.
60. Holzmann, C., Kilch, T., Kappel, S., Dorr, K., Jung, V., Stockle, M., Bogeski, I., and Peinelt, C. 2015. Differential redox regulation of Ca^{2+} signaling and viability in normal and malignant prostate cells. *Biophys J* 109:1410–1419.
61. Bogeski, I., Kummerow, C., Al-Ansary, D., Schwarz, E.C., Koehler, R., Kozai, D., Takahashi, N. et al. 2010. Differential redox regulation of ORAI ion channels: A mechanism to tune cellular calcium signaling. *Sci Signal* 3:ra24.
62. Paredes, R.M., Etzler, J.C., Watts, L.T., Zheng, W., and Lechleiter, J.D. 2008. Chemical calcium indicators. *Methods* 46:143–151.

12 Regulation and Role of Store-Operated Ca^{2+} Entry in Cellular Proliferation

Rawad Hodeify, Fang Yu, Raphael Courjaret, Nancy Nader, Maya Dib, Lu Sun, Ethel Adap, Satanay Hubrack, and Khaled Machaca

CONTENTS

12.1 INTRODUCTION

Ca^{2+} is a ubiquitous intracellular messenger that transduces a variety of cellular responses downstream of the activation of G-protein-coupled or tyrosine kinase receptors. Depending on the agonist and cellular context, Ca^{2+} can mediate different responses in the same cell [1]. The specific cellular response transduced downstream of the particular Ca^{2+} transient is encoded in the spatial and temporal dynamics of the Ca^{2+} signal, leading to the activation of a subset of Ca^{2+}-dependent effectors and the ensuing cellular response. As such, the duration, amplitude, frequency, and spatial localization of Ca^{2+} signals encode targeted signals that activate Ca^{2+}-sensitive effectors to define a particular cellular response. To generate and fine-tune those Ca^{2+} signals, cells use two main Ca^{2+} sources: entry of extracellular Ca^{2+} and Ca^{2+} release from intracellular stores. The primary intracellular Ca^{2+} store is the endoplasmic reticulum (ER), which can concentrate Ca^{2+} in the hundreds of μM range [2].

In contrast, cytoplasmic Ca^{2+} concentration is kept at rest at extremely low levels (~100 nM or lower), thus providing a low-noise background for detection of complex Ca^{2+} dynamics [3].

The Ca^{2+}-signaling machinery includes Ca^{2+} entry and extrusion pathways in the plasma membrane (PM), ER membrane Ca^{2+} release channels, and Ca^{2+} reuptake ATPases within the ER membrane [4]. These Ca^{2+} transport pathways, in addition to intracellular Ca^{2+} buffers and Ca^{2+} uptake and release through other intracellular organelles, primarily the mitochondria, combine to shape highly tuned and dynamic Ca^{2+} transients that regulate cellular functions [5].

Under physiological conditions in non-excitable cells, Ca^{2+} transients are typically initiated downstream of agonist stimulation through the activation of the PLC–IP_3 signal transduction cascade, which leads to the opening of intracellular Ca^{2+} channel inositol 1,4,5-trisphosphate receptors (IP_3Rs) to release Ca^{2+} from intracellular stores [6]. Ca^{2+} release depletes the stores and activates a Ca^{2+} influx pathway in the PM termed store-operated Ca^{2+} entry (SOCE). SOCE is mediated by two key players: ER transmembrane Ca^{2+} sensors represented by the STIM family of proteins and PM Ca^{2+} channels of the Orai family that link directly to STIMs (see Chapters 1 through 3). The N-terminus of STIM1 faces the ER lumen and consists of two EF-hand domains that detect luminal Ca^{2+} concentration. The loss of STIM1 Ca^{2+} binding upon store depletion leads to conformational changes in the protein and its aggregation into clusters that translocate and stabilize into ER–PM junctions with very close apposition (~20 nm) [7]. STIM1 within these ER–PM junctions binds to and recruits Orai1 through a diffusional trap mechanism, resulting in opening Orai1 channels and Ca^{2+} entry [8]. As such, the STIM–Orai clusters at ER–PM junctions define a specific microdomain at ER–PM junctions that also include the ER Ca-ATPase (SERCA) [9,10].

The tightly regulated remodeling of the Ca^{2+}-signaling machinery upon store depletion allows for specific Ca^{2+} signaling in the midrange between Ca^{2+} microdomains and global Ca^{2+} waves [10] (see Chapter 5). Spatially, Ca^{2+} signaling can occur in localized spatially restricted elementary Ca^{2+} release events that activate effectors located in the immediate proximity of the Ca^{2+} channel. Alternatively, Ca^{2+} signals/waves occur/spread through the entire cell resulting in a global spatially unrestricted signal.

We have recently described a SOCE-dependent Ca^{2+}-signaling modularity that signals in the midrange between these two spatial extremes [10]. Store depletion downstream of receptor activation and IP_3 generation results in a localized Ca^{2+} entry point source at the SOCE clusters that induces Ca^{2+} entry into the cytoplasm, which is readily taken up into the ER lumen through SERCA activity only to be released again through open IP_3Rs distally to the SOCE entry site and gate Ca^{2+}-activated Cl^- channels as downstream Ca^{2+} effectors. This mechanism, referred to as "Ca^{2+} teleporting," allows for specific activation of Ca^{2+} effectors that are distant from the point source Ca^{2+} channel without inducing a global Ca^{2+} wave, thus providing a novel module in the Ca^{2+}-signaling repertoire. A cartoon summary of Ca^{2+} teleporting is found in Figure 12.1.

The relationship between Ca^{2+} signaling and cellular proliferation is complex with Ca^{2+} transients detected at various stages of the cell cycle [2]. These transients are

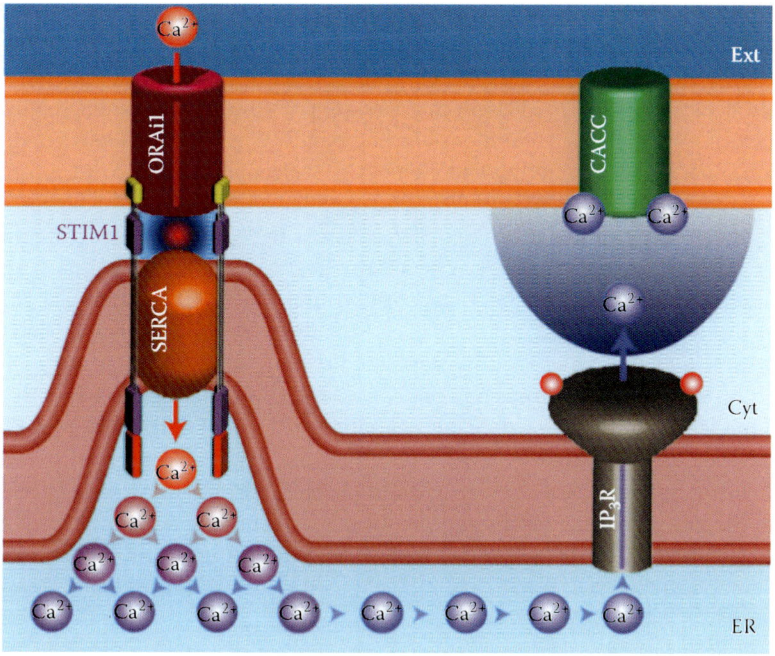

FIGURE 12.1 Cartoon illustrating Ca^{2+} teleporting that mediates midrange Ca^{2+} signaling. Receptor activation generates IP_3 that releases Ca^{2+} from stores. Following store depletion, the SOCE machinery forms spatially localized clusters that support Ca^{2+} entry. Ca^{2+} flowing through Orai1 is taken up by SERCA into the ER lumen. ER Ca^{2+} is then released through open IP_3Rs to activate Ca^{2+}-activated Cl channels (CaCC) that localize away from the point source SOCE clusters. See main text for more details. (Adapted from Courjaret, R., Dib, M., and Machaca, K., *J Cell Physiol*, 232, 1095, 2017.)

thought to activate a multitude of Ca^{2+} effectors downstream of the initial Ca^{2+} signal, which were shown to be important for cellular proliferation, including, for example, calmodulin (CaM) and Ca^{2+}–CaM-dependent protein kinase II (CaMKII).

However, there are a few cases where Ca^{2+} signals have been shown directly to be critical for cell cycle progression, including in mitosis for nuclear envelope breakdown and for chromosome disjunction [11]. In contrast, nuclear envelope breakdown during meiosis in vertebrate oocytes occurs independently of Ca^{2+}, but Ca^{2+} is required for the completion of meiosis I in vertebrate oocytes [12–14]. Interestingly, multiple Ca^{2+} signaling pathways are modified during M-phase of the cell cycle, with the best defined example being *Xenopus* oocyte maturation [15].

Several Ca^{2+} influx pathways have been implicated in cell proliferation and cell cycle progression, including TRP channels, voltage-gated Ca^{2+} channels (Ca_V), purinergic P2X receptors, ionotropic glutamate receptors, and SOCE [16]. Blockers of voltage-gated Ca^{2+} channels were shown to slow down cell growth, arguing for a role for these channels in cell cycle progression [16–18]. Experimental manipulation of the expression levels of members of the TRPC, TRPV, and TRPM families of cation channels, which are Ca^{2+} permeable, was linked to cell proliferation with differential effects depending on the particular channel studied [16]. However, some of these studies are difficult to interpret because channel knockdown or overexpression could have significantly broader effects on Ca^{2+} signaling than affecting Ca^{2+} influx through the specific channel in question, as it may lead to changes in expression of other Ca^{2+}-signaling pathways as a compensatory mechanism. Furthermore, the majority of TRP channels conduct cations with some Ca^{2+} permeability and are not Ca^{2+} selective like Orai1 or Ca_V channels, with the exception of TRPV5 and TRPV6 (see Chapter 13). Hence, changes in their expression is likely to affect the ionic balance across the cell membrane with effects on resting membrane potential, which may in turn affect cell proliferation.

The relationship between SOCE and cell proliferation is an intimate one that goes beyond the well-recognized roles of Ca^{2+} signaling in cellular growth and proliferation. SOCE is dramatically downregulated during the division phase of the cell cycle through mechanisms that have not been fully elucidated. This is in line with the significant remodeling of the Ca^{2+}-signaling machinery during M-phase, which has been well characterized during oocyte meiosis. Furthermore, there is mounting evidence from multiple neoplasms for an important role for SOCE in metastasis.

This chapter presents a brief overview of our current knowledge as to the mechanisms regulating SOCE during cell cycle from cellular proliferation to metastasis with an emphasis on SOCE regulation during cell division (mitosis and meiosis).

12.2 ROLE OF Ca^{2+} SIGNALS IN CELLULAR PROLIFERATION

The requirement for Ca^{2+} in cell proliferation has been known for decades with the optimum extracellular Ca^{2+} concentration to support proliferation being in the 0.5–1 mM range [16]. However, the requirement for Ca^{2+} during cell growth and proliferation is vague and difficult to define given the broad involvement of Ca^{2+} signaling in many aspects of cellular physiology and signaling. Also, bifurcating signaling pathways such as G-protein-coupled receptor activation can induce cell proliferation together with a Ca^{2+} signal without any direct link between Ca^{2+} and cell growth. Nonetheless, Ca^{2+} influx appears to support cellular proliferation since immortalized/neoplastic cells do not require as much Ca^{2+} in the extracellular medium as primary cells, and this is of importance for cell cycle studies and in the design of anticancer drugs.

Ca^{2+} release from intracellular stores through IP_3Rs or ryanodine receptors (RyR) has been implicated in proliferation. IP_3R activation is involved in stimulating the proliferation of stem cells, kidney cells, pancreatic beta cells, breast cancer, and vascular smooth muscle cells [19–26]. RyR are more likely to be involved in the regulation of cellular differentiation [25]. The activation of IP_3Rs also stimulates the proliferation of breast cancer cells [26]. Interestingly, breast cancer cells rely for their survival and proliferation on Ca^{2+} released via IP_3Rs to fuel mitochondrial respiration, making this Ca^{2+}-dependent cross talk between the ER and mitochondria a potential target for drug development [27].

Further strengthening a role for Ca^{2+} in proliferation, several Ca^{2+}-dependent downstream effectors have been implicated in cellular proliferation. A primary intracellular Ca^{2+}-binding effector, the ubiquitous cellular Ca^{2+}-sensing protein CaM, is required for cell cycle progression and proliferation [28,29]. Furthermore, CaM levels are highly regulated during the cell cycle with an increase at the G_1/S transition [30].

Several effectors downstream of CaM have been identified, including CaM kinases I and II that interfere with various steps of the cell cycle, and CaM-dependent protein phosphatase calcineurin [31–33]. In T cells, calcineurin dephosphorylates and activates the nuclear factor of activated T cells (NFAT), inducing its nuclear translocation, where it stimulates gene transcription in support of cell proliferation [34] (see Chapter 5). Several other effectors have been identified downstream of Ca^{2+} signals and the activation of CaM-dependent phosphorylation/dephosphorylation processes such as NF-κB and the cAMP response element [33].

Ca^{2+} influx through multiple pathways has also been implicated in cell proliferation. Ca_V channels, particularly T-type channels, have been described as regulating the proliferation of several cancer cells [35], and in the case of TRP channels, most knockdown experiments revealed that they are also required for cell proliferation [16]. SOCE is now probably the most extensively studied Ca^{2+} pathway involved in cell proliferation. Beyond the immune response, NFAT stimulation by SOCE is involved in the proliferation of other cell types such as neural progenitor cells [36,37], osteoblasts [38], kidney cells [39], and endothelial and endothelial progenitor cells involved in angiogenesis in cancer patients [40,41].

12.3 REMODELING OF THE Ca^{2+}-SIGNALING MACHINERY DURING MEIOSIS

As discussed earlier, not only are Ca^{2+} signals important for cellular proliferation and cell cycle progression, but Ca^{2+} transport pathways are themselves modulated during the cell cycle. This regulation of Ca^{2+} signaling during the cell cycle is presumed to support its progression although direct evidence for this remains scarce. An important case study is the remodeling of Ca^{2+}-signaling pathways during *Xenopus* oocyte meiosis, where multiple Ca^{2+}-signaling modules are modified to prepare the egg for fertilization and the egg-to-embryo transition.

Vertebrate oocytes arrest at prophase of meiosis I for prolonged periods of time [42,43]. Before oocytes acquire the competency to be fertilized, they undergo a maturation period during which they progress to metaphase of meiosis II in a process termed "oocyte maturation." Mature oocytes, typically referred to as eggs, complete meiosis at fertilization and transition to the mitotic cell cycle. Egg activation is highly Ca^{2+} dependent as Ca^{2+} mediates critical steps to initiate development, including the block of polyspermy and the completion of meiosis II to ensure a proper egg-to-embryo transition. In order to be effective, these events need to occur in a chronological fashion since polyspermy needs to be prevented before the completion of meiosis to ensure zygote viability. Throughout phylogeny in all sexually reproducing species investigated to date, Ca^{2+} has been shown to encode the egg-to-embryo transition through a species-specific Ca^{2+} transient at fertilization with well-defined spatial, temporal, and amplitude dynamics [44–47]. Eggs acquire the ability to produce this fertilization-specific Ca^{2+} transient only after oocyte maturation, owing to dramatic remodeling of the Ca^{2+}-signaling machinery during oocyte maturation that affects multiple Ca^{2+} transport proteins such as IP_3R, PMCA, and Orai/STIM [15].

The fertilization-specific Ca^{2+} signal in *Xenopus* is distinguished by a sustained elevated Ca^{2+} plateau that lasts for several minutes following a local Ca^{2+} rise at the site of sperm entry that spreads slowly (~9 μm/s), in the form of a Ca^{2+} wave, across the entire egg [48–51]. In contrast, Ca^{2+} transients in oocytes tend to have a saltatory mode of propagation with a more rapid wave speed (~20 μm/s) [52–54]. Given that the frog oocytes express only the type 1 IP_3R isoform and no RyR [55], these changes in Ca^{2+} release during oocyte maturation are thus due to changes in the regulation of the IP_3R during maturation as discussed in more detail in the following text. IP_3-dependent Ca^{2+} release plays a central role in mediating the Ca^{2+} transient at fertilization in vertebrate eggs.

In mammals, the slow oscillations that are maintained for long periods of time depend on Ca^{2+} influx from the extracellular space, presumably through the SOCE pathway [56–58]. Fertilization activates phospholipase Cγ (PLCγ) in *Xenopus* or delivers PLCζ in mammalian eggs, which increases IP_3 production and gates IP_3Rs to release Ca^{2+} from the ER [59–62]. It is clear that IP_3-dependent Ca^{2+} release properties are modulated during oocyte maturation when an increase in IP_3-dependent Ca^{2+}-release sensitivity is noted and conserved among different species [54,63–67].

IP_3-dependent Ca^{2+} release is sensitized during *Xenopus* oocyte maturation [54]. Threshold IP_3 concentrations lead to small and spatially separate elementary Ca^{2+} release events (puffs) in oocytes [68]; whereas in eggs, similar IP_3 concentrations lead to larger consolidated events referred to as single release events (SREs) (Figure 12.2b) [54]. These larger release events are likely due to the clustering of the smaller elementary Ca^{2+} puffs [54,69–71]. The ER remodels during oocyte maturation forming

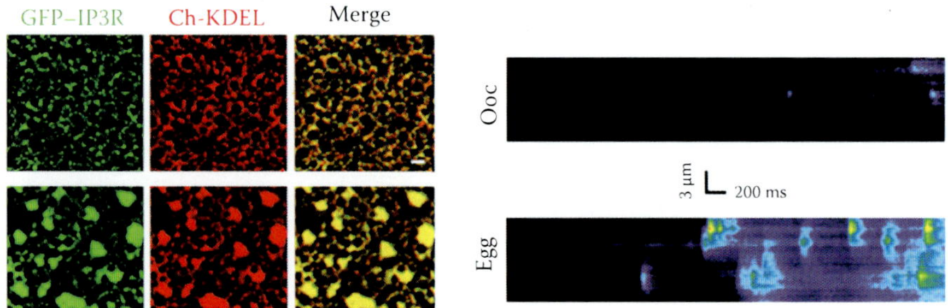

FIGURE 12.2 Clustering of elementary Ca^{2+} release events during oocyte maturation. (a) The ER remodels during oocyte maturation to form large patches in the egg. Oocytes were injected with GFP–IP_3R (50 ng/cell) and mCherry-KDEL (10 ng/cell). mCherry-KDEL marks the ER and shows dramatic remodeling of the ER in the animal hemisphere of the egg into large ER patches. IP_3Rs localize to these ER patches, as well as the reticular ER in the egg. Scale bar, 2 μm. (b) Elementary Ca^{2+} release events during oocyte maturation. *Xenopus* oocytes were injected with 10 μM caged IP_3 and 40 μM Oregon Green. Oocyte maturation was induced with progesterone, and both immature oocytes and fully mature eggs were imaged in line scan mode at 488 nm with the 405 nm laser at low intensity (0.2%) to continuously uncage cIP_3. The same region in the cell was scanned continuously in line scan mode with the x-axis representing time and the y-axis space. The single isolated Ca^{2+} puffs observed in the oocyte coalesce into larger release events referred to as single release events (SRE). (Adapted from Sun, L. et al., *PLoS One*, 6, e27928, 2011.)

large patches that are enriched in IP_3Rs as compared to the neighboring reticular ER (Figure 12.2a) [72]. Interestingly, we found that this clustering sensitizes IP_3Rs within ER patches, to respond to lower IP_3 concentrations as compared to IP_3Rs that localize to the reticular ER [72]. This sensitization appears to be due to increased Ca^{2+}-dependent cooperativity at subthreshold IP_3 concentrations due to the physical clustering of IP_3Rs within ER patches, because IP_3Rs freely exchange between the two ER compartments (i.e., patches and reticular ER) [72]. A similar sensitization of the IP_3R is observed during mitosis [73] and has been suggested to depend on cdk1 phosphorylation of IP_3R [74].

Sensitization of IP_3-dependent Ca^{2+} release during oocyte maturation takes place simultaneously with the activation of multiple kinase cascades that drive this differentiation pathway. This argued for a potential role for phosphorylation of IP_3R in modulating its sensitivity in both *Xenopus* and mouse eggs, especially that IP_3R was found to be phosphorylated specifically during oocyte maturation at maturation-promoting factor (MPF)/mitogen-activated protein kinase (MAPK) conserved sites [54,75,76]. However, direct evidence showing that phosphorylation at these residues modulates IP_3 sensitivity of the channels is lacking. Furthermore, IP_3R sensitization could also be related to the number of functional IP_3Rs, given their gating cooperativity. In that context, the number of functional IP_3Rs increases during *Xenopus* oocyte maturation (without a significant change in total IP_3R protein pool) as they translocate from annulate lamellae to the ER [77,78]. Annulate lamellae represent an oocyte-specific vesicular compartment to which IP_3Rs localize but are silenced presumably through protein–protein interactions [77,78].

Two additional transport pathways important in defining Ca^{2+} transient in the frog oocyte are the PM Ca^{2+} ATPase (PMCA) that possesses high affinity for Ca^{2+} and functions to decrease Ca^{2+} levels back to the resting state [79,80] and the SERCA pump that ensures Ca^{2+} sequestration into the ER lumen [81]. PMCA, which localizes to the cell membrane in immature *Xenopus* oocytes, is internalized into an intracellular vesicular pool during oocyte maturation [71]. Coupled to the continuous and constant SERCA-dependent Ca^{2+} reuptake and the increased sensitivity of IP_3-dependent Ca^{2+} release, PMCA internalization contributes to the sustained Ca^{2+} plateau after the slow rising Ca^{2+} phase at fertilization [15,71].

Finally, SOCE completely inactivates in the mature *Xenopus* egg compared to immature oocytes [82,83]. The regulation of SOCE during meiosis is covered in more detail in the next section.

12.4 SOCE INACTIVATION DURING M-PHASE

Immature *Xenopus* oocytes possess a robust SOC current that has similar biophysical properties to CRAC, the prototypical SOC channel activity originally characterized in immune cells [82,84,85] (see Chapter 1). However, during oocyte maturation, SOCE is inactivated completely: in mature eggs, SOCE is no longer activated when stores are depleted [82]. The inhibition of SOCE requires the activation of maturation-promoting factor (MPF, composed of CDK1 and cyclin B) [83]. This was shown by measuring both the SOC current and the levels of activation of multiple kinases that drive oocyte maturation at the single-oocyte level, to allow for direct correlation between the activity of various kinases and SOCE at the single-cell level [83]. Therefore, the fertilization-specific Ca^{2+} transient in *Xenopus* eggs is generated without contribution from SOCE. *Xenopus* eggs respond to sperm entry with a single sweeping Ca^{2+} transient that lasts several minutes [49,86]. This Ca^{2+} signal encodes subsequent events associated with egg activation in the following order: (1) fast block to polyspermy due to gating of Ca^{2+}-activated Cl^- channels that depolarize the cell membrane; (2) slow block to polyspermy due to cortical granule fusion; and (3) completion of meiosis due to calcineurin and CaMKII activation [15,87,88]. SOCE inactivation is likely to contribute to shaping the dynamics of the fertilization-specific Ca^{2+} signal and therefore promote the egg-to-embryo transition.

Interestingly, SOCE is downregulated but not completely inactivated during mammalian oocyte meiosis as store depletion in metaphase II eggs induces Ca^{2+} entry, which supports Ca^{2+} oscillations following fertilization [89]. In contrast to *Xenopus* eggs, mammalian oocytes respond at fertilization with multiple Ca^{2+} oscillations that can last for hours [57]. Maintenance of these Ca^{2+} oscillations depends on Ca^{2+} influx through SOCE, presumably to refill Ca^{2+} stores and provide a continuous Ca^{2+} source for the Ca^{2+} release waves. There is, hence, a correlation between the occurrence of SOCE during meiosis and the ability of the oocyte to support Ca^{2+} oscillations at fertilization. Some reports have argued that SOCE amplitude increases during oocyte maturation in both mouse and pig [90,91]. In contrast, we and others have shown that SOCE is downregulated but not completely inhibited during mouse oocyte meiosis, and that downregulated SOCE is required for the egg-to-embryo transition [92,93]. The reasons for these discrepancies remain unclear. Nonetheless, collectively the data suggest that SOCE downregulation during vertebrate oocyte maturation represents an important determinant of the remodeling of Ca^{2+} signaling in preparation for fertilization.

Furthermore, similar to what is observed in meiosis of frog oocytes, SOCE is also inhibited during mitosis of mammalian cells. In the late 1980s, Volpi and Berlin showed that histamine stimulation during interphase in HeLa cells produced an initial Ca^{2+} rise owing to Ca^{2+} release from internal stores, followed by an elevated plateau due to Ca^{2+} influx from the extracellular space [94]. By contrast, histamine stimulation in mitotic cells resulted only in the Ca^{2+} release phase, arguing that Ca^{2+} influx is inhibited during mitosis. This observation was made around the time when the initial ideas regarding SOCE were being formulated. The same group later argued that SOCE inhibition during mitosis occurs due to uncoupling of store depletion from SOCE, as thapsigargin (an agent that causes store depletion by blocking SERCA) activated SOCE in interphase but not mitotic cells [95]. More recent studies confirmed that SOCE is inactivated during mitosis in RBL-2H3, HeLa, and HEK293 cells [96–98].

Investigating SOCE levels throughout the cell cycle showed that there is a slight enhancement of SOCE during the G_1 and S phases, and dramatic downregulation during M-phase [98]. Consistent with this cell cycle–dependent modulation of SOCE activity, SOCE has been shown to control the G_1/S transition but is not necessary during S-phase or the G_2/M transition [99]. Furthermore, SOCE has emerged as an important player in cell proliferation, yet the mechanisms by which it controls distinct phases of the cell cycle remain elusive. Recent studies show that inactivation of

SOCE by silencing STIM1 in smooth muscle cells, cervical and breast cancer cells significantly inhibited proliferation by slowing down cell cycle progression [21,100]. This is discussed in more detail in Section 12.6.

Collectively, most current evidence argues that SOCE downregulation during M-phase is conserved and as such could be physiologically significant. Although this has not been directly addressed experimentally, one can speculate that tight regulation of Ca^{2+} signaling is required during M-phase owing to its important functional role at multiple steps throughout the process including nuclear-envelope breakdown, anaphase onset, and cell cleavage. Hence, SOCE inactivation might represent a safety mechanism that prevents sporadic Ca^{2+} signals from occurring during cell division, which may disrupt its normal progression.

12.5 MECHANISMS REGULATING SOCE INACTIVATION DURING M-PHASE

Aside from the role of Maturation Promoting Factor (MPF, Cdk1) in SOCE inhibition, very little was known regarding the mechanistic regulation of SOCE inactivation during M-phase. It has also been argued that SOCE inhibition during mitosis in COS-7 cells is the result of the microtubule-network remodeling that accompanies mitosis [101]. Laser scanning confocal microscopy to monitor cytosolic Ca^{2+} dynamics revealed that SOCE was progressively inhibited in mitosis and became virtually absent during metaphase [101]. Russa and colleagues used various cytoskeletal modifying drugs and immunofluorescence to assess the contribution of microtubule and actin filaments to SOCE. Nocodazole treatment caused microtubule reorganization and retraction from the cell periphery that mimicked the natural mitotic microtubule remodeling that was also accompanied by SOCE inhibition. Short exposure to paclitaxel, a microtubule-stabilizing drug, strengthened SOCE, whereas long exposure resulted in microtubule disruption and SOCE inhibition. Actin-modifying drugs (cytochalasin D, calyculin A) did not affect SOCE. These findings indicate that mitotic microtubule remodeling plays a significant role in the inhibition of SOCE during mitosis. However, recent studies investigating the behavior of STIM1 and Orai1 during M-phase have provided additional insights [96,97,102–104].

STIM1 is a phosphoprotein, as revealed in large-scale mass spectrometry studies with different findings as to the specific phosphorylated residues from immunoprecipitated STIM1, presumably due to the different cell types used with lack of careful control of the cell cycle stage [105]. Smyth and colleagues reported that during mitosis of HeLa and HEK293 cells, which were treated with 1.67 μM nocodazole for 12–16 h, STIM1 fails to move to peripheral junctions to form puncta and interact with Orai1 [97]. Furthermore, during mitosis, STIM1 becomes phosphorylated at multiple sites identified by mass spectrometry. Phosphorylation of STIM1 could also be detected by an anti-phospho-Ser/Thr-Pro MPM-2 antibody [106] in this study. STIM1 contains 10 minimal MPF–MAPK consensus sites (S/T-P), all located in the far C-terminus (Figure 12.3). Removal of these 10 residues by truncation at amino acid 482 abolished MPM-2 recognition of mitotic STIM1 [97]. The resulting truncated protein, when coexpressed with Orai1 in mitotic cells, partially rescues SOCE as measured by Ca^{2+} imaging and SOC current recording

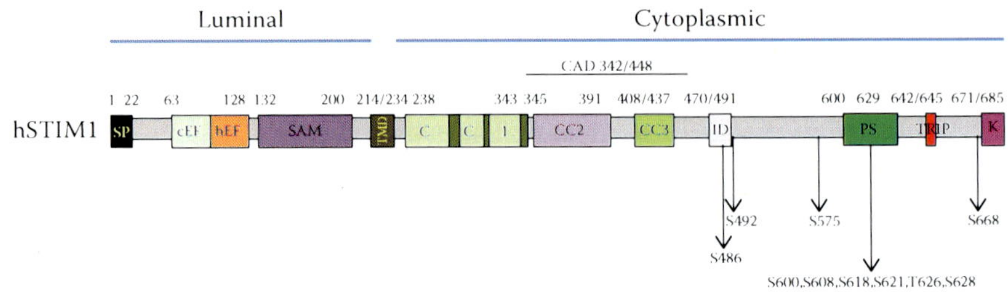

FIGURE 12.3 The molecular domains of human STIM1. ER STIM1 contains a luminal and a cytosolic domain. Indicated are the locations in the sequence of the N-terminal signal peptide (SP), a Ca^{2+}-binding canonical EF-hand domain (cEF), a non-Ca^{2+}-binding hidden EF-hand (hEF) domain, a sterile α-motif (SAM), and a single transmembrane domain (TMD), a C-terminal region of coiled-coil regions that includes CC1, CC2, and CC3. Downstream of CRAC activation domain (CAD) is an acidic inhibitory domain (ID) that mediates fast Ca^{2+}-dependent inactivation of Orai1. The C-terminal tail contains a Pro/Ser-rich domain (PS) in which multiple S/TP phosphorylation sites, a microtubule interacting domain (TRIP), and a Lys-rich domain responsible for phospholipid interaction at the plasma membrane are indicated.

by whole-cell patch–clamp technique. In addition, alanine substitution at two residues (Ser486 and Ser668) was sufficient to partially rescue SOCE, although to a lesser extent than the 482 deletion mutant [97]. Cotransfection with an Orai1 construct was necessary in these experiments because truncated STIM1 was not able to rescue SOCE in mitotic cells unless Orai1 was coexpressed. In fact, Smyth and colleagues reported that HEK293 cells expressing the truncated STIM1 did expand at a slightly but significantly slower rate, but no other significant alterations to the mitotic process were observed [97].

In a subsequent study, a STIM1 mutant retaining the full-length C-terminus, but lacking the 10 putative phosphorylation sites by alanine substitution (STIM1-10A), was able to rescue SOCE in mitotic cells expressing only endogenous Orai1 [96]. This would suggest that phosphorylation of STIM1 is the major underlying mechanism for shutting down SOCE during mitosis. When the cellular localization of STIM1-10A was examined by confocal microscopy in mitotic cells, the results were striking. STIM1-10A was incapable of dissociating from EB1 and accumulated in the spindle area. Mutation of the TRIP EB1-binding domain to TRNN rescued appropriate partitioning of STIM1 to the cell periphery. However, STIM1-10A supported SOCE in mitosis is likely not due to restoration of the EB1 interaction by STIM1-10A, because STIM1 activation of SOCE is independent of its EB1 interaction [107]. However, ER mislocalization driven by STIM-10A did not cause obvious mitotic defects. In addition, the phosphomimetic STIM1-10E mutant failed to inhibit SOCE activation in interphase cells [96].

Our group studied the mechanisms regulating SOCE inactivation during *Xenopus* oocyte meiosis with disparate results from what is observed in mitosis [103]. SOCE inactivation during meiosis is dependent on the kinase cascade that drives oocyte maturation, where activation of MPF was shown to be necessary and sufficient to inactivate SOCE in *Xenopus* oocytes [82,83]. Overexpression of human STIM1 and Orai1 by injection of *in vitro* transcribed mRNA into *Xenopus* oocytes greatly enhances SOCE, yet even this current induced by exogenous expression is inactivated during meiosis (Figure 12.4c) [103]. Associated with this inhibition of SOCE, STIM1 fails to cluster following Ca^{2+} store depletion (Figure 12.4b). Furthermore, meiosis was associated with inhibition of SOCE mediated by constitutively active STIM1 mutants [103], arguing that this inhibition is an active process. STIM1 is phosphorylated at multiple sites during meiosis as shown by a mobility shift on SDS-PAGE and by mass spectrometry. However, mutagenesis of all possible phosphorylation sites to alanines failed to rescue either STIM1 clustering or SOCE activation [103]. Interestingly, STIM1 clustering inhibition during meiosis required activation of MPF and was independent of the activity of the MAPK cascade, consistent with the requirement for MPF activation to inhibit endogenous SOCE during meiosis [83,103]. Our data suggest that STIM1 phosphorylation is not responsible for STIM1 clustering inhibition or SOCE inactivation during M-phase. Thus, while there are similarities between the regulation of STIM1 in meiosis and mitosis, there appear to be some differences as well. Indeed, it is likely that there is much more to be

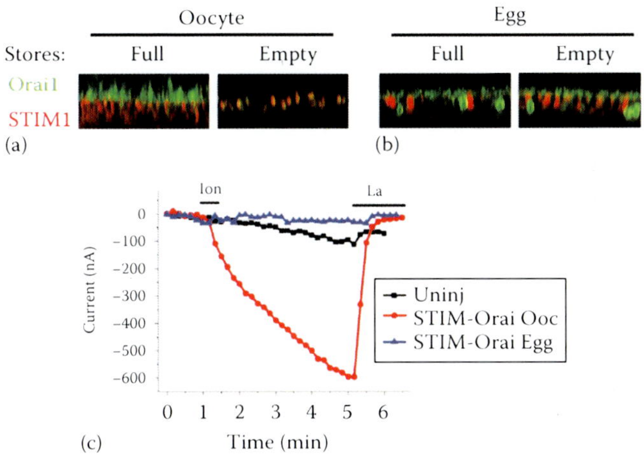

FIGURE 12.4 Meiosis is associated with inhibition of STIM1 clustering and Orai1 internalization. Oocyte were injected with mCherry-STIM1 and GFP-Orai1 (a) and matured into eggs (meiosis) (b). Images show the distribution of Orai1 and STIM1 before and after store depletion with TPEN (5 mM). Images show orthogonal section across a stack of confocal images representing a cross section across the plasma membrane and cortical cytoplasm. Whereas STIM1–Orai1 cocluster in oocytes (a), this clustering is inhibited in eggs (b). The ER remodels in eggs and forms large membrane patches to which STIM1 localizes. Orai1 is enriched in intracellular vesicles in eggs. Scale bar is 2 μm. (c) SOCE inactivates during meiosis. Control oocytes or oocytes and eggs injected with STIM1 and Orai1 were treated with ionomycin (10 μM) to deplete intracellular Ca^{2+} stores and activate SOCE. SOCE activates in oocytes but not in eggs overexpressing STIM1 and Orai1. (Adapted from Yu, F. et al., *Proc. Natl. Acad. Sci. USA*, 106, 17401, 2009.)

learned about the causes and consequences of STIM1 phosphorylation. More complex and sophisticated assays or models may be necessary to fully understand STIM1 phosphorylation.

In a follow-up study, substitution of the regulatory region of STIM1, which contains the 10 MPF–MAPK putative phosphorylation sites, with GFP-rescued clustering of STIM1 into puncta during meiosis, but still did not rescue SOCE [102]. These data argue that SOCE inactivation during meiosis is not only due to inhibition of STIM1 clustering in response to store depletion. Indeed, SOCE inactivation during meiosis is also associated with the removal of Orai1 from the cell membrane into an endosomal compartment (Figure 12.4b) [103,104]. Orai1 is enriched in the cell membrane of immature oocytes (Figure 12.4a) and continuously recycles between the cell membrane and an endosomal compartment through a Rho- and Rab5-dependent pathway [104]. However, in eggs, Orai1 is internalized into an endosomal compartment, which requires the activities of dynamin and caveolin in addition to Rab5. Orai1 possesses a consensus caveolin-binding domain in its N-terminal cytoplasmic region that when mutated inhibits the ability of Orai1 to be internalized during meiosis, supporting the role of caveolin-mediated endocytosis in Orai1 internalization [104]. Whether Orai1 trafficking is regulated during mitosis in a similar fashion to meiosis remains unclear (Figure 12.5). Nonetheless, even without the regulation of STIM1 by phosphorylation, Orai1 internalization would be sufficient to inhibit SOCE during M-phase, thus bringing into question the role of STIM1 phosphorylation.

An important difference between mitosis and meiosis studies is the experimental approaches used as they may affect both the results and interpretations. *Xenopus* oocyte meiosis provides an important advantage in that oocytes are physiologically arrested at prophase I and eggs at metaphase II of meiosis, as discussed earlier. This removes the need for pharmacological or other experimental interventions to arrest cells in M-phase. M-phase is a transient phase that involves dramatic remodeling of multiple aspects of cellular physiology as the cell prepares to divide. Interventions such as nocodazole treatment may modify physiological processes in ways that are not always predictable. Such treatments, however, are necessary to allow cell synchronization in mitosis. Hence, future research should focus on the regulation of SOCE and STIM1 phosphorylation in mitosis under more physiological conditions, ideally in primary cells. Nonetheless, understanding SOCE inhibition during M-phase will undoubtedly provide important clues regarding the basic mechanisms controlling SOCE activation and regulation.

12.6 SOCE AND CANCER

The level of Ca^{2+}-signaling remodeling in cancer cells is remarkable and is associated with dysregulation of several Ca^{2+} channels and pumps [108,109]. The expression levels of essential components of SOCE, including members of the STIM and Orai families (Orai1, Orai3, STIM1, and STIM2), are modulated in several types of tumors. SOCE affects hallmarks of cancer progression, including cell cycle progression, escaping apoptosis, tumorigenesis, metastasis, angiogenesis, and tumor immunity. The differential expression of these components seems to depend on the cancer type and tumor stage (reviewed in [110–116]).

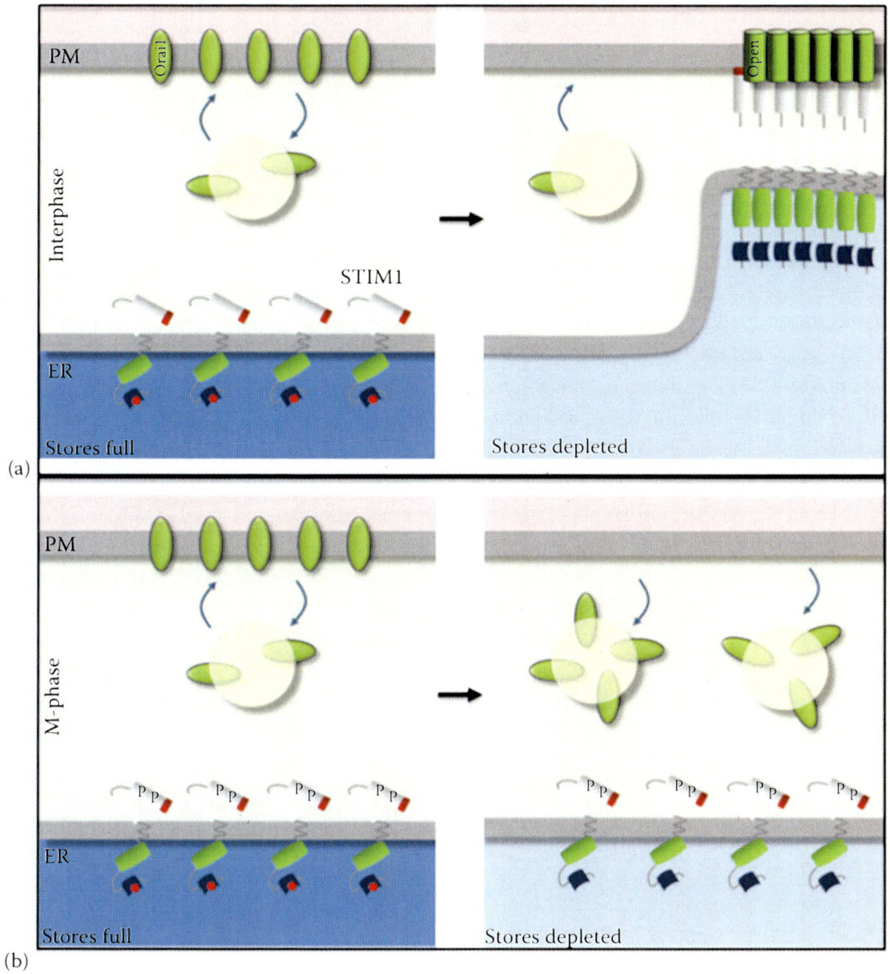

FIGURE 12.5 Working model of SOCE inactivation during M-phase. (a) Events mediating STIM1Orai1 coupling during interphase. Orai1 has been shown to recycle between an endosomal compartment (Endo) and the cell membrane in *Xenopus* oocytes; however, it is not known whether this also occurs in mammalian cells. (b) During M-phase, STIM1 is phosphorylated and cannot form large clusters (indicated by red X) in response to Ca^{2+}-store depletion. Orai1 internalizes into an endosomal compartment during *Xenopus* oocyte meiosis, but it is not known whether this also occurs during mitosis. The interphase panel (a) refers to stage VI *Xenopus* oocytes arrested in an interphase-like state at the G_2–M transition of the meiotic cell cycle. PM, plasma membrane; ER, endoplasmic reticulum.

12.6.1 STIM1 and Orai1 Role in Cell Progression, Proliferation, and Cell Death of Cancer Cells

In the past few years, several reports supported a role for SOCE key players, STIM1 and Orai1, in cell cycle progression, proliferation, apoptosis, and during tumor development and progression. Sabbioni and colleagues were the first to report STIM1 (initially known as GOK) deletion in human rhabdomyosarcoma and rhabdoid tumor cell lines, RD and G401, and to show that increasing levels of STIM1 caused growth arrest in these cells [117]. Feng et al. showed that Orai1 knockdown in MCF-7 cells suppressed cell proliferation measured colorimetrically, using colony formation assays, and inhibited tumor formation in nude mice [118]. Expression of two nonfunctional Orai1 mutants L273S and R91W triggers apoptosis resistance in prostate LNCaP cells measured by the TUNEL technique and by Hoechst staining. The rescue of Orai1 function by overexpression of wild-type (WT) Orai1 increased apoptosis levels [119]. In another study, Kondratska and colleagues showed high levels of Orai1 and STIM1 expression in the pancreatic adenocarcinoma Panc1 cell line. siRNA-mediated downregulation of Orai1 and/or STIM1-intensified apoptosis-induced thermotherapy [120]. Similarly, Orai1 overexpression in A549 lung cancer cells decreased SOCE, arrested cells in G_0/G_1 phase, induced p21 expression, decreased ERK1/2 and Akt phosphorylation, and inhibited EGF-induced proliferation [121]. Treating MDA-MB-231 cells with SOCE inhibitor SKF96365 blocked TGFβ-induced cell proliferation and cell cycle arrest measured by colony formation assay and flow cytometry, respectively. The effect of TGF on proliferation was shown to be mediated by a decrease in STIM1 levels [122].

In cervical cancer tissues from patients with early stage cervical cancer, Chen et al. reported elevated levels of STIM1 protein compared to noncancerous tissues, as analyzed by immunoblotting. On the other hand, siRNA-mediated STIM1 knockdown in cervical cancer SiHa cells inhibits cell proliferation by arresting the cell cycle at S and G_2/M phases as determined by flow cytometry of PI-stained cells. This effect of STIM1 on cell cycle progression was suggested to be due to STIM1-dependent modulation of p21 and Cdc25C involved in G_2/M checkpoint progression [21].

In another study, DU145 and PC3 prostate cancer cells stably expressing STIM1 and Orai1 showed slower growth rate determined by cell growth curve measured by counting cell number over time [123]. The induction of STIM1 and Orai1 levels, and subsequently SOCE, was accompanied by an increase in the percentage of cells in the G_0/G_1 phase and decrease in the G_2/M phase. The effect of STIM1 and Orai1 overexpression on cell cycle progression was accompanied by altered expression of cell cycle regulatory proteins, cyclin E2, cyclin D1, Wee1, and Myt1. Moreover, cells overexpressing STIM1 and/or Orai1 promoted cell senescence and induced expression of apoptosis inhibitors (DcR2, XIAP, and Bcl2). In another study, Bcl2 overexpression in human prostate cancer LNCaP cells inhibited SOCE and promoted apoptosis resistance [124,125]. Dubois and colleagues showed that heteromeric channels formed by Orai1–Orai3 are regulated by arachidonic acid, and independent of intracellular stores content, to promote cell proliferation and apoptosis resistance of prostate cancer cells. The effect on proliferation was linked to NFAT activation and expression of cyclin D1 controlling G_1/S transition [126].

In mouse melanoma cells, silencing of STIM1 caused a reduction in cell growth and increased cell death [127]. Also, pharmacological inhibition of Orai1, or siRNA-mediated silencing of Orai1 and/or STIM2, caused melanoma cells to grow faster but reduced their invading potential. In a similar study, Umemura et al. showed that inhibition of SOCE by inhibitor YM58483 or siRNA knockdown of STIM1 suppresses proliferation and invasion of melanoma cells [128]. In malignant B16BL6 melanoma cells, mitochondrial Ca^{2+} uptake is coupled to SOCE, which promotes PKB activity favoring cell survival [129]. In a recent study, Hooper and colleagues reported very small SOCE in invasive melanoma depending on PKC-mediated phosphorylation of Orai1 [130]. Collectively, these studies show a correlation between SOCE levels and cancer cell proliferation and the ability to resist apoptosis.

12.6.2 SOCE in Cell Motility, Metastasis, and Tumor Microenvironment

Yang et al. reported that serum-induced migration of human MDA-MB-231 breast cancer cells with STIM1 or Orai1 levels knocked down by siRNA was decreased by 60%–85%, measured by Boyden chamber assay without affecting cell proliferation [100]. The results from these *in vitro* studies were replicated in mouse models where tumor cells expressing luciferase reporter gene were injected into immune-deficient mice through the tail vein [131,132]. The metastasis of injected cells with STIM1 and Orai1 siRNA to the lungs was much less than the control siRNA-treated cells.

Analysis of cancer tissues from clear-cell renal-cell carcinoma (ccRCC) showed increased expression of Orai1 and STIM1 protein levels in comparison to adjacent normal tissues [133]. Pharmacologic inhibition of SOCE using SKF96365 or 2-APB commonly used blockers, and siRNA-mediated knockdown of Orai1 or STIM1, decreased the motility of ccRCC cell lines, Caki1 and ACHN, independent of proliferation, as measured by wound-healing assay in the presence of the antiproliferative drug Mitomycin C.

In cervical cancer tissues from patients with early stage cervical cancer, the levels of STIM1 were positively correlated with tumor size and metastasis to the lymph nodes [21]. Similarly, injection of cervical cancer cells selected for high STIM1 expression significantly enhanced tumor growth, local spread, and angiogenesis, whereas cells with short hairpin RNAs (shRNAs) targeting human STIM1 showed a significant decrease in tumor growth and angiogenesis.

In human prostate cancer tissues, STIM1 and Orai1 were expressed at significantly lower levels in hyperplasia and tumor tissues at advanced stages. However, the expression of STIM1 was higher in tumors with earlier histological grade than in hyperplasia tissues, suggesting that STIM1 can play a dual role depending on the cancer stage, where it favors malignant transformation of prostate cells at early stages and prohibit tumor growth at advanced stages [123]. The expression levels of STIM1 and Orai1 in the hyperplasic human prostate cell line BPH-1 were lower as compared to the metastatic LNCaP, DU145, and PC3 cell lines with a significantly greater SOCE. Overexpression of Orai1 and STIM1 in prostate cancer cell lines DU145 and PC3 promoted cell migration as tested using the wound-healing and Transwell assays. Furthermore, *in vivo*, immunodeficient mice (SCID) injected with DU145 cells expressing STIM1-YFP or Orai1-YFP showed retardation in tumor growth and decrease in expression of E-cadherin, demonstrating that STIM1 enhances epithelial–mesenchymal transition (EMT).

Colorectal cancer tissue collected from patients showed high expression of STIM1, with significant correlation between STIM1 levels, tumor growth, invasion depth, and metastasis to lymph nodes. STIM1 knockdown by 50%–90% using shRNA or SOCE inhibition using 2-APB and SKF96365 in three colon cancer cell lines (DLD-1, HCT116, and SW480) strongly inhibited cell motility, and this depends on COX-2 expression and prostaglandin synthesis [134].

Studies by Trebak and colleagues showed that in breast cancer MCF7 cells, SOCE is mediated by STIM1 and Orai3, rather than the ubiquitous STIM1–Orai1 pair [135,136]. Induction of estrogen-receptor-positive breast cancer MCF7 cells using breast growth factors activated Orai3-dependent Ca^{2+} influx [137]. Orai3 knockdown resulted in a significant decrease in tumor cell invasion measured using Boyden chamber invasion. In addition, Orai3 knockdown with shRNA-encoding lentiviruses significantly reduced the number of MCF7 colonies on agar. In primary glioblastoma cell lines, STIM1 and Orai1 are the key components of SOCE. Interestingly, using siRNA approaches to knockdown Orai3 in MCF7, or STIM1 and Orai1 in glioblastoma cells showed strong reduction in Matrigel invasion of these cells compared to nonmalignant primary astrocytes [138].

Collectively, these studies strongly argue for a ubiquitous role for Ca^{2+} influx through SOCE in mediating tumor cell metastasis in different cancers. This could be due to the role of SOCE in regulating the cytoskeleton in a polarized fashion in migrating cells, which is important for cell motility [139]. Given that over 90% of breast cancer deaths are associated with metastasis rather than growth of the primary tumor, SOCE may represent an attractive anticancer therapeutic target.

12.6.3 Orai1 and STIM1 in Antitumor Immunity

In addition to the role of SOCE in regulation metastasis, Orai1 and STIM1 have been implicated in regulating immune function in the context of its antitumor activity. The study by Xu et al. using Transwell migration assay reported higher recruitment of human leukemic monocyte macrophage cell line U937 when using culture supernatant from STIM1-knockdown DU145 and PC3 cells and decreased migration of U937 when using supernatant from DU145 and PC3 cells overexpressing STIM1 [123]. Real-time RT-PCR showed decreased expression of cytokines in U937 cells incubated with medium from DU145 and PC3 cells overexpressing STIM1 and/Orai1. In addition, the markers for EMT transition, VEGFA and MMP9 decreased in U937 macrophages, suggesting that STIM1 and Orai1 overexpression, and hence enhanced SOCE activity in prostate cancer cells, hinders EMT in macrophages and their recruitment to tumor sites.

Using Stim1$^{fl/fl}$Stim2$^{fl/fl}$ Cd4Cre (DKO) mice with CD4+ and CD8+ T cells lacking SOCE [140], Weidinger et al. examined injecting B16-OVA melanoma cells and MC-38 colon carcinoma cells in DKO and WT mice with depleted immunosuppressive Treg cells. DKO mice failed to control tumor growth suggesting that SOCE in

CD8+ T cells mediated by the STIM family favors antitumor immunity [141]. The loss of SOCE in DKO did not compromise the priming or migration of tumor antigen–specific CD8+ T cells, but it inhibited the ability of CTLs to control tumor engraftment and growth. The cytotoxicity of SOCE-deficient CTLs from DKO mice against cocultured EG7-OVA cells was diminished, and the ability of these cells to produce IFN-γ and TNF-α upon stimulation was also reduced, compared to wild-type CTLs.

Although several studies investigating the role of SOCE in cancer have been carried out over the past few years, much remains to be elucidated regarding the mechanisms by which SOCE affects cancer development. This complexity is due to the fact that SOCE regulation is likely to be cancer type, as well as stage specific. Nonetheless, accumulating evidence suggests that SOCE plays a critical role in cancer cell proliferation, metastasis, and antitumor immunity.

REFERENCES

1. Berridge, M. J., Lipp, P., and Bootman, M. D. (2000) The versatility and universality of calcium signalling. *Nature Reviews. Molecular Cell Biology* **1**, 11–21.
2. Demaurex, N. and Frieden, M. (2003) Measurements of the free luminal ER Ca^{2+} concentration with targeted "cameleon" fluorescent proteins. *Cell Calcium* **34**, 109–119.
3. Berridge, M. J. (2012) Calcium signalling remodelling and disease. *Biochemical Society Transactions* **40**, 297–309.
4. Clapham, D. E. (1995) Calcium signaling. *Cell* **80**, 259–268.
5. Berridge, M. J., Bootman, M. D., and Roderick, H. L. (2003) Calcium signalling: Dynamics, homeostasis and remodelling. *Nature Reviews. Molecular Cell Biology* **4**, 517–529.
6. Clapham, D. E. (2007) Calcium signaling. *Cell* **131**, 1047–1058.
7. Wu, M. M., Buchanan, J., Luik, R. M., and Lewis, R. S. (2006) Ca^{2+} store depletion causes STIM1 to accumulate in ER regions closely associated with the plasma membrane. *Journal of Cell Biology* **174**, 803–813.
8. Hogan, P. G., Lewis, R. S., and Rao, A. (2010) Molecular basis of calcium signaling in lymphocytes: STIM and ORAI. *Annual Review of Immunology* **28**, 491–533.
9. Hogan, P. G. (2015) The STIM1-ORAI1 microdomain. *Cell Calcium* **58**, 357–367.
10. Courjaret, R. and Machaca, K. (2014) Mid-range Ca^{2+} signalling mediated by functional coupling between store-operated Ca^{2+} entry and IP3-dependent Ca^{2+} release. *Nature Communications* **5**, 3916.
11. Arredouani, A., Yu, F., Sun, L., and Machaca, K. (2010) Regulation of store-operated Ca^{2+} entry during the cell cycle. *Journal of Cell Science* **123**, 2155–2162.
12. Sun, L. and Machaca, K. (2004) Ca^{2+} cyt negatively regulates the initiation of oocyte maturation. *Journal of Cell Biology* **165**, 63–75.
13. Sun, L., Hodeify, R., Haun, S., Charlesworth, A., MacNicol, A. M., Ponnappan, S., Ponnappan, U., Prigent, C., and Machaca, K. (2008) Ca^{2+} homeostasis regulates Xenopus oocyte maturation. *Biology of Reproduction* **78**, 726–735.
14. Tombes, R. M., Simerly, C., Borisy, G. G., and Schatten, G. (1992) Meiosis, egg activation, and nuclear envelope breakdown are differentially reliant on Ca^{2+}, whereas germinal vesicle breakdown is Ca^{2+} independent in the mouse oocyte. *Journal of Cell Biology* **117**, 799–811.
15. Machaca, K. (2007) Ca^{2+} signaling differentiation during oocyte maturation. *Journal of Cellular Physiology* **213**, 331–340.
16. Capiod, T. (2013) The need for calcium channels in cell proliferation. *Recent Patents on Anti-Cancer Drug Discovery* **8**, 4–17.

17. Nel, A. E., Dirienzo, W., Stefanini, G. F., Wooten, M. W., Canonica, G. W., Lattanze, G. R., Stevenson, H. C., Miller, P., Fudenberg, H. H., and Galbraith, R. M. (1986) Inhibition of T3 mediated T-cell proliferation by Ca^{2+}-channel blockers and inhibitors of Ca^{2+}/phospholipid-dependent kinase. *Scandinavian Journal of Immunology* **24**, 283–290.
18. Taylor, J. T., Huang, L., Pottle, J. E., Liu, K., Yang, Y., Zeng, X., Keyser, B. M., Agrawal, K. C., Hansen, J. B., and Li, M. (2008) Selective blockade of T-type Ca^{2+} channels suppresses human breast cancer cell proliferation. *Cancer Letters* **267**, 116–124.
19. Aguiari, G., Trimi, V., Bogo, M., Mangolini, A., Szabadkai, G., Pinton, P., Witzgall, R. et al. (2008) Novel role for polycystin-1 in modulating cell proliferation through calcium oscillations in kidney cells. *Cell Proliferation* **41**, 554–573.
20. Resende, R. R., Adhikari, A., da Costa, J. L., Lorencon, E., Ladeira, M. S., Guatimosim, S., Kihara, A. H., and Ladeira, L. O. (2010) Influence of spontaneous calcium events on cell-cycle progression in embryonal carcinoma and adult stem cells. *Biochimica et Biophysica Acta* **1803**, 246–260.
21. Chen, Y. F., Chiu, W. T., Chen, Y. T., Lin, P. Y., Huang, H. J., Chou, C. Y., Chang, H. C., Tang, M. J., and Shen, M. R. (2011) Calcium store sensor stromal-interaction molecule 1-dependent signaling plays an important role in cervical cancer growth, migration, and angiogenesis. *Proceedings of the National Academy of Sciences of the United States of America* **108**, 15225–15230.
22. Stanimirovic, D. B., Ball, R., Mealing, G., Morley, P., and Durkin, J. P. (1995) The role of intracellular calcium and protein kinase C in endothelin-stimulated proliferation of rat type I astrocytes. *Glia* **15**, 119–130.
23. Shawl, A. I., Park, K. H., and Kim, U. H. (2009) Insulin receptor signaling for the proliferation of pancreatic beta-cells: Involvement of Ca^{2+} second messengers, IP3, NAADP and cADPR. *Islets* **1**, 216–223.
24. Cheng, D., Zhu, X., Barchiesi, F., Gillespie, D. G., Dubey, R. K., and Jackson, E. K. (2011) Receptor for activated protein kinase C1 regulates cell proliferation by modulating calcium signaling. *Hypertension* **58**, 689–695.
25. Forostyak, O., Forostyak, S., Kortus, S., Sykova, E., Verkhratsky, A., and Dayanithi, G. (2016) Physiology of Ca^{2+} signalling in stem cells of different origins and differentiation stages. *Cell Calcium* **59**, 57–66.
26. Szatkowski, C., Parys, J. B., Ouadid-Ahidouch, H., and Matifat, F. (2010) Inositol 1,4,5-trisphosphate-induced Ca^{2+} signalling is involved in estradiol-induced breast cancer epithelial cell growth. *Molecular Cancer* **9**, 156.
27. Cardenas, C., Muller, M., McNeal, A., Lovy, A., Jana, F., Bustos, G., Urra, F. et al. (2016) Selective vulnerability of cancer cells by inhibition of Ca transfer from endoplasmic reticulum to mitochondria. *Cell Reports* **15**, 219–220.
28. Rasmussen, C. D. and Means, A. R. (1989) Calmodulin is required for cell-cycle progression during G1 and mitosis. *EMBO Journal* **8**, 73–82.
29. Hidaka, H., Sasaki, Y., Tanaka, T., Endo, T., Ohno, S., Fujii, Y., and Nagata, T. (1981) N-(6-aminohexyl)-5-chloro-1-naphthalenesulfonamide, a calmodulin antagonist, inhibits cell proliferation. *Proceedings of the National Academy of Sciences of the United States of America* **78**, 4354–4357.
30. Kahl, C. R. and Means, A. (2003) Regulation of cell cycle progression by calcium/calmodulin-dependent pathways. *Endocrine Reviews* **24**, 719–736.
31. Skelding, K. A., Rostas, J. A., and Verrills, N. M. (2011) Controlling the cell cycle: The role of calcium/calmodulin-stimulated protein kinases I and II. *Cell Cycle* **10**, 631–639.
32. Courjaret, R. and Machaca, K. (2012) STIM and Orai in cellular proliferation and division. *Frontiers in Bioscience* **4**, 331–341.

33. Pinto, M. C., Kihara, A. H., Goulart, V. A., Tonelli, F. M., Gomes, K. N., Ulrich, H., and Resende, R. R. (2015) Calcium signaling and cell proliferation. *Cellular Signalling* **27**, 2139–2149.
34. Shaw, P. J. and Feske, S. (2012) Regulation of lymphocyte function by ORAI and STIM proteins in infection and autoimmunity. *The Journal of Physiology* **590**, 4157–4167.
35. Dziegielewska, B., Gray, L. S., and Dziegielewski, J. (2014) T-type calcium channels blockers as new tools in cancer therapies. *Pflügers Archiv* **466**, 801–810.
36. Somasundaram, A., Shum, A. K., McBride, H. J., Kessler, J. A., Feske, S., Miller, R. J., and Prakriya, M. (2014) Store-operated CRAC channels regulate gene expression and proliferation in neural progenitor cells. *The Journal of Neuroscience* **34**, 9107–9123.
37. Li, M., Chen, C., Zhou, Z., Xu, S., and Yu, Z. (2012) A TRPC1-mediated increase in store-operated Ca^{2+} entry is required for the proliferation of adult hippocampal neural progenitor cells. *Cell Calcium* **51**, 486–496.
38. Hu, F., Pan, L., Zhang, K., Xing, F., Wang, X., Lee, I., Zhang, X., and Xu, J. (2014) Elevation of extracellular Ca^{2+} induces store-operated calcium entry via calcium-sensing receptors: A pathway contributes to the proliferation of osteoblasts. *PLoS One* **9**, e107217.
39. Madsen, C. P., Klausen, T. K., Fabian, A., Hansen, B. J., Pedersen, S. F., and Hoffmann, E. K. (2012) On the role of TRPC1 in control of Ca^{2+} influx, cell volume, and cell cycle. *American Journal of Physiology. Cell Physiology* **303**, C625–C634.
40. Kito, H., Yamamura, H., Suzuki, Y., Yamamura, H., Ohya, S., Asai, K., and Imaizumi, Y. (2015) Regulation of store-operated Ca^{2+} entry activity by cell cycle dependent up-regulation of Orai2 in brain capillary endothelial cells. *Biochemical and Biophysical Research Communications* **459**, 457–462.
41. Lodola, F., Laforenza, U., Bonetti, E., Lim, D., Dragoni, S., Bottino, C., Ong, H. L. et al. (2012) Store-operated Ca^{2+} entry is remodelled and controls *in vitro* angiogenesis in endothelial progenitor cells isolated from tumoral patients. *PLoS One* **7**, e42541.
42. Smith, L. D. (1989) The induction of oocyte maturation: Transmembrane signaling events and regulation of the cell cycle. *Development* **107**, 685–699.
43. Miyazaki, S. (1995) Calcium signalling during mammalian fertilization. *Ciba Foundation Symposium* **188**, 235–251.
44. Antoine, A. F., Faure, J. E., Cordeiro, S., Dumas, C., Rougier, M., and Feijo, J. A. (2000) A calcium influx is triggered and propagates in the zygote as a wavefront during *in vitro* fertilization of flowering plants. *Proceedings of the National Academy of Sciences of the United States of America* **97**, 10643–10648.
45. Miyazaki, S., Shirakawa, H., Nakada, K., and Honda, Y. (1993) Essential role of the inositol 1,4,5-trisphosphate receptor/Ca^{2+} release channel in Ca^{2+} waves and Ca^{2+} oscillations at fertilization of mammalian eggs. *Developmental Biology* **158**, 62–78.
46. Runft, L. L., Jaffe, L. A., and Mehlmann, L. (2002) Egg activation at fertilization: Where it all begins. *Developmental Biology* **245**, 237–254.
47. Stricker, S. A. (1999) Comparative biology of calcium signaling during fertilization and egg activation in animals. *Developmental Biology* **211**, 157–176.
48. Busa, W. B., Ferguson, J. E., Joseph, S. K., Williamson, J. R., and Nuccitelli, R. (1985) Activation of frog (*Xenopus laevis*) eggs by inositol trisphosphate. I. Characterization of Ca^{2+} release from intracellular stores. *Journal of Cell Biology* **101**, 677–682.
49. Fontanilla, R. A. and Nuccitelli, R. (1998) Characterization of the sperm-induced calcium wave in Xenopus eggs using confocal microscopy. *Biophysical Journal* **75**, 2079–2087.
50. Kline, D. and Nuccitelli, R. (1985) The wave of activation current in the Xenopus egg. *Developmental Biology* **111**, 471–487.

51. Nuccitelli, R., Yim, D. L., and Smart, T. (1993) The sperm-induced Ca^{2+} wave following fertilization of the Xenopus egg requires the production of Ins(1, 4, 5)P3. *Developmental Biology* **158**, 200–212.
52. Lechleiter, J. D. and Clapham, D. E. (1992) Molecular mechanisms of intracellular calcium excitability in *X. laevis* oocytes. *Cell* **69**, 283–294.
53. Callamaras, N., Marchant, J. S., Sun, X. P., and Parker, I. (1998) Activation and co-ordination of InsP3-mediated elementary Ca^{2+} events during global Ca^{2+} signals in Xenopus oocytes. *The Journal of Physiology* **509**(Pt 1), 81–91.
54. Machaca, K. (2004) Increased sensitivity and clustering of elementary Ca^{2+} release events during oocyte maturation. *Developmental Biology* **275**, 170–182.
55. Parys, J. B. and Bezprozvanny, I. (1995) The inositol trisphosphate receptor of Xenopus oocytes. *Cell Calcium* **18**, 353–363.
56. Igusa, Y. and Miyazaki, S. (1983) Effects of altered extracellular and intracellular calcium concentration on hyperpolarizing responses of the hamster egg. *Journal of Physiology* **340**, 611–632.
57. Kline, D. and Kline, J. T. (1992) Thapsigargin activates a calcium influx pathway in the unfertilized mouse egg and suppresses repetitive calcium transients in the fertilized egg. *Journal of Biological Chemistry* **267**, 17624–17630.
58. Mohri, T., Shirakawa, H., Oda, S., Sato, M. S., Mikoshiba, K., and Miyazaki, S. (2001) Analysis of Mn^{2+}/Ca^{2+} influx and release during Ca^{2+} oscillations in mouse eggs injected with sperm extract. *Cell Calcium* **29**, 311–325.
59. Sato, K., Iwao, Y., Fujimura, T., Tamaki, I., Ogawa, K., Iwasaki, T., Tokmakov, A. A., Hatano, O., and Fukami, Y. (1999) Evidence for the involvement of a Src-related tyrosine kinase in Xenopus egg activation. *Developmental Biology* **209**, 308–320.
60. Sato, K., Tokmakov, A. A., Iwasaki, T., and Fukami, Y. (2000) Tyrosine kinase-dependent activation of phospholipase Cgamma is required for calcium transient in Xenopus egg fertilization. *Developmental Biology* **224**, 453–469.
61. Tokmakov, A. A., Sato, K. I., Iwasaki, T., and Fukami, Y. (2002) Src kinase induces calcium release in Xenopus egg extracts via PLCgamma and IP3-dependent mechanism. *Cell Calcium* **32**, 11–20.
62. Saunders, C. M., Larman, M. G., Parrington, J., Cox, L. J., Royse, J., Blayney, L. M., Swann, K., and Lai, F. A. (2002) PLCzeta: A sperm-specific trigger of Ca^{2+} oscillations in eggs and embryo development. *Development* **129**, 3533–3544.
63. Chiba, K., Kado, R. T., and Jaffe, L. A. (1990) Development of calcium release mechanisms during starfish oocyte maturation. *Developmental Biology* **140**, 300–306.
64. Mehlmann, L. and Kline, D. (1994) Regulation of intracellular calcium in the mouse egg: Calcium release in response to sperm or inositol trisphosphate is enhanced after meiotic maturation. *Biology of Reproduction* **51**, 1088–1098.
65. Fujiwara, T., Nakada, K., Shirakawa, H., and Miyazaki, S. (1993) Development of inositol trisphosphate-induced calcium release mechanism during maturation of hamster oocytes. *Developmental Biology* **156**, 69–79.
66. Jones, K. T., Carroll, J., and Whittingham, D. G. (1994) Ionomycin, thapsigargin, ryanodine, and sperm induced Ca^{2+} release increase during meiotic maturation of mouse oocytes. *Journal of Biological Chemistry* **270**, 6671–6677.
67. Terasaki, M., Runft, L. L., and Hand, A. R. (2001) Changes in organization of the endoplasmic reticulum during Xenopus oocyte maturation and activation. *Molecular Biology of the Cell* **12**, 1103–1116.
68. Parker, I., Choi, J., and Yao, Y. (1996) Elementary events of InsP 3-induced Ca^{2+} liberation in Xenopus oocytes: Hot spots, puffs and blips. *Cell Calcium* **20**, 105–121.
69. Parys, J. B., McPherson, S. M., Mathews, L., Campbell, K. P., and Longo, F. J. (1994) Presence of inositol 1,4,5-trisphosphate receptor, calreticulin, and calsequestrin in eggs of sea urchins and *Xenopus laevis*. *Developmental Biology* **161**, 466–476.

70. Kume, S., Yamamoto, A., Inoue, T., Muto, A., Okano, H., and Mikoshiba, K. (1997) Developmental expression of the inositol 1,4,5-trisphosphate receptor and structural changes in the endoplasmic reticulum during oogenesis and meiotic maturation of *Xenopus laevis*. *Developmental Biology* **182**, 228–239.
71. El Jouni, W., Jang, B., Haun, S., and Machaca, K. (2005) Calcium signaling differentiation during Xenopus oocyte maturation. *Developmental Biology* **288**, 514–525.
72. Sun, L., Yu, F., Ullah, A., Hubrack, S., Daalis, A., Jung, P., and Machaca, K. (2011) Endoplasmic reticulum remodeling tunes IP3-dependent Ca^{2+} release sensitivity. *PLoS One* **6**, e27928.
73. Malathi, K., Kohyama, S., Ho, M., Soghoian, D., Li, X., Silane, M., Berenstein, A., and Jayaraman, T. (2003) Inositol 1,4,5-trisphosphate receptor (type 1) phosphorylation and modulation by Cdc2. *Journal of Cellular Biochemistry* **90**, 1186–1196.
74. Li, X., Malathi, K., Krizanova, O., Ondrias, K., Sperber, K., Ablamunits, V., and Jayaraman, T. (2005) Cdc2/cyclin B1 interacts with and modulates inositol 1,4,5-trisphosphate receptor (type 1) functions. *Journal of Immunology* **175**, 6205–6210.
75. Sun, L., Haun, S., Jones, R. C., Edmondson, R. D., and Machaca, K. (2009) Kinase-dependent regulation of IP3-dependent Ca^{2+} release during oocyte maturation. *Journal of Biological Chemistry* **284**, 20184–20196.
76. Lee, B., Vermassen, E., Yoon, S. Y., Vanderheyden, V., Ito, J., Alfandari, D., De Smedt, H., Parys, J. B., and Fissore, R. A. (2006) Phosphorylation of IP3R1 and the regulation of $[Ca^{2+}]i$ responses at fertilization: A role for the MAP kinase pathway. *Development* **133**, 4355–4365.
77. Boulware, M. J. and Marchant, J. S. (2005) IP3 receptor activity is differentially regulated in endoplasmic reticulum subdomains during oocyte maturation. *Current Biology* **15**, 765–770.
78. Boulware, M. J. and Marchant, J. S. (2008) Nuclear pore disassembly from endoplasmic reticulum membranes promotes Ca^{2+} signalling competency. *The Journal of Physiology* **586**, 2873–2888.
79. Guerini, D., Coletto, L., and Carafoli, E. (2005) Exporting calcium from cells. *Cell Calcium* **38**, 281–289.
80. Strehler, E. E., Filoteo, A. G., Penniston, J. T., and Caride, A. J. (2007) Plasma-membrane Ca^{2+} pumps: Structural diversity as the basis for functional versatility. *Biochemical Society Transactions* **35**, 919–922.
81. Strehler, E. E. and Treiman, M. (2004) Calcium pumps of plasma membrane and cell interior. *Current Molecular Medicine* **4**, 323–335.
82. Machaca, K. and Haun, S. (2000) Store-operated calcium entry inactivates at the germinal vesicle breakdown stage of Xenopus meiosis. *Journal of Biological Chemistry* **275**, 38710–38715.
83. Machaca, K. and Haun, S. (2002) Induction of maturation-promoting factor during Xenopus oocyte maturation uncouples Ca^{2+} store depletion from store-operated Ca^{2+} entry. *Journal of Cell Biology* **156**, 75–85.
84. Hartzell, H. C. (1996) Activation of different Cl currents in Xenopus oocytes by Ca liberated from stores and by capacitative Ca influx. *Journal of General Physiology* **108**, 157–175.
85. Yao, Y. and Tsien, R. Y. (1997) Calcium current activated by depletion of calcium stores in Xenopus oocytes. *Journal of General Physiology* **109**, 703–715.
86. Busa, W. B. and Nuccitelli, R. (1985) An elevated free cytosolic Ca^{2+} wave follows fertilization in eggs of the frog, *Xenopus laevis*. *Journal of Cell Biology* **100**, 1325–1329.
87. Liu, J. and Maller, J. L. (2005) Calcium elevation at fertilization coordinates phosphorylation of XErp1/Emi2 by Plx1 and CaMK II to release metaphase arrest by cytostatic factor. *Current Biology* **15**, 1458–1468.
88. Mochida, S. and Hunt, T. (2007) Calcineurin is required to release Xenopus egg extracts from meiotic M phase. *Nature* **449**, 336–340.

89. Martin-Romero, F. J., Lopez-Guerrero, A. M., Alvarez, I. S., and Pozo-Guisado, E. (2012) Role of store-operated calcium entry during meiotic progression and fertilization of mammalian oocytes. *International Review of Cell and Molecular Biology* **295**, 291–328.
90. Gomez-Fernandez, C., Lopez-Guerrero, A. M., Pozo-Guisado, E., Alvarez, I. S., and Martin-Romero, F. J. (2012) Calcium signaling in mouse oocyte maturation: The roles of STIM1, ORAI1 and SOCE. *Molecular Human Reproduction* **18**, 194–203.
91. Wang, C., Lee, K., Gajdocsi, E., Papp, A. B., and Machaty, Z. (2012) Orai1 mediates store-operated Ca^{2+} entry during fertilization in mammalian oocytes. *Developmental Biology* **365**, 414–423.
92. Lee, B., Palermo, G., and Machaca, K. (2013) Downregulation of store-operated Ca^{2+} entry during mammalian meiosis is required for the egg-to-embryo transition. *Journal of Cell Science* **126**, 1672–1681.
93. Cheon, B., Lee, H. C., Wakai, T., and Fissore, R. A. (2013) Ca^{2+} influx and the store-operated Ca^{2+} entry pathway undergo regulation during mouse oocyte maturation. *Molecular Biology of the Cell* **24**, 1396–1410.
94. Volpi, M. and Berlin, R. D. (1988) Intracellular elevations of free calcium induced by activation of histamine H1 receptors in interphase and mitotic HeLa cells: Hormone signal transduction is altered during mitosis. *Journal of Cell Biology* **107**, 2533–2539.
95. Preston, S. F., Sha'afi, R. I., and Berlin, R. D. (1991) Regulation of Ca^{2+} influx during mitosis: Ca^{2+} influx and depletion of intracellular Ca^{2+} stores are coupled in interphase but not mitosis. *Cell Regulation* **2**, 915–925.
96. Smyth, J. T., Beg, A. M., Wu, S., Putney, J. W., Jr., and Rusan, N. M. (2012) Phosphoregulation of STIM1 leads to exclusion of the endoplasmic reticulum from the mitotic spindle. *Current Biology* **22**, 1487–1493.
97. Smyth, J. T., Petranka, J. G., Boyles, R. R., DeHaven, W. I., Fukushima, M., Johnson, K. L., Williams, J. G., and Putney, J. W., Jr. (2009) Phosphorylation of STIM1 underlies suppression of store-operated calcium entry during mitosis. *Nature Cell Biology* **11**, 1465–1472.
98. Tani, D., Monteilh-Zoller, M. K., Fleig, A., and Penner, R. (2007) Cell cycle-dependent regulation of store-operated I_{CRAC} and Mg^{2+}-nucleotide-regulated MagNuM (TRPM7) currents. *Cell Calcium* **41**, 249–260.
99. Chen, Y. W., Chen, Y. F., Chen, Y. T., Chiu, W. T., and Shen, M. R. (2016) The STIM1-Orai1 pathway of store-operated Ca^{2+} entry controls the checkpoint in cell cycle G1/S transition. *Scientific Reports* **6**, 22142.
100. Yang, S., Zhang, J. J., and Huang, X. Y. (2009) Orai1 and STIM1 are critical for breast tumor cell migration and metastasis. *Cancer Cell* **15**, 124–134.
101. Russa, A. D., Ishikita, N., Masu, K., Akutsu, H., Saino, T., and Satoh, Y. (2008) Microtubule remodeling mediates the inhibition of store-operated calcium entry (SOCE) during mitosis in COS-7 cells. *Archives of Histology and Cytology* **71**, 249–263.
102. Yu, F., Sun, L., Courjaret, R., and Machaca, K. (2011) Role of the STIM1 C-terminal domain in STIM1 clustering. *Journal of Biological Chemistry* **286**, 8375–8384.
103. Yu, F., Sun, L., and Machaca, K. (2009) Orai1 internalization and STIM1 clustering inhibition modulate SOCE inactivation during meiosis. *Proceedings of the National Academy of Sciences of the United States of America* **106**, 17401–17406.
104. Yu, F., Sun, L., and Machaca, K. (2010) Constitutive recycling of the store-operated Ca^{2+} channel Orai1 and its internalization during meiosis. *Journal of Cell Biology* **191**, 523–535.
105. Pozo-Guisado, E. and Martin-Romero, F. J. (2013) The regulation of STIM1 by phosphorylation. *Communicative & Integrative Biology* **6**, e26283.
106. Westendorf, J. M., Rao, P. N., and Gerace, L. (1994) Cloning of cDNAs for M-phase phosphoproteins recognized by the MPM2 monoclonal antibody and determination of

the phosphorylated epitope. *Proceedings of the National Academy of Sciences of the United States of America* **91**, 714–718.

107. Grigoriev, I., Gouveia, S. M., van, D. V., Demmers, J., Smyth, J. T., Honnappa, S., Splinter, D. et al. (2008) STIM1 is a MT-plus-end-tracking protein involved in remodeling of the ER. *Current Biology* **18**, 177–182.
108. Monteith, G. R., Davis, F. M., and Roberts-Thomson, S. J. (2012) Calcium channels and pumps in cancer: Changes and consequences. *The Journal of Biological Chemistry* **287**, 31666–31673.
109. Prevarskaya, N., Ouadid-Ahidouch, H., Skryma, R., and Shuba, Y. (2014) Remodelling of Ca^{2+} transport in cancer: How it contributes to cancer hallmarks? *Philosophical Transactions of the Royal Society of London. Series B, Biological Sciences* **369**, 20130097.
110. Xie, J., Pan, H., Yao, J., Zhou, Y., and Han, W. (2016) SOCE and cancer: Recent progress and new perspectives. *International Journal of Cancer* **138**, 2067–2077.
111. Wong, H. S. and Chang, W. C. (2015) Correlation of clinical features and genetic profiles of stromal interaction molecule 1 (STIM1) in colorectal cancers. *Oncotarget* **6**, 42169–42182.
112. Stanisz, H., Vultur, A., Herlyn, M., Roesch, A., and Bogeski, I. (2016) The role of Orai/STIM calcium channels in melanocytes and melanoma. *Journal of Physiology* **594**, 2825–2835.
113. Vashisht, A., Trebak, M., and Motiani, R. K. (2015) STIM and Orai proteins as novel targets for cancer therapy. A review in the theme: Cell and molecular processes in cancer metastasis. *American Journal of Physiology. Cell Physiology* **309**, C457–C469.
114. Vanden Abeele, F., Shuba, Y., Roudbaraki, M., Lemonnier, L., Vanoverberghe, K., Mariot, P., Skryma, R., and Prevarskaya, N. (2003) Store-operated Ca^{2+} channels in prostate cancer epithelial cells: Function, regulation, and role in carcinogenesis. *Cell Calcium* **33**, 357–373.
115. Fiorio Pla, A., Kondratska, K., and Prevarskaya, N. (2016) STIM and ORAI proteins: Crucial roles in hallmarks of cancer. *American Journal of Physiology. Cell Physiology* **310**, C509–C519.
116. Bergmeier, W., Weidinger, C., Zee, I., and Feske, S. (2013) Emerging roles of store-operated Ca^{2+} entry through STIM and ORAI proteins in immunity, hemostasis and cancer. *Channels (Austin, TX)* **7**, 379–391.
117. Sabbioni, S., Barbanti-Brodano, G., Croce, C. M., and Negrini, M. (1997) GOK: A gene at 11p15 involved in rhabdomyosarcoma and rhabdoid tumor development. *Cancer Research* **57**, 4493–4497.
118. Feng, M., Grice, D. M., Faddy, H. M., Nguyen, N., Leitch, S., Wang, Y., Muend, S. et al. (2010) Store-independent activation of Orai1 by SPCA2 in mammary tumors. *Cell* **143**, 84–98.
119. Flourakis, M., Lehen'kyi, V., Beck, B., Raphael, M., Vandenberghe, M., Abeele, F. V., Roudbaraki, M. et al. (2010) Orai1 contributes to the establishment of an apoptosis-resistant phenotype in prostate cancer cells. *Cell Death Disease* **1**, e75.
120. Kondratska, K., Kondratskyi, A., Yassine, M., Lemonnier, L., Lepage, G., Morabito, A., Skryma, R., and Prevarskaya, N. (2014) Orai1 and STIM1 mediate SOCE and contribute to apoptotic resistance of pancreatic adenocarcinoma. *Biochimica et Biophysica Acta* **1843**, 2263–2269.
121. Hou, M. F., Kuo, H. C., Li, J. H., Wang, Y. S., Chang, C. C., Chen, K. C., Chen, W. C., Chiu, C. C., Yang, S., and Chang, W. C. (2011) Orai1/CRACM1 overexpression suppresses cell proliferation via attenuation of the store-operated calcium influx-mediated signalling pathway in A549 lung cancer cells. *Biochimica et Biophysica Acta* **1810**, 1278–1284.
122. Cheng, H., Wang, S., and Feng, R. (2016) STIM1 plays an important role in TGF-beta-induced suppression of breast cancer cell proliferation. *Oncotarget* **7**, 16866–16878.

123. Xu, Y., Zhang, S., Niu, H., Ye, Y., Hu, F., Chen, S., Li, X. et al. (2015) STIM1 accelerates cell senescence in a remodeled microenvironment but enhances the epithelial-to-mesenchymal transition in prostate cancer. *Scientific Reports* **5**, 11754.
124. Vanden Abeele, F., Skryma, R., Shuba, Y., Van Coppenolle, F., Slomianny, C., Roudbaraki, M., Mauroy, B., Wuytack, F., and Prevarskaya, N. (2002) Bcl-2-dependent modulation of Ca^{2+} homeostasis and store-operated channels in prostate cancer cells. *Cancer Cell* **1**, 169–179.
125. Skryma, R., Mariot, P., Bourhis, X. L., Coppenolle, F. V., Shuba, Y., Abeele, F. V., Legrand, G., Humez, S., Boilly, B., and Prevarskaya, N. (2000) Store depletion and store-operated Ca^{2+} current in human prostate cancer LNCaP cells: Involvement in apoptosis. *The Journal of Physiology* **527**(Pt 1), 71–83.
126. Dubois, C., Vanden Abeele, F., Lehen'kyi, V., Gkika, D., Guarmit, B., Lepage, G., Slomianny, C. et al. (2014) Remodeling of channel-forming ORAI proteins determines an oncogenic switch in prostate cancer. *Cancer Cell* **26**, 19–32.
127. Stanisz, H., Stark, A., Kilch, T., Schwarz, E. C., Muller, C. S., Peinelt, C., Hoth, M., Niemeyer, B. A., Vogt, T., and Bogeski, I. (2012) ORAI1 Ca^{2+} channels control endothelin-1-induced mitogenesis and melanogenesis in primary human melanocytes. *The Journal of Investigative Dermatology* **132**, 1443–1451.
128. Umemura, M., Baljinnyam, E., Feske, S., De Lorenzo, M. S., Xie, L. H., Feng, X., Oda, K. et al. (2014) Store-operated Ca^{2+} entry (SOCE) regulates melanoma proliferation and cell migration. *PLoS One* **9**, e89292.
129. Feldman, B., Fedida-Metula, S., Nita, J., Sekler, I., and Fishman, D. (2010) Coupling of mitochondria to store-operated Ca^{2+}-signaling sustains constitutive activation of protein kinase B/Akt and augments survival of malignant melanoma cells. *Cell Calcium* **47**, 525–537.
130. Hooper, R., Zhang, X., Webster, M., Go, C., Kedra, J., Marchbank, K., Gill, D. L., Weeraratna, A. T., Trebak, M., and Soboloff, J. (2015) Novel protein kinase C-mediated control of Orai1 function in invasive melanoma. *Molecular and Cellular Biology* **35**, 2790–2798.
131. Minn, A. J., Kang, Y., Serganova, I., Gupta, G. P., Giri, D. D., Doubrovin, M., Ponomarev, V., Gerald, W. L., Blasberg, R., and Massague, J. (2005) Distinct organ-specific metastatic potential of individual breast cancer cells and primary tumors. *The Journal of Clinical Investigation* **115**, 44–55.
132. Ponomarev, V., Doubrovin, M., Serganova, I., Vider, J., Shavrin, A., Beresten, T., Ivanova, A. et al. (2004) A novel triple-modality reporter gene for whole-body fluorescent, bioluminescent, and nuclear noninvasive imaging. *European Journal of Nuclear Medicine and Molecular Imaging* **31**, 740–751.
133. Kim, J. H., Lkhagvadorj, S., Lee, M. R., Hwang, K. H., Chung, H. C., Jung, J. H., Cha, S. K., and Eom, M. (2014) Orai1 and STIM1 are critical for cell migration and proliferation of clear cell renal cell carcinoma. *Biochemical and Biophysical Research Communications* **448**, 76–82.
134. Wang, J. Y., Sun, J., Huang, M. Y., Wang, Y. S., Hou, M. F., Sun, Y., He, H. et al. (2015) STIM1 overexpression promotes colorectal cancer progression, cell motility and COX-2 expression. *Oncogene* **34**, 4358–4367.
135. Motiani, R. K., Abdullaev, I. F., and Trebak, M. (2010) A novel native store-operated calcium channel encoded by Orai3: Selective requirement of Orai3 versus Orai1 in estrogen receptor-positive versus estrogen receptor-negative breast cancer cells. *Journal of Biological Chemistry* **285**, 19173–19183.
136. Zhang, W., Zhang, X., Gonzalez-Cobos, J. C., Stolwijk, J. A., Matrougui, K., and Trebak, M. (2015) Leukotriene-C4 synthase, a critical enzyme in the activation of store-independent Orai1/Orai3 channels, is required for neointimal hyperplasia. *The Journal of Biological Chemistry* **290**, 5015–5027.

137. Motiani, R. K., Zhang, X., Harmon, K. E., Keller, R. S., Matrougui, K., Bennett, J. A., and Trebak, M. (2013) Orai3 is an estrogen receptor alpha-regulated Ca^{2+} channel that promotes tumorigenesis. *The FASEB Journal* **27**, 63–75.
138. Motiani, R. K., Hyzinski-Garcia, M. C., Zhang, X., Henkel, M. M., Abdullaev, I. F., Kuo, Y. H., Matrougui, K., Mongin, A. A., and Trebak, M. (2013) STIM1 and Orai1 mediate CRAC channel activity and are essential for human glioblastoma invasion. *Pflügers Archiv* **465**, 1249–1260.
139. Tsai, F. C., Seki, A., Yang, H. W., Hayer, A., Carrasco, S., Malmersjo, S., and Meyer, T. (2014) A polarized Ca^{2+}, diacylglycerol and STIM1 signalling system regulates directed cell migration. *Nature Cell Biology* **16**, 133–144.
140. Oh-hora, M., Yamashita, M., Hogan, P. G., Sharma, S., Lamperti, E., Chung, W., Prakriya, M., Feske, S., and Rao, A. (2008) Dual functions for the endoplasmic reticulum calcium sensors STIM1 and STIM2 in T cell activation and tolerance. *Nature Immunology* **9**, 432–443.
141. Weidinger, C., Shaw, P. J., and Feske, S. (2013) STIM1 and STIM2-mediated Ca^{2+} influx regulates antitumour immunity by CD8(+) T cells. *EMBO Molecular Medicine* **5**, 1311–1321.
142. Courjaret, R., Dib, M., and Machaca, K. (2017) Store-operated Ca^{2+} entry in oocytes modulate the dynamics of IP_3-dependent Ca^{2+} release from oscillatory to tonic. *Journal of Cellular Physiology* **232**, 1095–1103.

13 TRPV5 and TRPV6 Calcium-Selective Channels

Ji-Bin Peng, Yoshiro Suzuki, Gergely Gyimesi, and Matthias A. Hediger

CONTENTS

13.1 INTRODUCTION

The epithelial calcium (Ca^{2+}) channels TRPV5 and TRPV6 are members of the transient receptor potential (TRP) channel family TRPV ("V" for vanilloid) subgroup. TRPV5 and TRPV6 play major roles in the maintenance of blood Ca^{2+} levels in higher organisms. Both channels exhibit similarities in many ways, as they share a high level (75%) of amino acid identity, comparable functional properties, and similar mechanisms of regulation. Also, they were discovered using similar cloning strategies [1,2]. Yet, their physiological contributions toward maintaining a systemic calcium balance are distinct. In addition, the following three key features distinguish TRPV5 and TRPV6 from other members of the TRP superfamily of cation channels: (1) high selectivity for Ca^{2+} over other cations, (2) apical membrane localization in Ca^{2+}-transporting epithelial tissues, and (3) responsiveness to 1,25-dihydroxyvitamin D_3 (1,25[OH]$_2D_3$) [3,4]. These features make TRPV5 and TRPV6 ideally suited to facilitate intestinal absorption and renal reabsorption of Ca^{2+}, serving as apical Ca^{2+} entry channels in transepithelial Ca^{2+} transport [5,6]. A major difference between the properties of TRPV5 and TRPV6 lies in their tissue distribution: TRPV5 is predominantly expressed in the distal convoluted tubules (DCT) and connecting tubules (CNT) of the kidney, with limited expression in extrarenal tissues [1,7]. In contrast, TRPV6 exhibits a broader expression pattern, showing prominent expression in the intestine with additional expression in the kidney [8–10], placenta, epididymis, exocrine tissues (i.e., pancreas, prostate, salivary gland, sweat gland), and a few other tissues [11–13]. Thus, while TRPV5 plays a key role in determining the level of urinary Ca^{2+} excretion, the physiological roles of TRPV6 are not limited to intestinal Ca^{2+} absorption. Much progress has recently been made in understanding the roles of TRPV5 and TRPV6 channels in the kidney [14], intestine [15], placenta [16], and epididymis [17]. However, their roles in other organs have as yet not been fully investigated.

In this chapter, we review the current status of our knowledge of the physiological and pathological roles of TRPV5 and TRPV6 and discuss a variety of techniques that have led to a deeper understanding of these channels. We review the identification strategies of TRPV5 and TRPV6 in searches for Ca^{2+} absorption channels, as well as specific techniques used to reveal their key features. These include radiotracer Ca^{2+} uptake and electrophysiology procedures, structure–function studies, methods to identify regulatory interacting partners, genetically engineered animals, strategies to study the role of TRPV6 in cancers, procedures for the development of small-molecule modulators of TRPV6 and TRPV5, the evaluation of variations/mutations in humans, and 3D structural determination. For additional information about TRPV5 and TRPV6, we would like to refer the interested reader to other comprehensive review articles [3–6,18].

13.2 Ca^{2+} TRANSPORT ACROSS EPITHELIA

Epithelia form barriers to separate cells from the internal and external environment, allowing specific exchange of nutrients and other substances across epithelial cells. Among the many electrolytes, Ca^{2+} is an important intracellular messenger and a major component of the mineral phase of bones and teeth. Thus, at the cellular level, the cytosolic Ca^{2+} concentration is kept low (around 100 nM at rest), whereas its concentration in the extracellular fluids (ECF), such as blood, is maintained at a much higher level and within a relatively narrow limit (around 1 mM) [19]. The bone, which contains roughly 99% of the Ca^{2+} in the human body, is subject to constant Ca^{2+} exchange with the ECF. Any reduction in the extracellular Ca^{2+} concentration ($[Ca^{2+}]_o$) in the ECF is being monitored by the G-protein-coupled Ca^{2+}-sensing receptor (CaSR) in the parathyroid gland, triggering secretion of the parathyroid hormone (PTH) [19]. The PTH increases bone resorption and renal reabsorption of Ca^{2+} and stimulates the hydroxylation of 25-hydroxyvitamin D_3 [$25(OH)_2D_3$] in the proximal tubule of the kidney to form $1,25(OH)_2D_3$ [20]. An important role of this activated form of vitamin D is the stimulation of intestinal Ca^{2+} absorption via the nuclear vitamin D receptor (VDR) [21]. Several organs and hormones work in concert to maintain stable levels of Ca^{2+} in the ECF. The intestinal and renal handling of Ca^{2+} is important as the intestine determines how much Ca^{2+} enters the body and the kidney determines how much Ca^{2+} is removed from the body.

Intestinal absorption and renal reabsorption of Ca^{2+} are similar processes, involving Ca^{2+} transport across the epithelia from the luminal side to the blood side [22,23]. In general, Ca^{2+} ions can cross the epithelia through the paracellular route, as well as the transcellular route (Figure 13.1). When $[Ca^{2+}]_o$ is higher than that of the plasma, luminal Ca^{2+} predominantly enters the intestine via the paracellular route through tight junctions between the epithelial cells (Figure 13.1a). However, this route does not operate in the absence of favorable transepithelial $[Ca^{2+}]$ gradients. Thus, when luminal $[Ca^{2+}]_o$ is lower than that in the plasma, Ca^{2+} will need to be actively absorbed across the epithelia via the transcellular route. For this to occur, Ca^{2+} first enters the cells across the apical membranes passively through TRPV6. This is followed by binding of Ca^{2+} to calbindin-D_{9K} and transfer to the basolateral membrane. Basolateral exit is then mediated by the plasma membrane Ca^{2+} ATPase (PMCA), that is, by primary active transport, at the expense of stored ATP [22].

Similar transport mechanisms also take place in the kidney during Ca^{2+} reabsorption (Figure 13.1b). Based on our current knowledge from animal studies, up to 95% of filtered Ca^{2+} is being reabsorbed mainly via the paracellular route in renal proximal tubules and loops of Henle, while about 5%–10% of Ca^{2+} reabsorption is delayed till the distal tubules, where reabsorption occurs via the transcellular route and in a regulated manner in order to fine-tune the whole-body calcium homeostasis. At this point, it is worthy to note that TRPV6 in the human kidney likely plays an additional role in transcellular Ca^{2+} transport in specific parts of the tubular system (see Section 13.8 for further details). Similar to the intestine, the transcellular route in the kidney involves passive Ca^{2+} entry across the apical membranes through TRPV5,

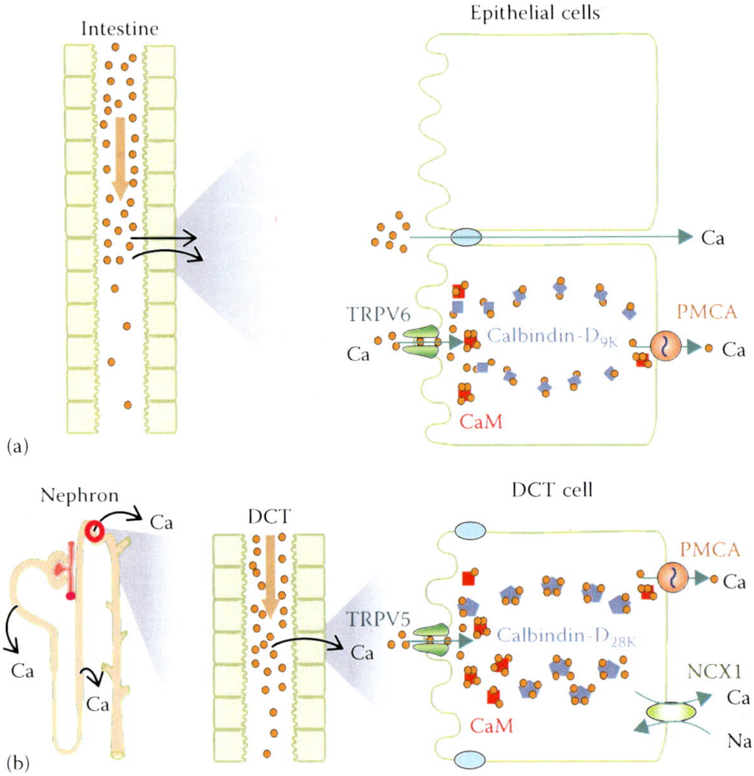

FIGURE 13.1 Mechanisms of Ca^{2+} transport in the intestine and kidney. (a) Intestinal Ca^{2+} absorption. At high luminal Ca^{2+} levels (mM range), Ca^{2+} will be absorbed predominantly via the paracellular route. At low luminal levels, Ca^{2+} is being absorbed via the transcellular route as follows: (1) luminal Ca^{2+} is first absorbed via TRPV6 expressed in the brush border membrane; (2) transport of Ca^{2+} to basolateral membrane is facilitated by calbindin-D_{9K} (calbindin also acts as an intracellular buffer, preventing second messenger signaling during the Ca^{2+} absorptive process); and (3) basolateral exit occurs mostly via the plasma Ca^{2+}-pump PMCA1b. During periods of hypocalcemia, 1,25-vitamin D_3 [1,25$(OH)_2D_3$] upregulates the expression of TRPV6 and calbindin via the nuclear vitamin D receptor (VDR/NR1I1) to stimulate intestinal Ca^{2+} absorption. (b) Renal reabsorption of Ca^{2+}. About 70% of filtered Ca^{2+} is being reabsorbed via the paracellular route in the proximal tubular and about 25% in the loop of Henle. About 5%–10% of filtered Ca^{2+} is reabsorbed in the distal tubules (distal convoluted tubule, DCT, and connecting tubule, CT) via the transcellular route as follows: (1) luminal absorption occurs through TRPV5; (2) transport of Ca^{2+} to the basolateral membrane is facilitated by calbindin-D_{28K}; and (3) basolateral exit occurs via the PMCA pump or the Na^+/Ca^+ exchanger NCX1 (SLC8A1). PTH, which is released following hypocalcemia, increases renal Ca^{2+} reabsorption by stimulation of the expression of TRPV5 and calbindin. The regulation of TRPV5 in the distal tubule plays an important role in the fine-tuning of the whole-body Ca^{2+} balance. Note that, in the human kidney, TRPV6 may play additional Ca^{2+}-absorptive roles in the tubular system.

transfer to the basolateral membrane after binding to calbindin-D_{28K}, and basolateral exit through the PMCA (primary active transport) or the Na^+/Ca^{2+} exchanger NCX1 (at the expense of the inwardly directed Na^+ gradient) [22,23].

Passive Ca^{2+} transport across the apical membrane via TRPV5 or TRPV6 is favored by the inwardly directed electrochemical gradient for Ca^{2+}, which is warranted by the low cytosolic Ca^{2+} levels of approximately 100 nM. As already noted, once Ca^{2+} ions have entered the epithelial cell, they bind to calbindins during Ca^{2+} uptake (Figure 13.1). Two types of calbindins with different molecular weight and number of Ca^{2+}-binding EF hands, calbindin-D_{9K} and calbindin-D_{28K}, are expressed in the Ca^{2+}-transporting epithelial cells in the mammalian intestine and kidney, respectively [24]. They belong to a large family of high-affinity Ca^{2+}-binding proteins ($K_d = 10^{-8} - 10^{-6}$ M) featuring EF-hand structural motifs [25]. Calbindins and likely other Ca^{2+}-binding proteins maintain the free cytosolic Ca^{2+} concentrations below toxic levels and help regulate TRPV5 and TRPV6 through their feedback mechanisms [26]. However, if apical Ca^{2+} entry exceeds the capacity of calbindins, Ca^{2+} will bind to calmodulin (CaM), resulting in a Ca^{2+} feedback inhibition of TRPV5 or TRPV6 channels (see Figure 13.1 on the right; also see Section 13.4.4).

In addition to the intestine and kidney, maternal–fetal transport of Ca^{2+} involves an active mechanism, as the serum [Ca^{2+}] is higher in fetuses than in mothers [27]. In humans, Ca^{2+} is needed for the rapid growth of fetal skeleton, especially in the third trimester. The syncytiotrophoblast layer is a specialized epithelium responsible for the exchange of Ca^{2+} along with other nutrients and gases between mother and fetus [28]. Ca^{2+} transport in human syncytiotrophoblasts has been the subject of many studies [29]. These Ca^{2+} channels are insensitive to L-type voltage-gated Ca^{2+} channel modulators but sensitive to Mg^{2+} and ruthenium red [30], resembling the properties of TRPV6.

Active Ca^{2+} transport also takes place in specialized organs to create local environments where [Ca^{2+}] needs to be controlled. For example, the luminal $[Ca^{2+}]_o$ is lower in the distal portion of the epididymis [31]. Recent studies revealed that this lowered [Ca^{2+}] is maintained by TRPV6-mediated Ca^{2+} transport into the epithelial cells of the epididymis and that this is important for preserving optimal motility of sperm and optimal male fertility [17]. Therefore, TRPV6 of the epididymis and the progesterone-regulated CatSper channel of sperm [32] both cooperate to ensure maximal motility of the sperm and fertility (see Section 13.5.4 for further details).

Another example is the observation that [Ca^{2+}] in the vestibular endolymph is only one-fourth of that of the perilymph (~1 mM). Indeed, $[Ca^{2+}]_o$ is important in regulating the hair-bundle stiffness and gating-spring integrity in hair cells of the inner ear [33]. Active transcellular Ca^{2+} transport by TRPV5 and TRPV6 is present in the epithelial cells of the inner ear and maintains the low [Ca^{2+}] of the vestibular endolymph [34,35].

The glandular tissue is also formed by epithelial cells. The release of content from secretory granules, which are loaded with Ca^{2+}, is accompanied by the loss of Ca^{2+} [36]. Given that exocrine glands secrete large amounts of fluid, a Ca^{2+} reuptake mechanism may be useful to supply Ca^{2+} released in the apical pole of exocrine cells back into new secretory vesicles.

Lastly, carcinomas are cancer cells that originate from epithelial cells, and altered Ca^{2+} homeostasis is a characteristic feature of carcinomas. This may be related to cancer cell proliferation, survival, and metastasis formation [37–39]. The development of specific small-molecule inhibitors targeting Ca^{2+} channels overexpressed in carcinomas may offer novel strategies for cancer treatment [40,41] (see Section 13.7).

13.3 IDENTIFICATION OF TRPV5 AND TRPV6 BY EXPRESSION CLONING

TRPV5 and TRPV6 were identified in 1999 by Bindels and colleagues and by our group from rabbit kidney cells and rat duodenum, respectively [1,2]. Both groups used expression cloning with radiotracer $^{45}Ca^{2+}$ uptake assays, screening intestinal or renal cDNA libraries for the ability of clones to induce Ca^{2+} uptake in complementary RNA (cRNA)–injected *Xenopus laevis* oocytes. This approach has proven very useful in the identification of novel transporters, channels, and receptors [42]. Before the epithelial Ca^{2+} channels were identified, calbindins were known to facilitate the diffusion of Ca^{2+} from the apical to basolateral membranes, followed by cellular exit via Na^+/Ca^{2+} exchangers and/or PMCAs. However, what mediated Ca^{2+} entry across the plasma membrane remained elusive for a long time [22,24]. To increase our chances for successfully cloning TRPV6, we fed rats a Ca^{2+}-deficient diet for 2 weeks to boost the expression of Ca^{2+}-reabsorptive gene products. Also, we isolated mRNA from the duodenum and cecum of the rat intestine, the sites where Ca^{2+} absorption primarily occurs. A twofold increase in radiotracer uptake was observed in oocytes injected with mRNA samples from the duodenum and cecum compared to control uninjected oocytes. We then fractionated the duodenal mRNA by preparative gel electrophoresis and tested the ability of the mRNA samples of different size fractions to stimulate Ca^{2+} uptake. An increase in Ca^{2+} uptake was observed in a 3 kb fraction. This fraction was used to make a cDNA library, and the library was divided into smaller pools of about 300 clones. Plasmid DNA was isolated from each pool and reverse-transcribed into capped cRNA. These cRNAs were then tested for their ability to stimulate Ca^{2+} uptake in injected oocytes. Once a positive pool was identified, it was divided into even smaller pools, and the earlier mentioned process was repeated in an iterative way, until a single positive clone was identified [2]. A similar approach was used to identify TRPV5 from rabbit kidney [1]. Because the distal tubule represents a very short portion of the nephron, cells from CT and cortical collecting ducts were isolated from rabbit kidney using a monoclonal antibody against these tubule segments to obtain enough mRNA for expression cloning of TRPV5 [1].

The identified rabbit TRPV5 and rat TRPV6 cDNAs encode proteins of 730 and 727 amino acid residues, respectively. *In vivo* evidence exists that the translation of TRPV6 protein may start from an upstream non-AUG codon. This does not change the TRPV6 function but may provide an additional scaffold for channel assembly [43]. TRPV5 was originally named epithelial Ca^{2+} channel (ECaC) [1] and TRPV6 Ca^{2+} transport protein subtype 1 (CaT1) [2]. The two proteins share 74% amino acid sequence identity. At the time of cloning, only two cloned transport proteins shared amino acid sequence similarity to TRPV5 and TRPV6: the capsaicin receptor TRPV1, a heat-activated cation channel that contributes to the sensations of pain and temperature [44], and OSM-9, a *Caenorhabditis elegans* protein involved in olfaction, mechanosensation, and olfactory adaptation [45]. The mammalian TRPV family now includes six members, whereby TRPV5 and TRPV6 share approximately 40%–45% amino acid sequence identity with the other four members [8]. While TRPV5 and TRPV6 represent specialized epithelial Ca^{2+} transport proteins, the other four members are nonselective cation channels that serve as sensors and can be activated by physical cues such as heat (TRPV1–TRPV3) or tonicity (TRPV4) [46].

13.4 Ca^{2+} TRANSPORT PROPERTIES UNCOVERED BY VARIOUS APPROACHES

13.4.1 FUNCTIONAL EXPRESSION IN *XENOPUS* OOCYTES

As both channels were identified by expression cloning using $^{45}Ca^{2+}$ uptake assays in cRNA-injected oocytes to measure channel activity, the same assays were also utilized to uncover the basic transport properties of TRPV5 and TRPV6 [1,2]. In these assays, oocytes were first injected with cRNAs encoding the Ca^{2+} channel of interest, and 2–3 days later, when the channel protein expression reached stable levels, $^{45}Ca^{2+}$ uptake assays and/or voltage-clamp experiments were performed. The uptake assay involved incubation of groups of oocytes (typically 7–9 oocytes/group) in an uptake solution containing $^{45}Ca^{2+}$. Uptake measurements were performed during the initial uptake period, when the influx of Ca^{2+} was still in the linear range, typically within 30 min. The transport process was then terminated by removing the uptake solution and washing the oocytes with ice-cold solution 3–6 times, in order to remove $^{45}Ca^{2+}$ nonspecifically absorbed to the oocyte surface. Individual oocytes were then lysed in 10% SDS solution, and the amount of $^{45}Ca^{2+}$ in the oocytes was determined using scintillation counting. The amount of Ca^{2+} that entered the oocyte was calculated based on the level radiation in counts per minute (CPM) in oocyte lysates, following normalization against results from standard solutions with known $^{45}Ca^{2+}$ values.

13.4.2 CHARACTERIZATION OF TRPV5 AND TRPV6 AT THE MACROSCOPIC LEVEL

The $^{45}Ca^{2+}$ uptake experiments revealed the properties of TRPV5 and TRPV6 at the macroscopic level. TRPV5- and TRPV6-mediated Ca^{2+} uptake exhibited saturation kinetics, with K_m values for Ca^{2+} in the submillimolar range [1,2,11,47]. The K_m values for Ca^{2+} were 0.2 mM for rabbit and 0.66 for rat TRPV5 [1,47]. Those for TRPV6 were 0.44 mM for rat and 0.25 mM for human TRPV6 [2,11]. $[Ca^{2+}]_o$ that is normally present in the kidney DCT and in the intestinal lumen varies from submillimolar to millimolar ranges. Thus, given the K_m values of renal TRPV5 and intestinal TRPV6 for Ca^{2+}, these channels would be well suited to facilitate optimal Ca^{2+} uptake across the apical membranes.

TRPV5 and TRPV6 were found to be sensitive to protons and metal ions such as La^{3+}, Gd^{3+}, Cu^{2+}, Pb^{2+}, and Cd^{2+} [1,2]. The pH sensitivity of TRPV5 and TRPV6 may contribute to the elevated serum Ca^{2+} level in milk–alkali syndrome and the increased Ca^{2+} excretion in metabolic acidosis.

Subsequent voltage-clamp techniques allowed further measurements of the charge fluxes through these channels, controlling the voltage across the plasma membrane of *Xenopus* oocytes expressing TRPV5 or TRPV6. The *Xenopus* oocyte with a diameter of ~1.2 mm is very suitable for a two-microelectrode voltage clamp, in order to determine transporter and channel characteristics, provided that the transport is accompanied by net electric charge movement. Two glass Ag–AgCl microelectrodes filled with 3 M KCl are inserted into an oocyte and connected to a commanding and recording system, allowing the voltage across the oocyte membrane to be controlled and the current flux to be measured accurately [48]. Employing this method, we demonstrated that Ca^{2+}, Ba^{2+}, and Sr^{2+}, but not Mg^{2+}, elicit measureable currents in oocytes expressing TRPV6 [2]. This approach also revealed that Na^+ can permeate through TRPV6 in the absence of extracellular Ca^{2+} [2]. As a follow-up to this observation, Na^+ currents were measured under divalent-free conditions with the patch–clamp technique. By measuring $^{45}Ca^{2+}$ influx in parallel, using voltage-clamped TRPV6-expressing oocytes, the influx of each Ca^{2+} ion was shown to be accompanied by the movement of two positive charges across the membrane [2]. This confirmed the hypothesis that Ca^{2+}-induced currents are solely due to Ca^{2+} passing through the channel, and not by movement of other ions.

13.4.3 Characterization of TRPV5 and TRPV6 by Patch Clamping

The properties of TRPV5 and TRPV6 at a microscopic level were investigated in great detail using the patch–clamp technique, which is discussed in more detail elsewhere in this book (see Chapter 1). Briefly, Na^+ currents of TRPV5 were detected when divalent cations (e.g., Ca^{2+} and Mg^{2+}) were removed from the extracellular solution [49]. Unitary Na^+ conductance of TRPV5 and TRPV6 was subsequently determined [50,51]. Both TRPV5 and TRPV6 turned out to be highly selective for Ca^{2+}, with Ca^{2+} to Na^+ permeability ratios (P_{Ca}:P_{Na}) of over 100 for both channels [50,51]. Together with site-directed mutagenesis studies, the key residue for Ca^{2+} permeation and Mg^{2+} blockade was identified as aspartate 542 in TRPV5 [52,53] (shown in Figure 13.2 as the site for Ca^{2+} permeation).

13.4.4 Feedback Control Mechanisms

Both TRPV5 and TRPV6 exhibit Ca^{2+}-dependent inactivation as feedback control mechanisms. However, differences exist between the two channels in this regard. The inactivation mechanism involves the intrinsic channel structure, interaction with Ca^{2+}-binding proteins, and $PI(4,5)P_2$ phospholipid. While the $PI(4,5)P_2$-mediated inactivation mechanism is similar between TRPV5 and TRPV6, there are differences in the intrinsic inactivation kinetics and interactions with Ca^{2+}-binding proteins. TRPV6 has a fast inactivation mechanism, compared to TRPV5 [54]. The first intracellular loop between transmembrane domains 2 and 3 is the molecular

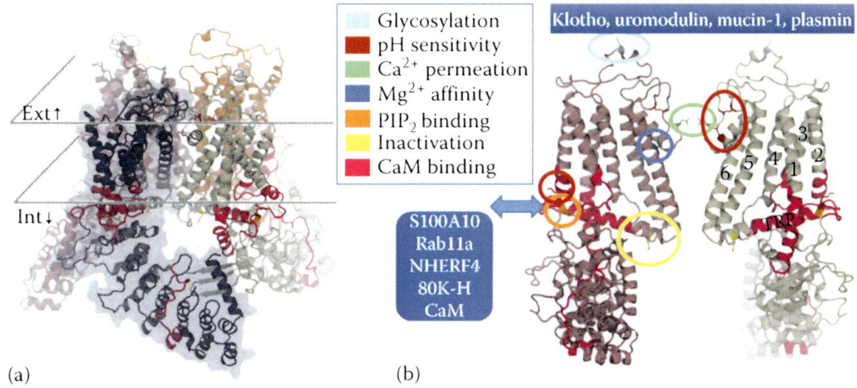

FIGURE 13.2 Model of the human TRPV5 channel based on the rat TRPV6 structure (PDB ID: 5IWK). (a) Tilted sideways view with one monomer of the tetrameric unit shaded in blue. Approximate position of the membrane is indicated. (b) Cutaway view of two oppositely positioned monomers. Transmembrane helices are numbered, and the conserved TRP helix is indicated. Residues or regions of functional importance are highlighted in stick representation and colored. Regions reported to bind CaM are colored in pink. Regions interacting with selected regulatory proteins are also indicated.

determinant of this mechanism [54]. The slower phase of Ca^{2+}-dependent inactivation involves Ca^{2+}/CaM binding (see Figure 13.2b). Flockerzi and colleagues showed that a CaM-binding site is present in the C-terminal of TRPV6 [55]. CaM binding mediates a slow phase of Ca^{2+}-dependent inactivation of TRPV6 [55–57]. The interaction between Ca^{2+}/CaM and TRPV6 increases dynamically as the intracellular Ca^{2+} level rises [56]. The CaM-binding site is not conserved between the two channels, and several CaM-binding sites were identified in TRPV5 [58,59]. Thus, the kinetics of the Ca^{2+}/CaM interaction with TRPV5 is likely more complex. In addition, the vitamin D–regulated Ca^{2+}-binding protein calbindin-D_{28K} binds Ca^{2+} that enters through TRPV5 [26]. This may further prevent inactivation of TRPV5. TRPV5 activity is also regulated by a Ca^{2+} sensor, 80K-H [60] (Figure 13.2b).

The common mechanism for TRPV5 and TRPV6 inactivation involves PI(4,5)P_2. Using the patch–clamp technique, Huang and colleagues demonstrated the regulation of TRPV5 by PI(4,5)P_2 [61]. They found that PI(4,5)P_2 activates TRPV5 and reduces Mg^{2+} blockade [61]. PI(4,5)P_2 is necessary for maintaining TRPV5 and TRPV6 activity and appears to prevent Ca^{2+}-dependent inactivation [61–64]. As the levels of PI(4,5)P_2 decrease, the activities of TRPV5 and TRPV6 also decline [61,63]. Ca^{2+} influx as a result of channel activation stimulates phospholipase C (PLC), which in turn hydrolyzes PI(4,5)P_2, leading to the reduction of channel activity [63]. As for the molecular site of PI(4,5)P_2 binding, Rohacs and colleagues found that arginine 599 in mouse or rat TRPV5 (corresponding to R606 in rabbit and human TRPV5) in the "TRP domain" is likely the PI(4,5)P_2-binding site [65] (Figure 13.2b). TRPV6 lacks a positively charged arginine residue between transmembrane domains 4 and 5 in TRPV1 that is important for high-affinity binding of PI(4,5)P_2 in TRPV1 [66,67]. Introducing this arginine residue into TRPV6 produced high-affinity binding of PI(4,5)P_2 in TRPV6 and abolished

Ca^{2+}-dependent inactivation of the channel [67]. Thus, the requirement of $PI(4,5)P_2$ for function and low-affinity binding of $PI(4,5)P_2$ are key features of TRPV6 (and likely of TRPV5 as well). Ca^{2+}-dependent hydrolysis of $PI(4,5)P_2$ by PLC underlies the Ca^{2+}-dependent inactivation mechanism of these channels.

13.4.5 Ca^{2+} Imaging

Ca^{2+} imaging, an approach to study Ca^{2+} signaling, has been utilized for screening small-molecule inhibitors of TRPV6 [68]. Binding of Ca^{2+} ions to a fluorescent dye results in either an increase in fluorescence intensity or changes in emission/excitation wavelengths. When cells are loaded with a Ca^{2+}-indicator dye, a rise in intracellular Ca^{2+}, either by release from internal Ca^{2+} stores or by influx through Ca^{2+}-permeant channels, can be captured with a camera. Bolanz and colleagues utilized Ca^{2+} imaging to study the inhibitory effect of tamoxifen on TRPV6 [69]. The same approach can also be utilized for high-throughput screening of inhibitors for TRPV6.

Employing the voltage-clamp approach, we found that Cd^{2+}, which is a potent blocker of TRPV5-mediated $^{45}Ca^{2+}$ uptake, is actually permeant through the TRPV5 and TRPV6 channels and generates current just like Ca^{2+} [47]. This forms the foundation for screening small-molecule modulators for TRPV5/6 using the FLIPR Tetra high-throughput cellular screening system (Molecular Devices), using the FLIPR Calcium 5 Assay Kit.

13.4.6 Additional Experimental Considerations

As shown in Figure 13.3, all of the aforementioned approaches have been utilized in evaluating channel function and regulation of TRPV5 and TRPV6. While radiotracer $^{45}Ca^{2+}$ uptake is a sensitive approach that directly measures Ca^{2+} influx, it cannot reveal detailed kinetics of channel activity. Two-electrode voltage-clamp studies reveal charge movement and voltage-dependence of membrane transport processes. However, its application is mostly limited to *Xenopus* oocytes. The patch–clamp approach offers detailed channel activity measurements, and it is also well suited to uncover the mechanisms of channel regulation. However, most patch–clamp experiments of TRPV5 and TRPV6 were done in divalent cation-free bathing solutions. Such experimental conditions might miss channel regulatory mechanisms that are Ca^{2+} dependent, for example. Furthermore, single-channel activities using Ca^{2+} as charge carrier have not been demonstrated for TRPV5 or TRPV6 so far. Nevertheless, data from this array of techniques are complementary and provide a well-documented understanding of the biophysical properties and regulatory aspects of TRPV5 and TRPV6.

13.4.7 Crystal Structure of Rat TRPV6

The recently reported crystal structure of rat TRPV6 at 3.25 Å resolution [70] (discussed in detail in Chapter 14) allows us to view the functionally important regions of the protein in the context of its 3D structure. In Figure 13.2, we show a structural model

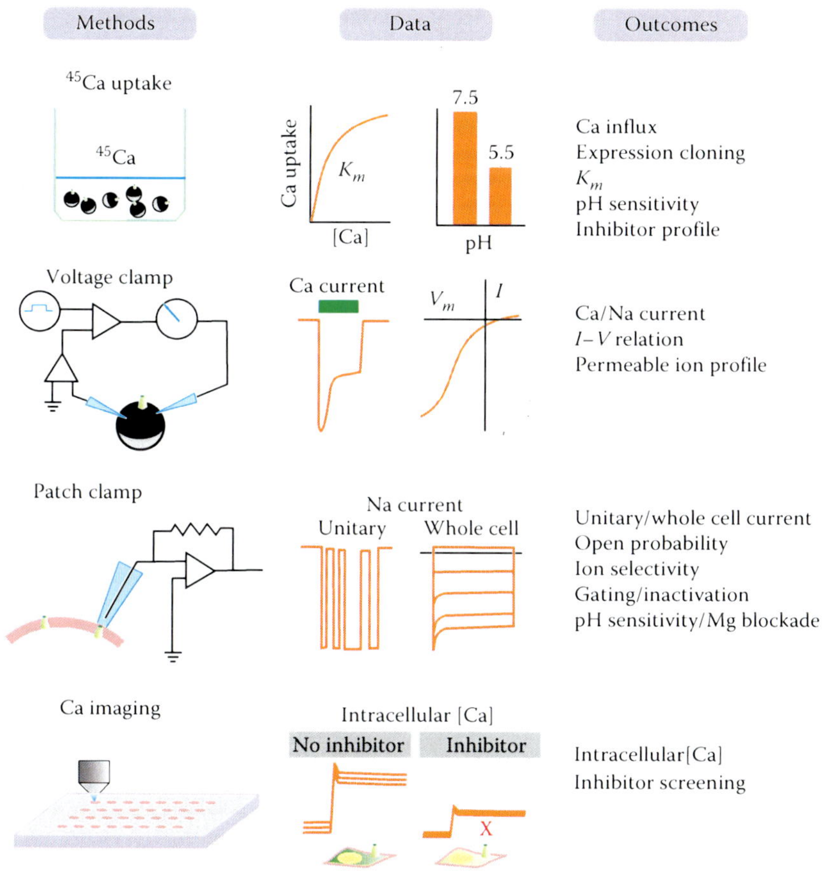

FIGURE 13.3 Major approaches to assess TRPV5 and TRPV6 function.

of the tetrameric human TRPV5 channel that we built using the Rosetta modeling suite [71–73], with the rat TRPV6 structure as a template. Residues shaping pore selectivity, gating, and inactivation are highlighted, as well as interaction sites with CaM. While some residues cluster near the pore, the TRP helix also appears to be an important structural integrator of various processes modulating TRPV5 channel function.

13.5 PHYSIOLOGICAL ROLES REVEALED USING GENETICALLY ENGINEERED ANIMAL MODELS

Genetically engineered mouse models were instrumental for exploring the physiological roles of TRPV5 and TRPV6. These include gene knockout (KO) models for TRPV5 [14] and TRPV6 [15], chemically induced gene mutations in TRPV5 [74], transgenic overexpression of TRPV6 in the small intestine [75], and a knock-in mouse model of TRPV6 [17]. These animal models and the findings from these models are summarized in the following sections.

13.5.1 TRPV5 and TRPV6 in Ca^{2+} Absorption and Reabsorption

Bindels and colleagues developed the TRPV5 gene KO mice with conditional potential. However, since TRPV5 expression is fairly restricted to the DCT and CNT, tissue-specific KO was unnecessary [14]. Our group developed a global TRPV6 KO mice [14]. The common features of the TRPV5 and TRPV6 KO mice include elevated $1,25(OH)_2D_3$ serum levels, increased urinary Ca^{2+} excretion, and a certain degree of bone deficiency [14,15]. Urinary Ca^{2+} excretion increased sixfold in TRPV5 KO mice [14]. Intestinal Ca^{2+} absorption decreased in KO mice [15] but increased in TRPV5 KO mice as duodenal TRPV6 and calbindin-D_{9K} mRNAs are upregulated due to elevated serum $1,25(OH)_2D_3$ levels [14]. Interestingly, the serum Ca^{2+} levels of the TRPV5 and TRPV6 KO mice were largely normal, although the elevated PTH levels in TRPV6 KO mice [15] and in older TRPV5 KO mice [76] are an indication of a subtle reduction in serum Ca^{2+} levels that may have occurred in these mice. The elevated levels of $1,25(OH)_2D_3$ are in further agreement with the overall deficiency in Ca^{2+} homeostasis in these mice. The increased urinary Ca^{2+} excretion in TRPV6 KO mice suggests that TRPV6 also plays a role in Ca^{2+} transport in the kidney. More recently, a chemical-induced S682P mutation of TRPV5 caused autosomal dominant hypercalciuria much like what was observed in the TRPV5 KO mice [74]. This confirms a key role of TRPV5 in regulating urinary Ca^{2+} excretion.

13.5.2 TRPV6 as a Central Component in Vitamin D–Regulated Active Ca^{2+} Absorption

While TRPV6 is considered the key apical channel for active Ca^{2+} absorption, a certain level of active Ca^{2+} absorption also occurs in the absence of TRPV6 [77]. Furthermore, the stimulating effect of $1,25(OH)_2D_3$ on intestinal Ca^{2+} absorption was still partially present in TRPV6 KO mice [78]. This raised the question as to what extent vitamin D–regulated transcellular Ca^{2+} absorption depends on TRPV6 expression. To address this question, Fleet and colleagues developed a transgenic mouse model in which intestinal epithelium–specific expression of human TRPV6 is driven by the villin gene promoter [75]. This approach directly demonstrated that TRPV6 functions as an apical Ca^{2+} transporter and intestinal absorption was elevated 300% in mice with intestine-specific transgenic expression of TRPV6 (TRPV6TG). In this study, TRPV6TG and VDRKO mice were also crossed to generate TRPV6TG–VDRKO mice.

The TRPV6TG mice on chow diet (0.72% calcium) exhibited elevated serum and urinary Ca^{2+} levels, poor growth, and soft tissue calcification. Lowering the calcium level in the diet to 0.25% normalized the serum Ca^{2+} levels. Femur bone mineral density of TRPV6TG mice was increased by 28% compared to normal mice. The $1,25(OH)_2D_3$ level was reduced by 56% likely due to the 98% lower CYP27B1 mRNA in the kidney of TRPV6TG mice. The reduced CYP27B1 mRNA level is likely a result of reduced PTH levels (46% lower). The phenotype of TRPV6TG mice is largely opposite to that of the TRPV6 KO mice, confirming the importance of TRPV6 in the intestine facilitating vitamin D–regulated transcellular Ca^{2+} absorption.

The key regulator of intestinal Ca^{2+} absorption is $1,25(OH)_2D_3$. TRPV6 expression is strongly regulated by vitamin D as the TRPV6 gene promoter has multiple vitamin-responsive elements [79]. In VDRKO mice, duodenal TRPV6 mRNA is reduced down to 5%–10% compared to wild-type mice [21,80]. VDRKO mice fed a diet with slightly reduced calcium content (0.5% calcium, standard commercial chow diet contained 0.72% calcium) for 10 weeks were hypocalcemic, with drastically elevated serum $1,25(OH)_2D_3$ and PTH levels, reduced bone mass, and decreased intestinal Ca^{2+} absorption [75].

In TRPV6TG–VDRKO mice, the serum Ca^{2+} remained elevated when kept on 0.5% calcium diet, serum $1,25(OH)_2D_3$ and PTH levels were very close to that of the wild type, bone mass was close to that in the TRPV6TG mice, and intestinal Ca^{2+} reabsorption remained elevated [75]. Thus, overexpression of TRPV6 in the intestine is sufficient to overcome the deficiency caused by the reduction of intestinal Ca^{2+} absorption in the absence of VDR. The results of this study also show that, once TRPV6 is restored, other components relevant to transcellular Ca^{2+} transport such as levels of calbindin-D_{9K}, $1,25(OH)_2D_3$, and PTH are normalized in the absence of VDR [75]. Taken together, given the results from studies of TRPV6 KO, VDRKO, and TRPV6TG mice (Figure 13.4), TRPV6 is likely a key element in vitamin D–regulated Ca^{2+} transport. Given the importance of Ca^{2+} absorption, it is also unlikely that TRPV6 is the only Ca^{2+} channel for intestinal Ca^{2+} absorption. The presence of a vitamin D–regulated active Ca^{2+} transport mechanism in the intestine in mice lacking TRPV6 might be due to physiological adaptation, that is, upregulation of alternative,

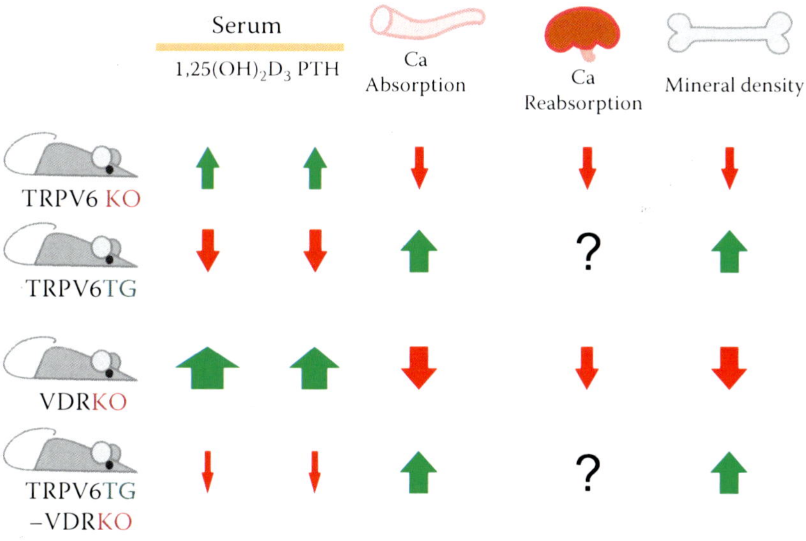

FIGURE 13.4 Ca^{2+} homeostasis in mice with genetically engineered TRPV6 and VDR. Shown are changes in Ca^{2+} homeostasis-related phenotypes including serum levels of $1,25(OH)_2D_3$ and PTH, intestinal Ca^{2+} absorption, renal Ca^{2+} reabsorption, and bone mineral density in TRPV6 KO, TRPV6TG, VDRKO, and TRPV6 TG–VDRKO mice. Arrow thickness indicates relative amplitude of change.

compensatory Ca^{2+} transport pathways. Also epigenetic and genetic alterations in the TRPV6 KO mice cannot be ruled out. The abovementioned studies suggest that an alternative transcellular Ca^{2+} entry pathway might exist in the intestine that is regulated by $1,25(OH)_2D_3$. Further studies will be required to identify this putative additional transport mechanism in the intestinal brush border membrane.

13.5.3 Role of TRPV6 in Maternal–Fetal Ca^{2+} Transport

The placenta transports Ca^{2+} from mother to fetus in order to maintain fetal bone mineralization [29,81]. It has been reported that blood $[Ca^{2+}]$ is higher in fetus compared to the mother during late pregnancy, when fetal bones are mineralized, suggesting that transcellular active Ca^{2+} transport is needed for fetal bone mineralization. Indeed, some Ca^{2+} deficiency symptoms in fetuses and neonates are thought to be derived from insufficient maternal–fetal Ca^{2+} transport in the placenta. However, the molecular identity of this process, as well as its mechanism of regulation, has, until recently, not been fully clarified. It was found that both TRPV5 and TRPV6 are expressed in the placenta [11,13,30,82,83]. However, TRPV6 mRNA levels are ~1000 times higher than those of TRPV5 in human placenta [8]. Specifically, TRPV6 is expressed in trophoblasts and syncytiotrophoblasts of the human placenta [13]. TRPV6 in human syncytiotrophoblasts is associated with cyclophilin B, a member of the immunophilin family [84]. The physiological meaning of this association is unclear, but cyclophilin B increased TRPV6 activity *in vitro*, suggesting that the two proteins may act together in placental Ca^{2+} transport.

To elucidate the role of TRPV6 in placental Ca^{2+} transport, we examined the maternal–fetal Ca^{2+} transport pathway in TRPV6 KO mice [16]. The maternal–fetal $^{45}Ca^{2+}$ transport activity was decreased by 40% in TRPV6 KO fetuses. TRPV6 KO mice exhibited a dramatic decrease in blood Ca^{2+} levels and ash weight in fetuses at 18 days post-fertilization (immediately before birth) [16]. These observations indicate that the fetal Ca^{2+} deficiency was caused by decreased maternal–fetal Ca^{2+} transport in the absence of TRPV6 [16].

A zebrafish loss-of-function TRPV6 mutation called "matt-und-schlapp" generated a 68% reduction of total body calcium content, with bone mineral defects, triggered by a marked reduction in Ca^{2+} uptake by the yolk sac and the gills [85]. It was found that the mutant possesses R304Stop before the transmembrane domain 1 (S1) of the TRPV6 gene. These results suggest that the role of TRPV6 in the Ca^{2+} transport in extraembryonic tissues involved in fetal bone formation is conserved among vertebrates.

13.5.4 Role of TRPV6 in Male Fertility

The reduced fertility in TRPV6 KO mice was another interesting observation [15]. The underlying mechanism was unclear until Weissgerber and colleagues observed that male fertility in mice depends on TRPV6-mediated Ca^{2+} absorption aimed at reducing the luminal Ca^{2+} levels in the epididymis [17]. This group generated a $Trpv6^{D541A/D541A}$ knock-in mutant mouse, whereby D541 corresponds to D542 in TRPV5, a key residue for Ca^{2+} permeation in TRPV6 [52]. The D541A mutation renders TRPV6 incapable of transporting Ca^{2+}. While female *Trpv6*$^{D541A/D541A}$ mice were normal in fertility, male homozygous knock-in mice rarely impregnated females despite normal copulatory behavior. There was more than 50% reduction in the number of caudal sperm measured by the "swim out" method in *Trpv6*$^{D541A/D541A}$ mice. Most capacitated caudal sperm was immotile or had impaired motility. The ability of capacitated caudal sperm from knock-in mice to fertilize oocytes *in vitro* was reduced to 14.4%. A closer examination revealed that TRPV6 is expressed in the apical membrane of epididymal epithelia. $[Ca^{2+}]_o$ determined by retrograde perfusion in the epididymal duct fluid of *Trpv6*$^{D541A/D541A}$ mice was 10 times that of wild-type littermates, and this matched the reduction of $^{45}Ca^{2+}$ uptake in the epididymis of *Trpv6*$^{D541A/D541A}$ mice. Thus, experimental results from this study indicate that blocking TRPV6 function in the epididymis increased $[Ca^{2+}]_o$ in the epididymal fluid, which is essential for sperm motility. This study shows that controlled $[Ca^{2+}]$ in the epididymal fluid by TRPV6 is critical in producing vital and healthy sperm cells. Raised $[Ca^{2+}]$ in the epididymal fluid in *Trpv6*$^{D541A/D541A}$ mice likely initiates premature motility of sperm cells and exhausts the energy that is needed for the motility after they exit the epididymis. The authors also examined this question in mice lacking Trpv6 and reached essentially the same conclusion [86].

Recent patch–clamp studies confirmed that the constitutive Ca^{2+} currents resemble that of TRPV6 in rat cauda epididymal principal cells [87]. The Ca^{2+}-activated chloride channel transmembrane member 16A (TMEM16A) is coexpressed with TRPV6 in the apical membrane of principal cells, and functional coupling between TMEM16A and TRPV6 likely plays a role in Ca^{2+} absorption in the epididymis [87].

13.6 PROTEINS THAT REGULATE TRPV5 AND TRPV6

A number of proteins have been identified that interact and/or regulate TRPV5 and TRPV6 (see examples in Figure 13.2b). These proteins were identified because (1) they interact with TRPV5 or TRPV6 as detected by yeast two-hybrid screening, (2) they are expressed in the same tissues or cells, or (3) they are relevant to Ca^{2+} absorption or reabsorption in a physiological or pathological context [3]. Most of these proteins were identified for TRPV5. However, since TRPV6 and TRPV5 are structurally and functionally similar, they generally have analogous roles in regulating TRPV6, provided that they are adequately coexpressed.

13.6.1 Proteins That Interact with TRPV5 and TRPV6

Several TRPV5-interacting proteins were identified by Bindels and colleagues using the yeast two-hybrid approach, with the C-terminal region of TRPV5 as bait. A group of TRPV5-interacting protein, such as S100A10 [88], Rab11a [89], and NHERF2 [90–92], regulates the trafficking and plasma membrane abundance of TRPV5. Another group of TRPV5-interacting protein, including calbindin-D_{28K} [26], CaM [58,59,93], 80H-K [60], BSPRY [94], and FKBP52 [95], also regulates TRPV5 activity. An interesting observation is that TRPV5 interacts with most of its binding partners through the region lying between 598 and 608 (human TRPV5 numbering) within the conserved "TRP domain" (Figure 13.2b). These include S100A10 [88],

Rab11a [89], 80K-H [60], NHERF4 [96], and one CaM [59]. It is worth mentioning that the TRP domain that follows the sixth transmembrane domain also contains a PI(4,5)P_2-binding site [65] and an intracellular pH sensor [97,98]. Histidine 711 (H712 in rabbit TRPV5), a critical residue involved in the constitutive internalization of TRPV5 [99], is also located in close proximity to the sites that interact with other proteins. It is very likely that TRPV5 function, membrane stability, and/or trafficking are affected when different binding partners interact with this channel. It is unclear whether all of the binding partners identified by the yeast two-hybrid approach or other *in vitro* approaches are physiologically relevant. Also, it is unlikely that all TRPV5-binding proteins regulate TRPV5 simultaneously. Some of the interactions may occur more often and some may occur at specific developmental or physiological stages. As already noted earlier, given that most of the interaction sites are conserved in TRPV5 and TRPV6 channels, it is possible that some of these interactions with TRPV5 are also relevant for TRPV6 in cells that express both TRPV6 and the interacting proteins.

13.6.2 Proteins That Are Physiologically Relevant to TRPV5 and TRPV6

Some proteins that regulate TRPV5 and TRPV6 were identified on the basis of physiological or pathological relevance. Because TRPV5 is fairly highly expressed in the DCT and CNT, proteins that are expressed in the DCT and associated with calcium homeostasis are possible candidates that regulate the TRPV5 channel.

An example is the study of the effect of the antiaging hormone Klotho on TRPV5 by Chang and colleagues [100]. These investigators showed that Klotho acts on the *N*-glycan moiety of the first extracellular loop of TRPV5 (see Figure 13.2b), thereby stimulating TRPV5-mediated Ca^{2+} uptake and stabilizing TRPV5 in the plasma membrane. This discovery revealed novel mechanisms of ion channel regulation [101–103]. This effect is dependent of the *N*-glycosylation state of TRPV5, since the *N*-glycosylation mutant $TRPV5^{N358Q}$ was not activated by Klotho. Subsequently, tissue transglutaminase was found to inhibit TRPV5 through a reduction of the pore diameter in a glycan-dependent manner [104]. Both, Klotho and tissue transglutaminase, act on TRPV5 through its *N*-glycan moiety, but in the opposite direction.

The novel regulation of TRPV5 by Klotho inspired a search for more relevant proteins that regulate TRPV5 from the extracellular side. As TRPV5 in the distal tubule faces proteins in the tubular fluid, some tubular proteins for which there was already evidence that they affect Ca^{2+} reabsorption were tested for their effects on TRPV5 function. Plasmin, a serine protease that is present in the nephrotic syndrome, suppresses TRPV5 through activating the protease-activated receptor-1 (PAR-1) [105]. The activation of the PAR-1/PLC/PKC pathway leads to altered CaM binding to TRPV5 and subsequent reduction in the pore size and open probability [105].

Wolf and colleagues found that the urinary protein uromodulin (Tamm–Horsfall glycoprotein/THP) and mucin-1 act extracellularly and decrease caveolin-mediated endocytosis of TRPV5 and thus enhance overall channel activity [106,107]. Because mutations of the uromodulin gene are associated with kidney stone disease, and mucin-1 protein is decreased in patients with hypercalciuric nephrolithiasis, it is tempting to speculate that uromodulin and mucin-1 prevent kidney stone formation in part through enhancing TRPV5 activity.

Other candidate proteins that may regulate TRPV5 function were identified because their gene mutations result in diseases associated with hypercalciuria. Studies of regulation of TRPV5 by the serine/threonine protein kinase WNK4, for example, were initiated because hypercalciuria was observed in pseudohypoaldosteronism type II (PHAII, also known as familial hyperkalemia and hypertension or Gordon's syndrome) in patients with the $WNK4^{Q565E}$ mutation [108]. Jiang and colleagues found that WNK4 increases the activity of the TRPV5 channel by increasing its forward trafficking to the plasma membrane via the secretory pathway in *Xenopus* oocytes [109,110]. This regulation is enhanced and stabilized by the sodium–hydrogen exchange regulatory cofactor NHERF2 [92]. However, diseases causing mutations in WNK4 did not abolish the regulation of TRPV5 by WNK4. WNK3, another member of the WNK family, enhances both TRPV5 and TRPV6 [111].

Genetic mutations of the inositol polyphosphate-5-phosphatase gene OCRL are associated with Dent's disease and Lowe syndrome (oculocerebrorenal syndrome), which often manifest hypercalciuria [112–114]. OCRL appears to inhibit TRPV6 by reducing the $PI(4,5)P_2$ levels and affecting its trafficking [115]. OCRL mutations associated with hypercalciuria impaired the inhibitory effect of OCRL on TRPV6. This suggests that hyperabsorption of Ca^{2+} in the intestine due to elevated TRPV6 activity may contribute to hypercalciuria in patients with these OCRL mutations. The OCRL-mediated regulation of TRPV6 likely overrides other conditions (e.g., proteinuria) in Dent-2 disease in the kidney.

Additionally, as a follow-up to the hypercalciuria phenotype observed in mice lacking the serine protease called tissue kallikrein, it was found that tissue kallikrein stimulates Ca^{2+} reabsorption via PKC-dependent plasma membrane accumulation of TRPV5 and that its genetic defect delays retrieval of TRPV5 from the plasma membrane [116]. Furthermore, the activity of TRPV5 may be regulated by the CaSR. The activation of CaSR increases TRPV5-mediated currents and elevates $[Ca^{2+}]_i$ in cells coexpressing TRPV5 and CaSR [117], providing an interesting type of regulation. Finally, TRPV5 and TRPV6 may be degraded in part through the ubiquitin E3 ligase Nedd4-2 and Nedd4, as demonstrated *in vitro* [118].

Thus, a variety of proteins have been found to regulate TRPV5 and TRPV6 using various approaches. It is unclear at this stage whether all of them are relevant *in vivo*. Therefore, the physiological significance of these regulatory pathways is awaiting further evaluation.

13.7 POTENTIAL USE OF TRPV6 IN THERAPY AND DEVELOPMENT OF CHEMICAL MODULATORS

13.7.1 Evaluation of TRPV6 as a Therapeutic Target

Ca^{2+} regulates a wide array of physiological functions including cell differentiation, proliferation, muscle contraction, neurotransmission, and fertilization [37–39,119,120]. Several Ca^{2+} channels, pumps, Ca^{2+}-binding proteins, and regulatory proteins work in concert to fine-tune cellular Ca^{2+} homeostasis. Thus, intracellular

Ca^{2+} homeostasis plays a critical role in the regulation of growth, proliferation, survival, and apoptosis of normal and malignant cells. TRPV6 is a highly selective Ca^{2+}channel that is, in general, thought to be constitutively active [51]. As already noted, TRPV6 mediates Ca^{2+} uptake in the duodenum where it is located within the apical membrane of intestinal enterocytes, and there is additional expression in epithelial tissues that bear tumors frequently [12]. Animal models indicate that alterations in TRPV6 expression and function lead to pathophysiological Ca^{2+} conditions, disruptions of renal transport, hypercalciuria, renal stone formation, preeclampsia, and osteoporosis [6]. Expression of TRPV6 is elevated in prostate, breast, colon, esophageal, and cervical tumor tissues, as well as in tumor cell lines, and increased expression stimulates the metastasis of cancer cells and confers chemotherapy resistance [4,12,121–129].

13.7.1.1 TRPV6 in Prostate and Breast Cancer

TRPV6 expression levels were shown to be upregulated in tissue samples originating from prostate, breast, thyroid, colon, ovarian, and pancreatic tumors [12,126]. Prostate and breast cancer are among the most lethal tumors in humans [130,131], yet the mechanisms underlying the development of aggressive tumor phenotypes are not well understood.

TRPV6 expression levels in prostate cancer are linked to tumor progression and have been proposed as a prognostic marker for tumor progression [132]. In the healthy prostate tissue and in benign prostate hyperplasia, TRPV6 is expressed at low levels, while high expression levels arise in prostate adenocarcinoma. TRPV6 mRNA levels were shown to significantly correlate with the Gleason grade score of prostate adenocarcinoma, and TRPV6 was found to be expressed in lymph node metastases of prostate origin [13,126,132]. Increased expression of TRPV6 mRNA is present in the human prostate cancer cell lines (LNCaP and PC-3) as compared to the normal and benign epithelial cells (PrEC and BPH-1) [126]. The androgen receptor (AR) agonist dihydrotestosterone was shown to inhibit TRPV6 expression in LNCaP cells while the AR antagonist bicalutamide (Casodex) increased TRPV6 expression [126,133,134]. TRPV6 expression has not been identified in the androgen-insensitive prostate cancer cell lines DU-145 and PC-3 [132].

Downregulation of TRPV6 in LNCaP cells using specific siRNA induces growth inhibition and S-phase cell cycle arrest and inhibits proliferating cell nuclear antigen expression [135]. The overexpression of TRPV6 in the LNCaP cell line led to increased resistance to apoptosis induced either by the SERCA Ca^{2+} pump inhibitor thapsigargin or DNA cross-linking by the chemotherapy drug cisplatin.

Recent studies by Raphael and colleagues indicate that TRPV6 is likely a tumor promoter in the prostate, functioning as a component of store-operated Ca^{2+} entry (SOCE) in these cells [127]. Being upregulated *de novo*, it significantly contributes to prostate cancer cell survival through proliferation and apoptosis resistance, promotes the formation of bone metastases, and potentiates tumorigenesis *in vivo*.

TRPV6 is also highly expressed in breast adenocarcinoma tissues and breast cancer cell lines (MDA-MB-231, MDA-MB-468, T47D, SKBR3, MCF-7) [69,136,137]. *In vitro* studies using the T47D cell line showed that TRPV6 can be regulated by estrogen, progesterone, tamoxifen, and 1α,25-dihydroxyvitamin D3 (vitamin D3), indicating that these agents affect breast cancer cell proliferation [136]. A recent study revealed that TRPV6 is mainly overexpressed in invasive breast cancer cells and that selective silencing of TRPV6 inhibited MDA-MB-231 migration and invasion, as well as MCF-7 migration [138].

TRPV1 agonist capsaicin has also been reported to upregulate TRPV6 protein expression levels and Ca^{2+} influx in AGS gastric cancer cell line and confers TRPV6-dependent proapoptotic activity [139]. A similar role of TRPV6 in capsaicin-induced apoptosis has been reported in small-cell lung carcinoma cell line [140].

Taken together, these data indicate that inhibition of TRPV6 may offer a promising therapeutic strategy for the treatment of prostate and breast cancers. As alluded to earlier mentioned text, selective TRPV6 blockers might also be useful for the treatment of hypercalciuria in kidney stone patients [6]. The development of specific TRPV6 inhibitors is therefore of great interest as it might open the door toward the development of more efficient therapeutics for cancer treatment.

Here we present an overview of the initiatives taken toward the discovery of TRPV6 modulators.

13.7.2 Development of Chemical Modulators of TRPV6

13.7.2.1 Development of Antiproliferative Compound TH-1177

Strategies to inhibit Ca^{2+} entry into electrically non-excitable cells by small-molecule compounds were addressed by Gray and colleagues in the early 2000s [40]. These investigators synthesized compounds such as **TH-1177** and its stereoisomer **TH-1211** that share similarities with known Ca^{2+} channel blockers derived from the molecules dihydropyridine, benzothiazepine, and phenylalkylamine (Figure 13.5a; compounds **1** and **2**). *In vitro* testing revealed that **TH-1177** was effective in blocking Ca^{2+} influx in prostate cancer cells. Efforts toward the identification of the molecular target of **TH-1177** revealed that the 25B splice variant of the $Ca_v3.2$ T-type Ca^{2+} channel (also known as CACNA1H) is inhibited by **TH-1177**, whereas its stereoisomer **TH-1211** had no effect. Moreover, these investigators postulated that increased expression levels and activity of T-type Ca^{2+} channels are likely responsible for Ca^{2+} influx and stimulation of proliferation in cancer cells. Indeed, **TH-1177** reduced the proliferative rate of human prostate cancer cells LNCaP and PC-3 *in vitro*, with a potency that correlates with its inhibition of Ca^{2+} influx. Furthermore, in xenograft studies, administration of **TH-1177** to nude mice inoculated with PC-3 human prostate cancer cells led to a significant dose-dependent increase in longevity. Paradoxically, T-type Ca^{2+} channels are actually known to open during membrane depolarization and thus are expected to mediate Ca^{2+} influx into cells after an action potential or in response to depolarizing signals, whereas the epithelial Ca^{2+} channel TRPV6 channels would seem more appropriate to facilitate Ca^{2+} entry into cancer cells, as they are open at the negative resting potentials of these cells. Thus, while **TH-1177** is well established as an inhibitor of Ca^{2+} entry in prostate cancer cells, it is reasonable to speculate that it might inhibit TRPV6 in these cells as well.

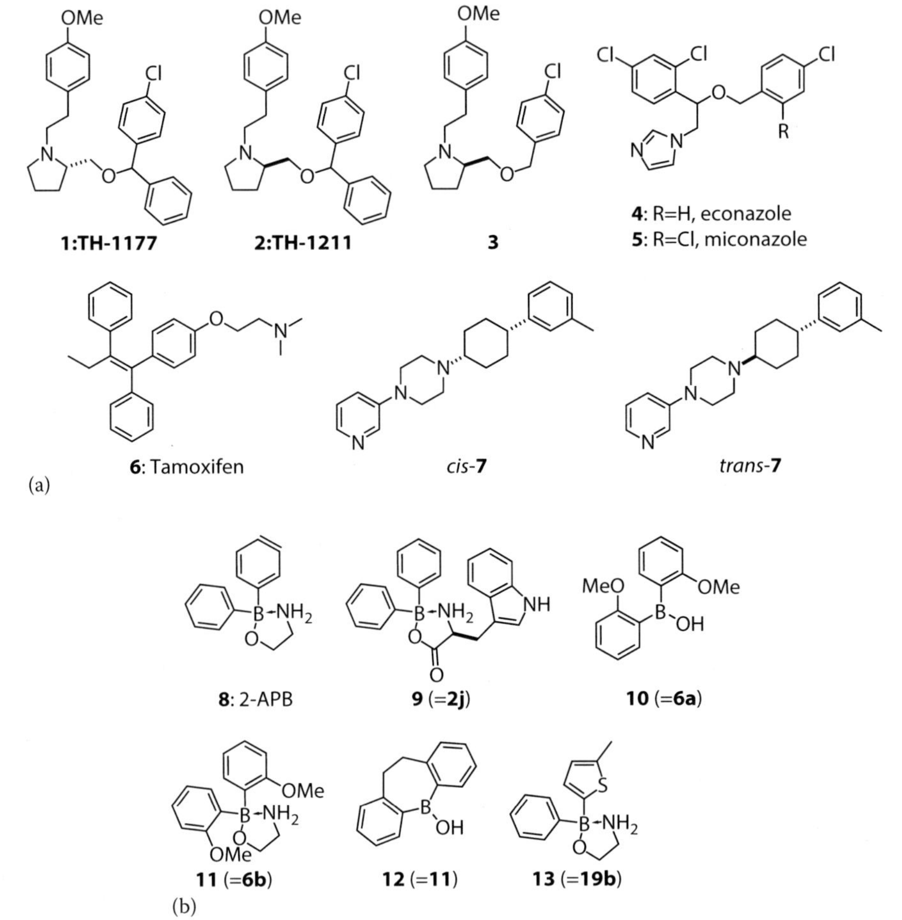

FIGURE 13.5 Chemical structures of known TRPV6 inhibitors: (a) heterocyclic compounds and (b) 2-APB analogs.

13.7.2.2 Development of TRPV6 Inhibitor "Compound #03"

To address the earlier question and to develop additional inhibitors of TRPV6 and also TRPV5, Landowski and colleagues synthesized compounds similar to **TH-1177**, (i.e., compounds **#02**, **#03**, **#05**, **#06**, and **#09**) [41]. Together with the antifungal agents econazole and miconazole that weakly inhibit TRPV6 [16,141,142] and **TH-1177**, these newly synthesized compounds were tested for their inhibitory effect on TRPV6 [41] in *Xenopus* oocytes expressing TRPV6 or TRPV5 and measuring $^{45}Ca^{2+}$ influx. Figure 13.5a shows the chemical structure of "compound **#03**" (compound **3**), as well as those of econazole and miconazole (compounds **4** and **5**). The compounds were also tested using LNCaP prostate cancer and T47D breast cancer cells, studying their effects on proliferation. In LNCaP cells, TRPV6 is believed to be the most abundantly expressed Ca^{2+} entry channel, and it was shown that there is a negligible expression of TRPV5 in these cells [126]. Indeed, **TH-1177** inhibited TRPV6 (and also TRPV5)-mediated Ca^{2+} uptake in oocytes, suggesting that the molecular target of **TH-1177** in prostate cancer cells is TRPV6. However, since it is well established that **TH-1177** mediates inhibition of Ca^{2+} entry via T-type Ca^{2+} channels and given that this compound also inhibits Ca^{2+} entry into cells that do not express TRPV6 [143], it is possible that, in prostate cancer cells, **TH-1177** inhibits both T-type and TRPV6 Ca^{2+} channels.

Compound **#03** of the study by Landowski and colleagues was found to be the most potent and selective inhibitor of TRPV6, significantly inhibiting prostate cancer cell proliferation *in vitro*, with an IC_{50} value of 0.44 μM [41]. Likewise, TRPV6-specific siRNAs were able to reduce proliferation. Indeed, it has been demonstrated that TRPV6 controls Ca^{2+} entry and proliferation in LNCaP cells (see Section 13.7.2.5 for further details).

13.7.2.3 Effects of Estrogen-Receptor Blocker Tamoxifen on TRPV6 Function

Looking for additional inhibitors, Bolanz et al. [69] demonstrated that the estrogen-receptor prodrug tamoxifen (Figure 13.5a, compound **6**) that is currently used for treatment of breast cancer also inhibits TRPV6 in *Xenopus* oocytes, with an IC_{50} of 7.5 μM based on Ca^{2+} uptake studies. This finding is interesting as it may help explain why certain ER-negative breast tumors favorably respond toward tamoxifen treatment.

13.7.2.4 Ligand-Based Virtual Screening (LBVS) to Develop Improved TRPV6 Inhibitors

All of the abovementioned agents are still relatively weak TRPV6 inhibitors, and several of them, including econazole and miconazole, are rather unspecific. Thus, to push forward the development of more effective TRPV6 inhibitors, Reymond and colleagues employed the LBVS approach [68]. For this, the publicly available ZINC database was exploited (Department of Pharmaceutical Chemistry at the University of California, San Francisco), a curated collection of commercially available chemical compounds prepared especially for virtual screening, allowing commercially available compounds to be tested without having to synthesize them first. Specific LBVS tools that have been developed by Reymond and colleagues were used to enable efficient exploitation of the ZINC database [68]. The strategy was to initially search for analogs of the abovementioned TRPV6 inhibitors, that is, **TH-1177**, compound **#03**, econazole, miconazole, and tamoxifen. Starting with these inhibitors, a pharmacophore shape similarity search for analogs in the ZINC database was performed, followed by purchase of selected virtual hits, to form small, focused libraries of a few hundred compounds. These were then experimentally tested using HEK293 cells stably transfected with TRPV6, measuring Ca^{2+} or Cd^{2+} uptake using the fluorophore calcium-5 and the FLIPR Tetra microplate fluorescence reader, in order to identify positive hits. Hits were used for a second round of LBVS, and the resulting improved compounds hits were further optimized by chemical synthesis and structure–activity relationship studies.

The most effective inhibitor *cis*-**22a** (IC_{50} value of 0.12 μM) was further investigated (Figure 13.5a, compound *cis*-**7**). It showed high selectivity against other Ca^{2+} channels and related TRP targets. The antiproliferative activity of *cis*-**22a** on TRPV6 expressing T47D human breast cancer cells was determined and compared to that of SKOV3 ovarian carcinoma cells that do not express TRPV6. Treatment of T47D cells with *cis*-**22a** decreased cellular proliferation by 20% at 5 μM (IC_{50} = 25 μM), a concentration sufficient to block TRPV6-mediated Ca^{2+} influx, while SKOV3 cells were unaffected. By contrast, the less potent diastereomer, *trans*-**22a** (Figure 13.5a, compound *trans*-**7**), had no significant effect on proliferation in these cells lines. The selective but significantly smaller effect of *cis*-**22a** compared to siRNA knockdown (resulting in 50% reduction in cell growth) may suggest that TRPV6 expression affects cell growth by mechanisms other than reduced Ca^{2+} uptake. In addition, it is possible that the TRPV6 inhibitor only works if its expression in the plasma membrane is activated as part of SOCE, as recently described by Prevarskaya and colleagues [127]. Therefore, further studies are needed to evaluate the beneficial effects of this compound on tumor growth and metastasis formation in prostate and breast cancer.

13.7.2.5 Analogs of 2-Aminoethyl Diphenylborinate (2-APB) as Potential Modulators of TRPV6 Function

2-APB is known to influence SOCE via Orai Ca^{2+} channels with different effects in different cell lines. Recent studies revealed that 2-APB also affects TRP channels, including TRPV6.

As already alluded, Prevarskaya's group demonstrated that TRPV6 translocates to the plasma membrane via an Orai1-mediated mechanism, controlling cancer cell survival [127]. According to their proposal, TRPV6 switches in prostate cancer cells from its constitutive activity to a store-operated mode. This switch is proposed to occur through STIM1, Orai1, and TRPC1-evoked translocation of TRPV6 to the plasma membrane, involving the Ca^{2+}/Annexin I/S100A11 pathway. As TRPV6 is virtually absent in healthy prostate, it is upregulated *de novo* in prostate cancer cells, where it changes its constitutively active role, supplying intracellular Ca^{2+} in a store-operated mode that is used in cancer cells to increase cell survival.

To investigate whether 2-APB analogs modulate TRPV6 function, a SAR study was conducted by Lochner and colleagues [144] (see Figure 13.5b for the chemical structures of relevant analogs) in an attempt to develop synthetic analogs that are more potent and more specific for TRPV6, compared to SOCE. The synthesis of a small library of 2-APB analogs was made, and the effects of the compounds were tested in HEK293 cells. To test the effect on human TRPV6, transiently expressing HEK293 cells were used. To test the effect on SOCE, measurements were performed in HEK293 cells in the absence of TRPV6 and the compounds were administered together with 1 μM thapsigargin, in order to activate SOCE. Several 2-APB analogs were discovered that were more potent inhibitors of TRPV6 than 2-APB. However, in general, they inhibited SOCE as well. Analog **19b** (Figure 13.5b; compound **13**) was an exception as it exhibits an approximately 2.5-fold higher selectivity for TRPV6 compared to SOCE. It appears that if steric bulk around the central boron atom or rigidity is increased, for example, as in compounds **2j**, **6a**, **6b**, and **11** (compounds

9–12 in Figure 13.5b), TRPV6 inhibitory activity is lost, while SOCE inhibitory activity is retained. The most beneficial effects were achieved when only one of the phenyl rings of 2-APB was replaced by a 2-methylthiophene (compound **19b**).

Interestingly, 2-APB and its amino alcohol variants turned out to be rather unstable compounds since they rapidly hydrolyze and transesterify in solution, which complicates the analysis of their pharmacological effects and their use as therapeutic compounds.

Taken together, further studies are needed to evaluate the therapeutic potential of 2-APB analogs and to determine whether such compounds are suitable for developing therapeutic applications, targeting specifically the TRPV6 Ca^{2+} channel. An additional issue to be addressed may be cellular acidification caused by 2-APB at high concentrations [145].

13.8 TRPV5 AND TRPV6 MUTATIONS AND KIDNEY STONE DISEASES

Kidney stone disease is a common condition with an incidence of approximately 5%–10% in each population. Although it is known that the environmental factors such as Ca^{2+} intake are important for the stone formation, 40% of stone-forming patients have family history, suggesting that there should be genetic factors as well. Several lines of evidence including ones from a stone-forming rat (GHS rat) suggest that kidney stone disease is a polygenic disease. Hypercalciuria (higher urine Ca^{2+} level) is a common risk factor for stone formation [6].

There are three types of hypercalciuria: (1) absorptive hypercalciuria caused by excess amount of intestinal Ca^{2+} absorption, (2) resorptive hypercalciuria that is derived from abnormal bone metabolism, and (3) renal leak hypercalciuria caused by a decreased Ca^{2+} reabsorption in the kidney. However, a hypercalciuria patient may have abnormal Ca^{2+} absorption and reabsorption and bone resorption at the same time. Since TRPV5 is suggested to be the molecular identity of the apical Ca^{2+} entry pathway for final renal Ca^{2+} reabsorption, and TRPV5 KO mice exhibited renal leak hypercalciuria-like phenotypes [14], it was reasonable to predict that TRPV5 mutations cause kidney stone disease with a renal Ca^{2+} leak. However, Muller et al. reported nine families with no association between TRPV5 gene and autosomal dominant hypercalciuria [146]. This is probably because the renal leak mutations are normally recessive mutations. However, TRPV5 cannot be excluded as a candidate gene for hypercalciuria. The same group also investigated 20 renal leak hypercalciuric patients, and 4 nonsynonymous polymorphisms were found. However, no association study was done in this report [147]. Later on, other groups investigated 365 kidney stone patients in Taiwan and found that the frequency of R154H (rs4236480) polymorphism in the TRPV5 gene was higher in the patients with multiple stones [148], indicating that the TRPV5 gene is somehow linked to renal stone formation.

Recently, the association between kidney stones and a novel TRPV5 gene mutation (L530R) was identified [149]. A genome-wide association study (GWAS) of 2636 Icelanders with kidney stones led to the identification of defects of the following genes that are responsible for kidney stone formation: (1) ALPL (tissue nonspecific alkali phosphatase for maintaining urine pyrophosphate that inhibits stone formation), (2) SLC34A1 (Na/Pi cotransporter for renal phosphate reabsorption), (3) claudin-14 (CLDN14), (4) CaSR, and (5) TRPV5. Claudin-14 and CaSR are thought to be important for renal Ca^{2+} handling to excrete excess Ca^{2+} via the paracellular pathway. With respect to the L530R mutation in the TRPV5 gene, it might represent a loss-of-function mutation, because it is localized in the pore loop of the TRPV5 channel. Thus, so far only one mutation of the TRPV5 gene has been identified that is associated with kidney stone disease, most likely resulting in renal leak hypercalciuria.

There are no further reports thus far showing an association of TRPV5 gene mutations with other human diseases. However, with respect to single nucleotide polymorphisms, 154H and 563T were more frequently distributed among African population compared to that of 154R or 563A. Renkema and colleagues showed that there were no functional differences in these polymorphisms with respect to Na^+ currents, Ca^{2+} currents, and $[Ca^{2+}]_i$-dependent inactivation in patch–clamp recordings [147]. On the contrary, when using the in *Xenopus* oocyte expression system, an increased $^{45}Ca^{2+}$ uptake in the 563T variant was reported [150]. Computational modeling based on the structure of TRPV1 revealed that the A563T variation induces structural, dynamical, and electrostatic changes in the TRPV5 pore, providing structural insights into the altered function of this variant [151]. The 154H variant exhibited the lowest Ca^{2+} uptake activity among five TRPV5 variants; however, the data did not reach statistical significance compared to the reference 154R variant [150]. In preliminary experiments, we found decreased ^{45}Ca uptake in the 154H variant but not in A563T, without changing membrane surface expression of TRPV5 proteins (Y. Suzuki et al., unpublished data). Interestingly, there appears to be an association between R154H and bone density of the femoral neck and the tibial epiphysis in adults, most likely due to impaired osteoclastic bone resorption (A. Pasch, Y. Suzuki, and M.A. Hediger, unpublished data). These results and also data from TRPV5 KO mice [152] suggest that variants of the TRPV5 gene are risk factors for impaired bone quality. Genetic variants of TRPV5 might contribute to differences in bone quality and strength among different populations.

Given the numerous evidence, including the study of TRPV6 KO mice that lack TRPV6 function results in impaired intestinal Ca^{2+} absorption [15], TRPV6 is believed to be a key component of apical Ca^{2+} entry during intestinal absorption of Ca^{2+} from food and drinking water. Therefore, it is reasonable to speculate that genetic variations of the TRPV6 gene are associated with absorptive hypercalciuria, one of the major risk factors of kidney stone formation. Indeed, this is the case: a gain-of-function haplotype of TRPV6 gene has been identified from sequencing of 170 Ca^{2+}-stone formers in Switzerland [153]. The frequency of this haplotype was statistically higher in the Ca^{2+}-stone formers compared to normal individuals, suggesting that the haplotype might be a risk factor for kidney stone patients with absorptive hypercalciuria. Another line of evidence that stems from molecular evolution highlights the functional significance of this haplotype: it has been reported that this very same haplotype reflects a positive selection during human evolution [154,155] and that this "derived TRPV6 form" provided an evolutionary advantage for the survival of humans in response to certain past environmental changes, when compared to the ancestral form. An improvement of $[Ca^{2+}]_i$-dependent inactivation has been suggested in the derived TRPV6 form, compared to the ancestral form [154], which is consistent with the observed increase in ^{45}Ca uptake in heterologous expression systems [155].

It is interesting to note that TRPV6 KO mice exhibit hypercalciuria [15]. This suggests that TRPV6 might also be involved in renal Ca^{2+} reabsorption. In fact, TRPV6 mRNA level appears to be higher than that of TRPV5 in the human kidney [8]. Also, prominent TRPV6 protein expression has been detected in both the proximal tubule and distal tubule in human kidney [10]. African-Americans exhibit lower urinary Ca^{2+} excretion and lower incidence of kidney stone disease [156]. The high prevalence of the ancestral haplotype of TRPV6 gene in African descendants suggest that TRPV6 may play a role in increased Ca^{2+} reabsorption in African populations.

13.9 CONCLUDING REMARKS

A variety of approaches have been utilized to achieve a better understanding of Ca^{2+} transport in epithelial cells, starting with expression cloning and moving all the way to in-depth cellular and biophysical analysis, generation of genetically engineered mouse models of TRPV5 and TRPV6, 3D structural studies, and chemical approaches. The remarkable Ca^{2+}-selectivity and distinct Ca^{2+}-dependent inactivation mechanisms involving $PI(4,5)P_2$ are examples of what patch–clamp techniques could achieve in understanding these channels. Genetically engineered animal models not only provided the expected roles of TRPV5 and TRPV6 in intestinal absorption and renal reabsorption of Ca^{2+} but also revealed unique roles of the TRPV6 channel in maternal–fetal Ca^{2+} transport, male fertility, and auditory system. The growing evidence for the involvement TRPV6 in cancer proliferation and metastasis formation generates pharmaceutical interest. New approaches, such as ligand-based virtual screening, have been utilized to design inhibitors as a first step toward therapeutic developments. The novel crystal structure of TRPV6 will be instrumental to our understanding of the Ca^{2+} transport mechanisms and regulation and help refine the chemical design of TRPV6 modulators. As the involvement of genetic mutations/variations of TRPV5 and TRPV6 in human health is emerging, the integration of multidisciplinary approaches will further advance our knowledge and help develop new strategies for the treatment of kidney stone disease and cancer.

ACKNOWLEDGMENTS

We thank our colleagues for their contributions to the research projects related to this book chapter. Our research was supported by the National Institute of Diabetes and Digestive and Kidney Diseases (R01DK072154), the American Heart Association National Center (0430125N), the Greater Southeast Affiliate (09GRNT2160024), and the Swiss National Center for Competence in Research, NCCR TransCure (www.transcure.org). Gergely Gyimesi acknowledges the COFUND Marie Curie international fellowship program, IFP TransCure.

REFERENCES

1. Hoenderop, J.G., van der Kemp, A.W., Hartog, A., van de Graaf, S.F., van Os, C.H., Willems, P.H., and Bindels, R.J. 1999. Molecular identification of the apical Ca^{2+} channel in 1,25-dihydroxyvitamin D_3-responsive epithelia. *J. Biol. Chem.* **274**:8375–8378.
2. Peng, J.B., Chen, X.Z., Berger, U.V., Vassilev, P.M., Tsukaguchi, H., and Brown, E.M., and Hediger, M.A. 1999. Molecular cloning and characterization of a channel-like transporter mediating intestinal calcium absorption. *J. Biol. Chem.* **274**:22739–22746.
3. Na, T. and Peng, J.B. 2014. TRPV5: A Ca^{2+} channel for the fine-tuning of Ca^{2+} reabsorption. *Handb. Exp. Pharmacol.* **222**:321–357.
4. Fecher-Trost, C., Weissgerber, P., and Wissenbach, U. 2014. TRPV6 channels. *Handb. Exp. Pharmacol.* **222**:359–384.
5. Hoenderop, J.G., Nilius, B., and Bindels, R.J. 2005. Calcium absorption across epithelia. *Physiol. Rev.* **85**:373–422.
6. Suzuki, Y., Landowski, C.P., and Hediger, M.A. 2008. Mechanisms and regulation of epithelial Ca^{2+} absorption in health and disease. *Annu. Rev. Physiol.* **70**:257–271.
7. Bindels, R.J. 2000. Molecular pathophysiology of renal calcium handling. *Kidney Blood Press. Res.* **23**:183–184.
8. Peng, J.B., Brown, E.M., and Hediger, M.A. 2001. Structural conservation of the genes encoding CaT1, CaT2, and related cation channels. *Genomics* **76**:99–109.
9. Song, Y., Peng, X., Porta, A., Takanaga, H., Peng, J.B., Hediger, M.A., Fleet, J.C., and Christakos, S. 2003. Calcium transporter 1 and epithelial calcium channel messenger ribonucleic acid are differentially regulated by 1,25 dihydroxyvitamin D_3 in the intestine and kidney of mice. *Endocrinology* **144**:3885–3894.
10. Wu, Y., Miyamoto, T., Li, K., Nakagomi, H., Sawada, N., Kira, S., Kobayashi, H. et al. 2011. Decreased expression of the epithelial Ca^{2+} channel TRPV5 and TRPV6 in human renal cell carcinoma associated with vitamin D receptor. *J. Urol.* **186**:2419–2425.
11. Peng, J.B., Chen, X.Z., Berger, U.V., Weremowicz, S., Morton, C.C., Vassilev, P.M., Brown, E.M., and Hediger, M.A. 2000. Human calcium transport protein CaT1. *Biochem. Biophys. Res. Commun.* **278**:326–332.
12. Zhuang, L., Peng, J.B., Tou, L., Takanaga, H., Adam, R.M., Hediger, M.A., and Freeman, M.R. 2002. Calcium-selective ion channel, CaT1, is apically localized in gastrointestinal tract epithelia and is aberrantly expressed in human malignancies. *Lab. Invest.* **82**:1755–1764.
13. Wissenbach, U., Niemeyer, B.A., Fixemer, T., Schneidewind, A., Trost, C., Cavalie, A., Reus, K., Meese, E., Bonkhoff, H., and Flockerzi, V. 2001. Expression of CaT-like, a novel calcium-selective channel, correlates with the malignancy of prostate cancer. *J. Biol. Chem.* **276**:19461–19468.
14. Hoenderop, J.G., van Leeuwen, J.P., van der Eerden, B.C., Kersten, F.F., van der Kemp, A.W., Merillat, A.M., Waarsing, J.H. et al. 2003. Renal Ca^{2+} wasting, hyperabsorption, and reduced bone thickness in mice lacking TRPV5. *J. Clin. Invest.* **112**:1906–1914.
15. Bianco, S.D., Peng, J.B., Takanaga, H., Suzuki, Y., Crescenzi, A., Kos, C.H., Zhuang, L. et al. 2007. Marked disturbance of calcium homeostasis in mice with targeted disruption of the Trpv6 calcium channel gene. *J. Bone Miner. Res.* **22**:274–285.
16. Suzuki, Y., Kovacs, C.S., Takanaga, H., Peng, J.B., Landowski, C.P., and Hediger, M.A. 2008. Calcium channel TRPV6 is involved in murine maternal-fetal calcium transport. *J. Bone Miner. Res.* **23**:1249–1256.
17. Weissgerber, P., Kriebs, U., Tsvilovskyy, V., Olausson, J., Kretz, O., Stoerger, C., Vennekens, R. et al. 2011. Male fertility depends on Ca^{2+} absorption by TRPV6 in epididymal epithelia. *Sci. Signal.* **4**:ra27.

18. Stoerger, C. and Flockerzi, V. 2014. The transient receptor potential cation channel subfamily V member 6 (TRPV6): Genetics, biochemical properties, and functions of exceptional calcium channel proteins. *Biochem. Cell Biol.* **92**:441–448.
19. Brown, E.M. 1991. Extracellular Ca^{2+} sensing, regulation of parathyroid cell function, and role of Ca^{2+} and other ions as extracellular (first) messengers. *Physiol. Rev.* **71**:371–411.
20. Khundmiri, S.J., Murray, R.D., and Lederer, E. 2016. PTH and vitamin D. *Compr. Physiol.* **6**:561–601.
21. Van Cromphaut, S.J., Dewerchin, M., Hoenderop, J.G., Stockmans, I., Van, H.E., Kato, S., Bindels, R.J. et al. 2001. Duodenal calcium absorption in vitamin D receptor-knockout mice: Functional and molecular aspects. *Proc. Natl. Acad. Sci. USA* **98**:13324–13329.
22. Bronner, F. 1998. Calcium absorption—A paradigm for mineral absorption. *J. Nutr.* **128**:917–920.
23. Friedman, P.A. 2000. Mechanisms of renal calcium transport. *Exp. Nephrol.* **8**:343–350.
24. Christakos, S., Gill, R., Lee, S., and Li, H. 1992. Molecular aspects of the calbindins. *J. Nutr.* **122**:678–682.
25. Kawasaki, H., Nakayama, S., and Kretsinger, R.H. 1998. Classification and evolution of EF-hand proteins. *Biometals* **11**:277–295.
26. Lambers, T.T., Mahieu, F., Oancea, E., Hoofd, L., de Lange, F., Mensenkamp, A.R., Voets, T. et al. 2006. Calbindin-D28K dynamically controls TRPV5-mediated Ca^{2+} transport. *EMBO J.* **25**:2978–2988.
27. Schauberger, C.W. and Pitkin, R.M. 1979. Maternal–perinatal calcium relationships. *Obstet. Gynecol.* **53**:74–76.
28. Smith, C.H., Moe, A.J., and Ganapathy, V. 1992. Nutrient transport pathways across the epithelium of the placenta. *Annu. Rev. Nutr.* **12**:183–206.
29. Stulc, J. 1997. Placental transfer of inorganic ions and water. *Physiol. Rev.* **77**:805–836.
30. Moreau, R., Daoud, G., Bernatchez, R., Simoneau, L., Masse, A., and Lafond, J. 2002. Calcium uptake and calcium transporter expression by trophoblast cells from human term placenta. *Biochim. Biophys. Acta* **1564**:325–332.
31. Jenkins, A.D., Lechene, C.P., and Howards, S.S. 1980. Concentrations of seven elements in the intraluminal fluids of the rat seminiferous tubules, rate testis, and epididymis. *Biol. Reprod.* **23**:981–987.
32. Carlson, A.E., Westenbroek, R.E., Quill, T., Ren, D., Clapham, D.E., Hille, B., Garbers, D.L., and Babcock, D.F. 2003. CatSper1 required for evoked Ca^{2+} entry and control of flagellar function in sperm. *Proc. Natl. Acad. Sci. USA* **100**:14864–14868.
33. Marquis, R.E. and Hudspeth, A.J. 1997. Effects of extracellular Ca^{2+} concentration on hair-bundle stiffness and gating-spring integrity in hair cells. *Proc. Natl. Acad. Sci. USA* **94**:11923–11928.
34. Yamauchi, D., Nakaya, K., Raveendran, N.N., Harbidge, D.G., Singh, R., Wangemann, P., and Marcus, D.C. 2010. Expression of epithelial calcium transport system in rat cochlea and vestibular labyrinth. *BMC Physiol.* **10**:1.
35. Nakaya, K., Harbidge, D.G., Wangemann, P., Schultz, B.D., Green, E.D., Wall, S.M., and Marcus, D.C. 2007. Lack of pendrin. *Am. J. Physiol. Renal Physiol.* **292**:F1314–F1321.
36. Nicaise, G., Maggio, K., Thirion, S., Horoyan, M., and Keicher, E. 1992. The calcium loading of secretory granules. A possible key event in stimulus-secretion coupling. *Biol. Cell* **75**:89–99.
37. Monteith, G.R., McAndrew, D., Faddy, H.M., and Roberts-Thomson, S.J. 2007. Calcium and cancer: Targeting Ca^{2+} transport. *Nat. Rev. Cancer* **7**:519–530.
38. Prevarskaya, N., Skryma, R., and Shuba, Y. 2011. Calcium in tumour metastasis: New roles for known actors. *Nat. Rev. Cancer* **11**:609–618.
39. Roderick, H.L. and Cook, S.J. 2008. Ca^{2+} signalling checkpoints in cancer: Remodelling Ca^{2+} for cancer cell proliferation and survival. *Nat. Rev. Cancer* **8**:361–375.

40. Gray, L.S., Perez-Reyes, E., Gomora, J.C., Haverstick, D.M., Shattock, M., McLatchie, L., Harper, J., Brooks, G., Heady, T., and Macdonald, T.L. 2004. The role of voltage gated T-type Ca^{2+} channel isoforms in mediating "capacitative" Ca^{2+} entry in cancer cells. *Cell Calcium* **36**:489–497.
41. Landowski, C.P., Bolanz, K.A., Suzuki, Y., and Hediger, M.A. 2011. Chemical inhibitors of the calcium entry channel TRPV6. *Pharm. Res.* **28**:322–330.
42. Romero, M.F., Kanai, Y., Gunshin, H., and Hediger, M.A. 1998. Expression cloning using *Xenopus laevis* oocytes. *Methods Enzymol.* **296**:17–52.
43. Fecher-Trost, C., Wissenbach, U., Beck, A., Schalkowsky, P., Stoerger, C., Doerr, J., Dembek, A. et al. 2013. The in vivo TRPV6 protein starts at a non-AUG triplet, decoded as methionine, upstream of canonical initiation at AUG. *J. Biol. Chem.* **288**:16629–16644.
44. Caterina, M.J., Schumacher, M.A., Tominaga, M., Rosen, T.A., Levine, J.D., and Julius, D. 1997. The capsaicin receptor: A heat-activated ion channel in the pain pathway. *Nature* **389**:816–824.
45. Colbert, H.A., Smith, T.L., and Bargmann, C.I. 1997. OSM-9, a novel protein with structural similarity to channels, is required for olfaction, mechanosensation, and olfactory adaptation in Caenorhabditis elegans. *J. Neurosci.* **17**:8259–8269.
46. Vennekens, R., Owsianik, G., and Nilius, B. 2008. Vanilloid transient receptor potential cation channels: An overview. *Curr. Pharm. Des.* **14**:18–31.
47. Peng, J.B., Chen, X.Z., Berger, U.V., Vassilev, P.M., Brown, E.M., and Hediger, M.A. 2000. A rat kidney-specific calcium transporter in the distal nephron. *J. Biol. Chem.* **275**:28186–28194.
48. Mackenzie, B. 1999. Selected techniques in membrane transport. In *Biomembrane Transport*. L.J. Van Winkle, ed. Academic Press, San Diego, CA, pp. 327–342.
49. Vennekens, R., Hoenderop, J.G., Prenen, J., Stuiver, M., Willems, P.H., Droogmans, G., Nilius, B., and Bindels, R.J. 2000. Permeation and gating properties of the novel epithelial Ca^{2+} channel. *J. Biol. Chem.* **275**:3963–3969.
50. Nilius, B., Vennekens, R., Prenen, J., Hoenderop, J.G., Bindels, R.J., and Droogmans, G. 2000. Whole-cell and single channel monovalent cation currents through the novel rabbit epithelial Ca^{2+} channel ECaC. *J. Physiol.* **527**(Pt 2):239–248.
51. Yue, L., Peng, J.B., Hediger, M.A., and Clapham, D.E. 2001. CaT1 manifests the pore properties of the calcium-release-activated calcium channel. *Nature* **410**:705–709.
52. Nilius, B., Vennekens, R., Prenen, J., Hoenderop, J.G., Droogmans, G., and Bindels, R.J. 2001. The single pore residue Asp542 determines Ca^{2+} permeation and Mg^{2+} block of the epithelial Ca^{2+} channel. *J. Biol. Chem.* **276**:1020–1025.
53. Dodier, Y., Banderali, U., Klein, H., Topalak, O., Dafi, O., Simoes, M., Bernatchez, G., Sauve, R., and Parent, L. 2004. Outer pore topology of the ECaC-TRPV5 channel by cysteine scan mutagenesis. *J. Biol. Chem.* **279**:6853–6862.
54. Nilius, B., Prenen, J., Hoenderop, J.G., Vennekens, R., Hoefs, S., Weidema, A.F., Droogmans, G., and Bindels, R.J. 2002. Fast and slow inactivation kinetics of the Ca^{2+} channels ECaC1 and ECaC2 (TRPV5 and TRPV6). Role of the intracellular loop located between transmembrane segments 2 and 3. *J. Biol. Chem.* **277**:30852–30858.
55. Niemeyer, B.A., Bergs, C., Wissenbach, U., Flockerzi, V., and Trost, C. 2001. Competitive regulation of CaT-like-mediated Ca^{2+} entry by protein kinase C and calmodulin. *Proc. Natl. Acad. Sci. USA* **98**:3600–3605.
56. Derler, I., Hofbauer, M., Kahr, H., Fritsch, R., Muik, M., Kepplinger, K., Hack, M.E. et al. 2006. Dynamic but not constitutive association of calmodulin with rat TRPV6 channels enables fine tuning of Ca^{2+}-dependent inactivation. *J. Physiol.* **577**:31–44.
57. Lambers, T.T., Weidema, A.F., Nilius, B., Hoenderop, J.G., and Bindels, R.J. 2004. Regulation of the mouse epithelial Ca^{2+} channel TRPV6 by the Ca^{2+}-sensor calmodulin. *J. Biol. Chem.* **279**:28855–28861.

58. de Groot T., Kovalevskaya, N.V., Verkaart, S., Schilderink, N., Felici, M., van der Hagen, E.A., Bindels, R.J., Vuister, G.W., and Hoenderop, J.G. 2011. Molecular mechanisms of calmodulin action on TRPV5 and modulation by parathyroid hormone. *Mol. Cell Biol.* **31**:2845–2853.
59. Kovalevskaya, N.V., Bokhovchuk, F.M., and Vuister, G.W. 2012. The TRPV5/6 calcium channels contain multiple calmodulin binding sites with differential binding properties. *J. Struct. Funct. Genomics* **13**:91–100.
60. Gkika, D., Mahieu, F., Nilius, B., Hoenderop, J.G., and Bindels, R.J. 2004. 80K-H as a new Ca^{2+} sensor regulating the activity of the epithelial Ca^{2+} channel transient receptor potential cation channel V5 (TRPV5). *J. Biol. Chem.* **279**:26351–26357.
61. Lee, J., Cha, S.K., Sun, T.J., and Huang, C.L. 2005. PIP2 activates TRPV5 and releases its inhibition by intracellular Mg^{2+}. *J. Gen. Physiol.* **126**:439–451.
62. Thyagarajan, B., Benn, B.S., Christakos, S., and Rohacs, T. 2009. Phospholipase C-mediated regulation of transient receptor potential vanilloid 6 channels: Implications in active intestinal Ca^{2+} transport. *Mol. Pharmacol.* **75**:608–616.
63. Thyagarajan, B., Lukacs, V., and Rohacs, T. 2008. Hydrolysis of phosphatidylinositol 4,5-bisphosphate mediates calcium-induced inactivation of TRPV6 channels. *J. Biol. Chem.* **283**:14980–14987.
64. Rohacs, T. and Nilius, B. 2007. Regulation of transient receptor potential (TRP) channels by phosphoinositides. *Pflugers Arch.* **455**:157–168.
65. Rohacs, T., Lopes, C.M., Michailidis, I., and Logothetis, D.E. 2005. $PI(4,5)P_2$ regulates the activation and desensitization of TRPM8 channels through the TRP domain. *Nat. Neurosci.* **8**:626–634.
66. Poblete, H., Oyarzun, I., Olivero, P., Comer, J., Zuniga, M., Sepulveda, R.V., Baez-Nieto, D., Gonzalez, L.C., Gonzalez-Nilo, F., and Latorre, R. 2015. Molecular determinants of phosphatidylinositol 4,5-bisphosphate ($PI(4,5)P_2$) binding to transient receptor potential V1 (TRPV1) channels. *J. Biol. Chem.* **290**:2086–2098.
67. Velisetty, P., Borbiro, I., Kasimova, M.A., Liu, L., Badheka, D., Carnevale, V., and Rohacs, T. 2016. A molecular determinant of phosphoinositide affinity in mammalian TRPV channels. *Sci. Rep.* **6**:27652.
68. Simonin, C., Awale, M., Brand, M., van Deursen, R., Schwartz, J., Fine, M., Kovacs, G. et al. 2015. Optimization of TRPV6 calcium channel inhibitors using a 3D ligand-based virtual screening method. *Angew. Chem. Int. Ed. Engl.* **54**:14748–14752.
69. Bolanz, K.A., Kovacs, G.G., Landowski, C.P., and Hediger, M.A. 2009. Tamoxifen inhibits TRPV6 activity via estrogen receptor-independent pathways in TRPV6-expressing MCF-7 breast cancer cells. *Mol. Cancer Res.* **7**:2000–2010.
70. Saotome, K., Singh, A.K., Yelshanskaya, M.V., and Sobolevsky, A.I. 2016. Crystal structure of the epithelial calcium channel TRPV6. *Nature* **534**:506–511.
71. Alford, R.F., Koehler, L.J., Weitzner, B.D., Duran, A.M., Tilley, D.C., Elazar, A., and Gray, J.J. 2015. An integrated framework advancing membrane protein modeling and design. *PLoS Comput. Biol.* **11**:e1004398.
72. Mandell, D.J., Coutsias, E.A., and Kortemme, T. 2009. Sub-angstrom accuracy in protein loop reconstruction by robotics-inspired conformational sampling. *Nat. Methods* **6**:551–552.
73. Kuhlman, B. and Baker, D. 2000. Native protein sequences are close to optimal for their structures. *Proc. Natl. Acad. Sci. USA* **97**:10383–10388.
74. Loh, N.Y., Bentley, L., Dimke, H., Verkaart, S., Tammaro, P., Gorvin, C.M., Stechman, M.J. et al. 2013. Autosomal dominant hypercalciuria in a mouse model due to a mutation of the epithelial calcium channel, TRPV5. *PLoS One* **8**:e55412.
75. Cui, M., Li, Q., Johnson, R., and Fleet, J.C. 2012. Villin promoter-mediated transgenic expression of transient receptor potential cation channel, subfamily V, member 6 (TRPV6) increases intestinal calcium absorption in wild-type and vitamin D receptor knockout mice. *J. Bone Miner. Res.* **27**:2097–2107.

76. van Abel, M., Huybers, S., Hoenderop, J.G., van der Kemp, A.W., van Leeuwen, J.P., and Bindels, R.J. 2006. Age-dependent alterations in Ca^{2+} homeostasis: Role of TRPV5 and TRPV6. *Am. J. Physiol. Renal Physiol.* **291**:F1177–F1183.
77. Benn, B.S., Ajibade, D., Porta, A., Dhawan, P., Hediger, M., Peng, J.B., Jiang, Y. et al. 2008. Active intestinal calcium transport in the absence of transient receptor potential vanilloid type 6 and calbindin-D_{9k}. *Endocrinology* **149**:3196–3205.
78. Kutuzova, G.D., Sundersingh, F., Vaughan, J., Tadi, B.P., Ansay, S.E., Christakos, S., and Deluca, H.F. 2008. TRPV6 is not required for 1alpha, 25-dihydroxyvitamin D_3-induced intestinal calcium absorption in vivo. *Proc. Natl. Acad. Sci. USA* **105**:19655–19659.
79. Meyer, M.B., Watanuki, M., Kim, S., Shevde, N.K., and Pike, J.W. 2006. The human transient receptor potential vanilloid type 6 distal promoter contains multiple vitamin D receptor binding sites that mediate activation by 1,25-dihydroxyvitamin D_3 in intestinal cells. *Mol. Endocrinol.* **20**:1447–1461.
80. Song, Y., Kato, S., and Fleet, J.C. 2003. Vitamin D receptor (VDR) knockout mice reveal VDR-independent regulation of intestinal calcium absorption and ECaC2 and calbindin D_{9k} mRNA. *J. Nutr.* **133**:374–380.
81. Pitkin, R.M. 1985. Calcium metabolism in pregnancy and the perinatal period: A review. *Am. J. Obstet. Gynecol.* **151**:99–109.
82. Bernucci, L., Henriquez, M., Diaz, P., and Riquelme, G. 2006. Diverse calcium channel types are present in the human placental syncytiotrophoblast basal membrane. *Placenta* **27**:1082–1095.
83. Moreau, R., Hamel, A., Daoud, G., Simoneau, L., and Lafond, J. 2002. Expression of calcium channels along the differentiation of cultured trophoblast cells from human term placenta. *Biol. Reprod.* **67**:1473–1479.
84. Stumpf, T., Zhang, Q., Hirnet, D., Lewandrowski, U., Sickmann, A., Wissenbach, U., Dorr, J., Lohr, C., Deitmer, J.W., and Fecher-Trost, C. 2008. The human TRPV6 channel protein is associated with cyclophilin B in human placenta. *J. Biol. Chem.* **283**:18086–18098.
85. Vanoevelen, J., Janssens, A., Huitema, L.F., Hammond, C.L., Metz, J.R., Flik, G., Voets, T., and Schulte-Merker, S. 2011. Trpv5/6 is vital for epithelial calcium uptake and bone formation. *FASEB J.* **25**:3197–3207.
86. Weissgerber, P., Kriebs, U., Tsvilovskyy, V., Olausson, J., Kretz, O., Stoerger, C., Mannebach, S. et al. 2012. Excision of Trpv6 gene leads to severe defects in epididymal Ca^{2+} absorption and male fertility much like single D541A pore mutation. *J. Biol. Chem.* **287**:17930–17941.
87. Gao, D.Y., Zhang, B.L., Leung, M.C., Au, S.C., Wong, P.Y., and Shum, W.W. 2016. Coupling of TRPV6 and TMEM16A in epithelial principal cells of the rat epididymis. *J. Gen. Physiol.* **148**:161–182.
88. Lewit-Bentley, A., Rety, S., Sopkova-de Oliveira, S.J., and Gerke, V. 2000. S100-annexin complexes: Some insights from structural studies. *Cell Biol. Int.* **24**:799–802.
89. van de Graaf, S.F., Chang, Q., Mensenkamp, A.R., Hoenderop, J.G., and Bindels, R.J. 2006. Direct interaction with Rab11a targets the epithelial Ca^{2+} channels TRPV5 and TRPV6 to the plasma membrane. *Mol. Cell Biol.* **26**:303–312.
90. Embark, H.M., Setiawan, I., Poppendieck, S., van de Graaf, S.F., Boehmer, C., Palmada, M., Wieder, T. et al. 2004. Regulation of the epithelial Ca^{2+} channel TRPV5 by the NHE regulating factor NHERF2 and the serum and glucocorticoid inducible kinase isoforms SGK1 and SGK3 expressed in *Xenopus* oocytes. *Cell Physiol. Biochem.* **14**:203–212.
91. Palmada, M., Poppendieck, S., Embark, H.M., van de Graaf, S.F., Boehmer, C., Bindels, R.J., and Lang, F. 2005. Requirement of PDZ domains for the stimulation of the epithelial Ca^{2+} channel TRPV5 by the NHE regulating factor NHERF2 and the serum and glucocorticoid inducible kinase SGK1. *Cell Physiol. Biochem.* **15**:175–182.

92. Jing, H., Na, T., Zhang, W., Wu, G., Liu, C., and Peng, J.B. 2011. Concerted actions of NHERF2 and WNK4 in regulating TRPV5. *Biochem. Biophys. Res. Commun.* **404**:979–984.
93. Holakovska, B., Grycova, L., Bily, J., and Teisinger, J. 2011. Characterization of calmodulin binding domains in TRPV2 and TRPV5 C-tails. *Amino Acids* **40**:741–748.
94. van de Graaf, S.F., van der Kemp, A.W., van den Berg, D., van Oorschot, M., Hoenderop, J.G., and Bindels, R.J. 2006. Identification of BSPRY as a novel auxiliary protein inhibiting TRPV5 activity. *J. Am. Soc. Nephrol.* **17**:26–30.
95. Gkika, D., Topala, C.N., Hoenderop, J.G., and Bindels, R.J. 2006. The immunophilin FKBP52 inhibits the activity of the epithelial Ca^{2+} channel TRPV5. *Am. J. Physiol. Renal Physiol.* **290**:F1253–F1259.
96. van de Graaf, S.F., Hoenderop, J.G., van der Kemp, A.W., Gisler, S.M., and Bindels, R.J. 2006. Interaction of the epithelial Ca^{2+} channels TRPV5 and TRPV6 with the intestine- and kidney-enriched PDZ protein NHERF4. *Pflugers Arch.* **452**:407–417.
97. Yeh, B.I., Kim, Y.K., Jabbar, W., and Huang, C.L. 2005. Conformational changes of pore helix coupled to gating of TRPV5 by protons. *EMBO J.* **24**:3224–3234.
98. Cha, S.K., Jabbar, W., Xie, J., and Huang, C.L. 2007. Regulation of TRPV5 single-channel activity by intracellular pH. *J. Membr. Biol.* **220**:79–85.
99. de Groot T., Verkaart, S., Xi, Q., Bindels, R.J., and Hoenderop, J.G. 2010. The identification of Histidine 712 as a critical residue for constitutive TRPV5 internalization. *J. Biol. Chem.* **285**:28481–28487.
100. Chang, Q., Hoefs, S., van der Kemp, A.W., Topala, C.N., Bindels, R.J., and Hoenderop, J.G. 2005. The beta-glucuronidase klotho hydrolyzes and activates the TRPV5 channel. *Science* **310**:490–493.
101. Lu, P., Boros, S., Chang, Q., Bindels, R.J., and Hoenderop, J.G. 2008. The beta-glucuronidase klotho exclusively activates the epithelial Ca^{2+} channels TRPV5 and TRPV6. *Nephrol. Dial. Transplant.* **23**:3397–3402.
102. Leunissen, E.H., Nair, A.V., Bull, C., Lefeber, D.J., van Delft, F.L., Bindels, R.J., and Hoenderop, J.G. 2013. The epithelial calcium channel TRPV5 is regulated differentially by klotho and sialidase. *J. Biol. Chem.* **288**:29238–29246.
103. Wolf, M.T., An, S.W., Nie, M., Bal, M.S., and Huang, C.L. 2014. Klotho up-regulates renal calcium channel transient receptor potential vanilloid 5 (TRPV5) by intra- and extracellular *N*-glycosylation-dependent mechanisms. *J. Biol. Chem.* **289**:35849–35857.
104. Boros, S., Xi, Q., Dimke, H., van der Kemp, A.W., Tudpor, K., Verkaart, S., Lee, K.P., Bindels, R.J., and Hoenderop, J.G. 2012. Tissue transglutaminase inhibits the TRPV5-dependent calcium transport in an *N*-glycosylation-dependent manner. *Cell Mol. Life Sci.* **69**:981–992.
105. Tudpor, K., Lainez, S., Kwakernaak, A.J., Kovalevskaya, N.V., Verkaart, S., van Genesen, S., van der Kemp, A., Navis, G., Bindels, R.J., and Hoenderop, J.G. 2012. Urinary plasmin inhibits TRPV5 in nephrotic-range proteinuria. *J. Am. Soc. Nephrol.* **23**:1824–1834.
106. Wolf, M.T., Wu, X.R., and Huang, C.L. 2013. Uromodulin upregulates TRPV5 by impairing caveolin-mediated endocytosis. *Kidney Int.* **84**:130–137.
107. Nie, M., Bal, M.S., Yang, Z., Liu, J., Rivera, C., Wenzel, A., Beck, B.B., Sakhaee, K., Marciano, D.K., and Wolf, M.T. 2016. Mucin-1 increases renal TRPV5 activity in vitro, and urinary level associates with calcium nephrolithiasis in patients. *J. Am. Soc. Nephrol.* **27**:3447–3458.
108. Mayan, H., Munter, G., Shaharabany, M., Mouallem, M., Pauzner, R., Holtzman, E.J., and Farfel, Z. 2004. Hypercalciuria in familial hyperkalemia and hypertension accompanies hyperkalemia and precedes hypertension: Description of a large family with the Q565E WNK4 mutation. *J. Clin. Endocrinol. Metab.* **89**:4025–4030.
109. Jiang, Y., Ferguson, W.B., and Peng, J.B. 2007. WNK4 enhances TRPV5-mediated calcium transport: Potential role in hypercalciuria of familial hyperkalemic hypertension caused by gene mutation of WNK4. *Am. J. Physiol. Renal Physiol.* **292**:F545–F554.

110. Jiang, Y., Cong, P., Williams, S.R., Zhang, W., Na, T., Ma, H.P., and Peng, J.B. 2008. WNK4 regulates the secretory pathway via which TRPV5 is targeted to the plasma membrane. *Biochem. Biophys. Res. Commun.* **375**:225–229.
111. Zhang, W., Na, T., and Peng, J.B. 2008. WNK3 positively regulates epithelial calcium channels TRPV5 and TRPV6 via a kinase-dependent pathway. *Am. J. Physiol. Renal Physiol.* **295**:F1472–F1484.
112. Hoopes, R.R., Jr., Shrimpton, A.E., Knohl, S.J., Hueber, P., Hoppe, B., Matyus, J., Simckes, A. et al. 2005. Dent disease with mutations in OCRL1. *Am. J. Hum. Genet.* **76**:260–267.
113. Shrimpton, A.E., Hoopes, R.R., Jr., Knohl, S.J., Hueber, P., Reed, A.A., Christie, P.T., Igarashi, T. et al. 2009. OCRL1 mutations in Dent 2 patients suggest a mechanism for phenotypic variability. *Nephron Physiol.* **112**:27–36.
114. Sliman, G.A., Winters, W.D., Shaw, D.W., and Avner, E.D. 1995. Hypercalciuria and nephrocalcinosis in the oculocerebrorenal syndrome. *J. Urol.* **153**:1244–1246.
115. Wu, G., Zhang, W., Na, T., Jing, H., Wu, H., and Peng, J.B. 2012. Suppression of intestinal calcium entry channel TRPV6 by OCRL, a lipid phosphatase associated with Lowe syndrome and Dent disease. *Am. J. Physiol. Cell Physiol.* **302**:C1479–C1491.
116. Gkika, D., Topala, C.N., Chang, Q., Picard, N., Thebault, S., Houillier, P., Hoenderop, J.G., and Bindels, R.J. 2006. Tissue kallikrein stimulates Ca^{2+} reabsorption via PKC-dependent plasma membrane accumulation of TRPV5. *EMBO J.* **25**:4707–4716.
117. Topala, C.N., Schoeber, J.P., Searchfield, L.E., Riccardi, D., Hoenderop, J.G., and Bindels, R.J. 2009. Activation of the Ca^{2+}-sensing receptor stimulates the activity of the epithelial Ca^{2+} channel TRPV5. *Cell Calcium* **45**:331–339.
118. Zhang, W., Na, T., Wu, G., Jing, H., and Peng, J.B. 2010. Down-regulation of intestinal apical calcium entry channel TRPV6 by ubiquitin E3 ligase Nedd4-2. *J. Biol. Chem.* **285**:36586–36596.
119. Berridge, M.J., Lipp, P., and Bootman, M.D. 2000. The versatility and universality of calcium signalling. *Nat. Rev. Mol. Cell Biol.* **1**:11–21.
120. Berridge, M.J., Bootman, M.D., and Roderick, H.L. 2003. Calcium signalling: Dynamics, homeostasis and remodelling. *Nat. Rev. Mol. Cell Biol.* **4**:517–529.
121. Fan, H., Shen, Y.X., and Yuan, Y.F. 2014. Expression and prognostic roles of TRPV5 and TRPV6 in non-small cell lung cancer after curative resection. *Asian Pac. J. Cancer Prev.* **15**:2559–2563.
122. Giusti, L., Cetani, F., Da, V.Y., Pardi, E., Ciregia, F., Donadio, E., Gargini, C. et al. 2014. First evidence of TRPV5 and TRPV6 channels in human parathyroid glands: Possible involvement in neoplastic transformation. *J. Cell Mol. Med.* **18**:1944–1952.
123. Huhn, S., Bevier, M., Pardini, B., Naccarati, A., Vodickova, L., Novotny, J., Vodicka, P., Hemminki, K., and Forsti, A. 2014. Colorectal cancer risk and patients' survival: Influence of polymorphisms in genes somatically mutated in colorectal tumors. *Cancer Causes Control* **25**:759–769.
124. Jiang, Y., Gou, H., Zhu, J., Tian, S., and Yu, L. 2016. Lidocaine inhibits the invasion and migration of TRPV6-expressing cancer cells by TRPV6 downregulation. *Oncol. Lett.* **12**:1164–1170.
125. Kim, S.Y., Hong, C., Wie, J., Kim, E., Kim, B.J., Ha, K., Cho, N.H., Kim, I.G., Jeon, J.H., and So, I. 2014. Reciprocal positive regulation between TRPV6 and NUMB in PTEN-deficient prostate cancer cells. *Biochem. Biophys. Res. Commun.* **447**:192–196.
126. Peng, J.B., Zhuang, L., Berger, U.V., Adam, R.M., Williams, B.J., Brown, E.M., Hediger, M.A., and Freeman, M.R. 2001. CaT1 expression correlates with tumor grade in prostate cancer. *Biochem. Biophys. Res. Commun.* **282**:729–734.
127. Raphael, M., Lehen'kyi, V., Vandenberghe, M., Beck, B., Khalimonchyk, S., Vanden Abeele, F., Farsetti, L. et al. 2014. TRPV6 calcium channel translocates to the plasma membrane via Orai1-mediated mechanism and controls cancer cell survival. *Proc. Natl. Acad. Sci. USA* **111**:E3870–E3879.

128. Skrzypski, M., Khajavi, N., Mergler, S., Szczepankiewicz, D., Kolodziejski, P.A., Metzke, D., Wojciechowicz, T., Billert, M., Nowak, K.W., and Strowski, M.Z. 2015. TRPV6 channel modulates proliferation of insulin secreting INS-1E beta cell line. *Biochim. Biophys. Acta* **1853**:3202–3210.
129. Zhang, S.S., Xie, X., Wen, J., Luo, K.J., Liu, Q.W., Yang, H., Hu, Y., and Fu, J.H. 2016. TRPV6 plays a new role in predicting survival of patients with esophageal squamous cell carcinoma. *Diagn. Pathol.* **11**:14.
130. Cooperberg, M.R., Moul, J.W., and Carroll, P.R. 2005. The changing face of prostate cancer. *J. Clin. Oncol.* **23**:8146–8151.
131. Schroder, F.H. 2010. Prostate cancer around the world. An overview. *Urol. Oncol.* **28**:663–667.
132. Fixemer, T., Wissenbach, U., Flockerzi, V., and Bonkhoff, H. 2003. Expression of the Ca^{2+}-selective cation channel TRPV6 in human prostate cancer: A novel prognostic marker for tumor progression. *Oncogene* **22**:7858–7861.
133. Bodding, M., Fecher-Trost, C., and Flockerzi, V. 2003. Store-operated Ca^{2+} current and TRPV6 channels in lymph node prostate cancer cells. *J. Biol. Chem.* **278**:50872–50879.
134. Vanden Abeele, F., Roudbaraki, M., Shuba, Y., Skryma, R., and Prevarskaya, N. 2003. Store-operated Ca^{2+} current in prostate cancer epithelial cells. Role of endogenous Ca^{2+} transporter type 1. *J. Biol. Chem.* **278**:15381–15389.
135. Lehen'kyi, V., Flourakis, M., Skryma, R., and Prevarskaya, N. 2007. TRPV6 channel controls prostate cancer cell proliferation via Ca^{2+}/NFAT-dependent pathways. *Oncogene* **26**:7380–7385.
136. Bolanz, K.A., Hediger, M.A., and Landowski, C.P. 2008. The role of TRPV6 in breast carcinogenesis. *Mol. Cancer Ther.* **7**:271–279.
137. Peters, A.A., Simpson, P.T., Bassett, J.J., Lee, J.M., Da, S.L., Reid, L.E., Song, S. et al. 2012. Calcium channel TRPV6 as a potential therapeutic target in estrogen receptor-negative breast cancer. *Mol. Cancer Ther.* **11**:2158–2168.
138. Dhennin-Duthille, I., Gautier, M., Faouzi, M., Guilbert, A., Brevet, M., Vaudry, D., Ahidouch, A., Sevestre, H., and Ouadid-Ahidouch, H. 2011. High expression of transient receptor potential channels in human breast cancer epithelial cells and tissues: Correlation with pathological parameters. *Cell Physiol. Biochem.* **28**:813–822.
139. Chow, J., Norng, M., Zhang, J., and Chai, J. 2007. TRPV6 mediates capsaicin-induced apoptosis in gastric cancer cells—Mechanisms behind a possible new "hot" cancer treatment. *Biochim. Biophys. Acta* **1773**:565–576.
140. Lau, J.K., Brown, K.C., Dom, A.M., Witte, T.R., Thornhill, B.A., Crabtree, C.M., Perry, H.E. et al. 2014. Capsaicin induces apoptosis in human small cell lung cancer via the TRPV6 receptor and the calpain pathway. *Apoptosis* **19**:1190–1201.
141. Nilius, B., Prenen, J., Vennekens, R., Hoenderop, J.G., Bindels, R.J., and Droogmans, G. 2001. Pharmacological modulation of monovalent cation currents through the epithelial Ca^{2+} channel ECaC1. *Br. J. Pharmacol.* **134**:453–462.
142. Schwarz, E.C., Wissenbach, U., Niemeyer, B.A., Strauss, B., Philipp, S.E., Flockerzi, V., and Hoth, M. 2006. TRPV6 potentiates calcium-dependent cell proliferation. *Cell Calcium* **39**:163–173.
143. Cove-Smith, A., Mulgrew, C.J., Rudyk, O., Dutt, N., McLatchie, L.M., Shattock, M.J., and Hendry, B.M. 2013. Anti-proliferative actions of T-type calcium channel inhibition in Thy1 nephritis. *Am. J. Pathol.* **183**:391–401.
144. Hofer, A., Kovacs, G., Zappatini, A., Leuenberger, M., Hediger, M.A., and Lochner, M. 2013. Design, synthesis and pharmacological characterization of analogs of 2-aminoethyl diphenylborinate (2-APB), a known store-operated calcium channel blocker, for inhibition of TRPV6-mediated calcium transport. *Bioorg. Med. Chem.* **21**:3202–3213.

145. Chokshi, R., Fruasaha, P., and Kozak, J.A. 2012. 2-Aminoethyl diphenyl borinate (2-APB) inhibits TRPM7 channels through an intracellular acidification mechanism. *Channels (Austin)* **6**:362–369.
146. Muller, D., Hoenderop, J.G., Vennekens, R., Eggert, P., Harangi, F., Mehes, K., Garcia-Nieto, V. et al. 2002. Epithelial Ca^{2+} channel (ECAC1) in autosomal dominant idiopathic hypercalciuria. *Nephrol. Dial. Transplant.* **17**:1614–1620.
147. Renkema, K.Y., Lee, K., Topala, C.N., Goossens, M., Houillier, P., Bindels, R.J., and Hoenderop, J.G. 2009. TRPV5 gene polymorphisms in renal hypercalciuria. *Nephrol. Dial. Transplant.* **24**:1919–1924.
148. Khaleel, A., Wu, M.S., Wong, H.S., Hsu, Y.W., Chou, Y.H., and Chen, H.Y. 2015. A single nucleotide polymorphism (rs4236480) in TRPV5 calcium channel gene is associated with stone multiplicity in calcium nephrolithiasis patients. *Mediators Inflamm.* **2015**:375427.
149. Oddsson, A., Sulem, P., Helgason, H., Edvardsson, V.O., Thorleifsson, G., Sveinbjornsson, G., Haraldsdottir, E. et al. 2015. Common and rare variants associated with kidney stones and biochemical traits. *Nat. Commun.* **6**:7975.
150. Na, T., Zhang, W., Jiang, Y., Liang, Y., Ma, H.P., Warnock, D.G., and Peng, J.B. 2009. The A563T variation of the renal epithelial calcium channel TRPV5 among African Americans enhances calcium influx. *Am. J. Physiol. Renal Physiol.* **296**:F1042–F1051.
151. Wang, L., Holmes, R.P., and Peng, J.B. 2016. Molecular modeling of the structural and dynamical changes in calcium channel TRPV5 induced by the African-specific A563T variation. *Biochemistry* **55**:1254–1264.
152. van der Eerden, B.C., Hoenderop, J.G., de Vries, T.J., Schoenmaker, T., Buurman, C.J., Uitterlinden, A.G., Pols, H.A., Bindels, R.J., and van Leeuwen, J.P. 2005. The epithelial Ca^{2+} channel TRPV5 is essential for proper osteoclastic bone resorption. *Proc. Natl. Acad. Sci. USA* **102**:17507–17512.
153. Suzuki, Y., Pasch, A., Bonny, O., Mohaupt, M.G., Hediger, M.A., and Frey, F.J. 2008. Gain-of-function haplotype in the epithelial calcium channel TRPV6 is a risk factor for renal calcium stone formation. *Hum. Mol. Genet.* **17**:1613–1618.
154. Akey, J.M., Swanson, W.J., Madeoy, J., Eberle, M., and Shriver, M.D. 2006. TRPV6 exhibits unusual patterns of polymorphism and divergence in worldwide populations. *Hum. Mol. Genet.* **15**:2106–2113.
155. Hughes, D.A., Tang, K., Strotmann, R., Schoneberg, T., Prenen, J., Nilius, B., and Stoneking, M. 2008. Parallel selection on TRPV6 in human populations. *PLoS One* **3**:e1686.
156. Sarmina, I., Spirnak, J.P., and Resnick, M.I. 1987. Urinary lithiasis in the black population: An epidemiological study and review of the literature. *J. Urol.* **138**:14–17.

14 Determining the Crystal Structure of TRPV6

Kei Saotome, Appu K. Singh, and Alexander I. Sobolevsky

CONTENTS

14.1 INTRODUCTION

Calcium ions play important roles in many physiological processes, including neurotransmitter release, excitation–contraction coupling, cell motility, and gene expression [1]. Cellular calcium levels are precisely tuned by various channels and transporters. Transient receptor potential (TRP) channels, which are generally non-selective cation channels, conduct Ca^{2+} in response to disparate activators, including sensory stimuli such as temperature, touch, and pungent chemicals [2]. Members 5 and 6 of vanilloid subfamily (TRPV5 and TRPV6, previously named ECaC1 and ECaC2/CaT1, respectively) are uniquely Ca^{2+}-selective ($P_{Ca}/P_{Na} > 100$) [3,4] TRP channels, both of which were identified in 1999 by expression cloning strategies utilizing cDNA libraries from rabbit kidney [5] and rat duodenum [6], respectively. While TRPV5 expression is mainly restricted to the kidney, TRPV6 is expressed in various tissues including the stomach, small intestine, prostate, esophagus, colon, and placenta. Genetic knockout of TRPV5 or TRPV6 in mice suggests the importance of these channels for Ca^{2+} homeostasis. TRPV5 knockout mice showed defects in renal Ca^{2+} reabsorption and reduced bone thickness [7], and the knockout of TRPV6 resulted in defective intestinal calcium absorption, decreased bone mineral density, reduced fertility, and hypocalcemia when challenged with a low Ca^{2+} diet [8]. Further support for the role of these channels in Ca^{2+} absorption and homeostasis stems from the robust regulation of their expression by the calciotropic hormone vitamin D [9–12] (see Chapter 13).

TRPV6 has been shown to be aberrantly expressed in numerous cancer types, including carcinomas of the colon, prostate, breast, and thyroid [13–18]. The correlation between TRPV6 expression and tumor malignancy and its potential contribution to cancer cell survival has highlighted TRPV6 as a target for cancer diagnosis and treatment [15,17,19,20]. Indeed, a selective inhibitor of TRPV6 activity derived from northern short-tailed shrew venom [21] has entered phase I clinical trials in patients with advanced solid tumors of tissues known to express TRPV6, including the pancreas and ovary [22].

Structurally, TRPV5 and TRPV6 share ~75% sequence identity with each other and are ~25% identical to the founding member of the TRPV subfamily TRPV1. The transmembrane (TM) domain has the same topology as tetrameric K^+ channels [23], with six TM helices (S1–S6) and a pore-forming re-entrant loop between the S5 and S6. Importantly, this loop contains a conserved aspartate residue that is critical for the calcium permeability of TRPV5 and TRPV6 [24,25], suggesting that this residue at least in part comprises the selectivity filter. Flanking the TM domain are relatively large intracellular N- and C-termini. The N-terminus, which includes six ankyrin repeats [26], is critical for proper channel assembly and function [27,28], while the C-terminus contains domains involved in Ca^{2+}/calmodulin-dependent inactivation [29–31] (see Chapter 13).

To help understand the functional mechanisms of TRPV5/6 and potentially inform rational drug design, we sought to obtain a high-resolution structure of an intact channel. Until several years ago, the only viable method of obtaining such a structure was x-ray crystallography. However, producing well-diffracting crystals can be a notoriously difficult and resource-consuming process because membrane proteins typically have low expression and purification yields, poor stability in detergent, and inherent flexibility [32]. However, structural biologists are now able to circumvent this major bottleneck, owing to recent advances in single-particle cryo-electron microscopy (cryo-EM), which have facilitated the determination of membrane protein structures at near-atomic resolutions without prior crystallization [33]. These advances have had a particularly profound effect on the TRP channel field, as atomic-level cryo-EM structures have been determined for TRPV1 [34–36]; TRPV2 [37,38]; and ankyrin subfamily member [39]. The ability to computationally select specific conformational states from a heterogeneous cryo-EM sample can be especially powerful when studying mechanisms of gating, as exemplified by studies of TRPV1 in various ligand-induced conformations [34–36]. Cryo-EM will surely continue to be exploited with great effect to elucidate structures of TRP channels and other membrane proteins.

As yet, there are several benefits that may make obtaining an x-ray structure desirable over cryo-EM. First, crystallographers can use true statistical approaches such as the Free R value [40] to evaluate the accuracy of atomic models against experimental data, while analogous methods in cryo-EM [41–43] are still relatively nascent. Second, the resolution of a cryo-EM map usually varies widely across a single reconstruction, with more flexible regions, typically in the periphery, being less resolved or completely absent. For example, in TRPV1, while the TM domain is well resolved, the first two ankyrin repeats at the distal N-terminus are missing from the electron density maps [34–36], presumably due to their flexibility (Figure 14.1a).

In x-ray structures, the resolution obtained from the diffraction data is more representative of the structure as a whole, and peripheral or flexible regions may be stabilized by crystal contacts and thus adequately resolved (Figure 14.1b). Third and perhaps most importantly, the position of anomalous scatterers, such as selenium atoms in selenomethionine-labeled protein, sulfur atoms in native cysteine or methionine side chains, or heavy atoms bound to the protein, can be accurately identified with little ambiguity. Anomalous scattering can therefore be utilized to robustly aid or validate sequence registry (Figure 14.1c), which is especially important for low-resolution structures and/or regions with poor electron density. Using anomalous scattering to identify bound ions is particularly useful for studying ion channel structures, as ion binding at specific locations is vital for understanding permeation and ion channel block.

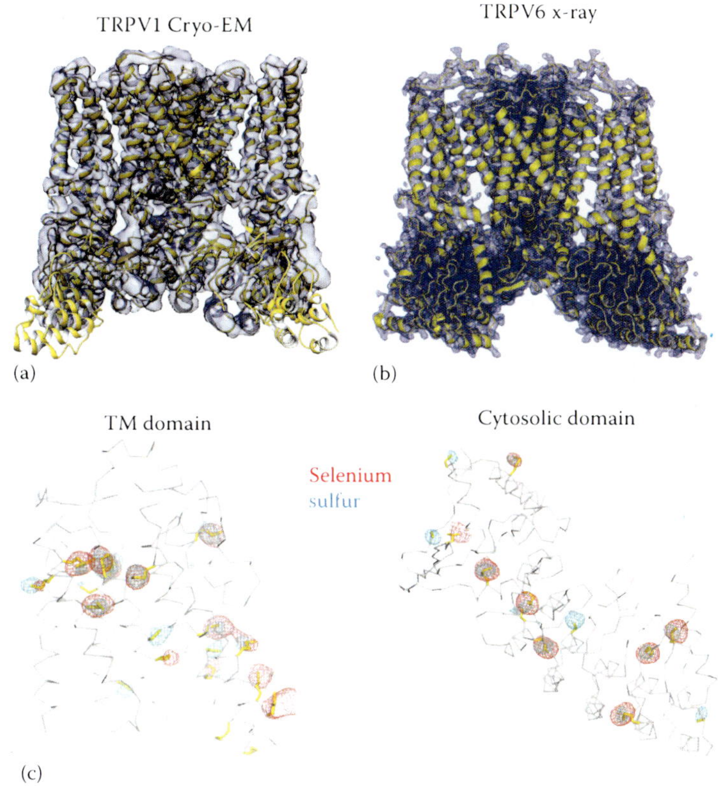

FIGURE 14.1 Crystallographic analysis of TRPV6. (a) Cartoon model of TRPV1 (yellow, PDB ID: 3J5P) with cryo-EM density (blue surface, EMD-5778) superimposed. Note the peripheral regions of the model, consisting of the first two ankyrin repeats, do not have corresponding density. (b) Cartoon model of TRPV6 (yellow, PDB ID: 5IWK) with $2F_O–F_C$ map (blue mesh, contoured at 1.0 σ) superimposed. (c) Anomalous difference Fourier maps for selenium (red mesh, contoured at 3.2 σ) and sulfur (teal mesh, contoured at 3.0 σ) superimposed onto the ribbon model of TM (left) and cytosolic (right) domains of TRPV6. Side chains of cysteines and methionines are shown as yellow sticks. *(Continued)*

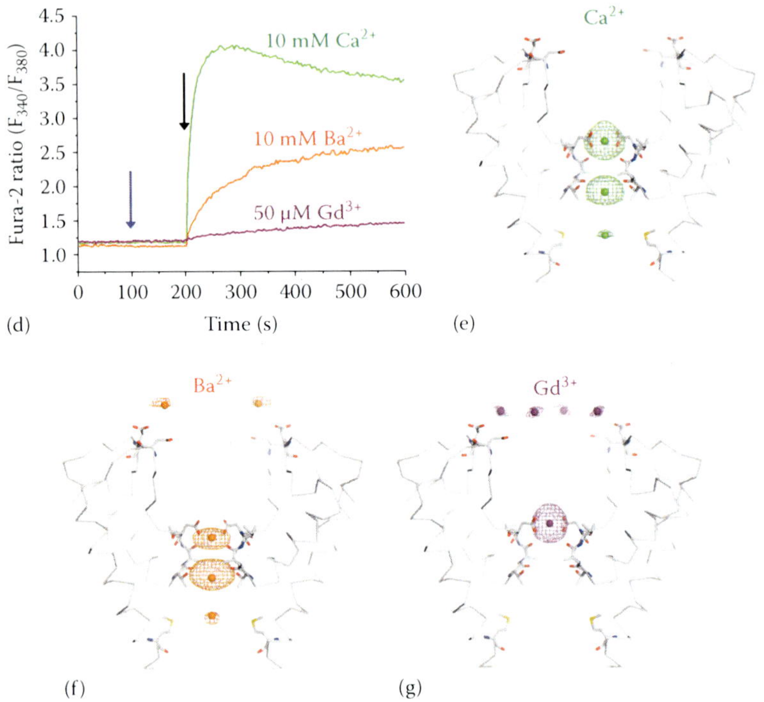

FIGURE 14.1 (*Continued*) Crystallographic analysis of TRPV6. (d) Ratiometric Fura-2 measurements of cation permeation and a block of rat TRPV6 expressed in HEK cells. To study Ca^{2+} influx (green trace) and Ba^{2+} influx (orange trace), 10 mM of the corresponding ion was added at the time indicated by a black arrow. To study the Gd^{3+} block of Ca^{2+} influx (purple trace), 50 μM Gd^{3+} and 5 mM Ca^{2+} were added at the times indicated by the blue and black arrows, respectively. (e–g) Ribbon models of the TRPV6 pore bound to Ca^{2+} (e, green spheres), Ba^{2+} (f, orange spheres), or Gd^{3+} (g, purple spheres). Only two of four subunits are shown with the front and back subunits removed for clarity. Residues important for cation-binding are shown as sticks. Anomalous difference Fourier maps generated from diffraction data collected at 1.75 Å wavelength x-rays for Ca^{2+} (e, green mesh, contoured at 2.3 σ) and Ba^{2+} (f, orange mesh, contoured at 3.5 σ) and at 1.56 Å wavelength x-rays for Gd^{3+} (g, purple mesh, contoured at 8.0 σ) are overlaid onto the models.

For TRPV6, we used these techniques to identify binding sites for the permeant cations Ca^{2+} and Ba^{2+}, as well as the channel blocker Gd^{3+} (Figure 14.1d through g). Methods to unambiguously identify specific atoms or small labels in cryo-EM electron density maps have yet to be developed. We were motivated by each of these factors as we attempted to crystallize TRPV5/6 in the midst of the cryo-EM "resolution revolution."

In 2016, we reported the crystal structure of intact rat TRPV6 at 3.25 Å resolution [44]. To our knowledge, this represented the first crystal structure of a TRP channel and the second crystal structure of a naturally occurring Ca^{2+}-selective channel, after the structure of the calcium release-activated calcium (CRAC) channel Orai reported in 2012 [45]. A detailed description of the TRPV6 structure can be found elsewhere [44]. In this chapter we will focus on the multiyear journey taken to determine the structure, in which >150 constructs were purified and subjected to crystallization screening, and thousands of crystals were tested for diffraction. We will summarize the methods used to screen constructs and precrystallization conditions, express and purify protein, grow and optimize crystals, and collect and analyze diffraction data. Finally, we will briefly compare the structural bases of Ca^{2+}-selective permeation in TRPV6 and Orai.

14.2 PRECRYSTALLIZATION SCREENING OF PROTEIN EXPRESSION AND BIOCHEMICAL BEHAVIOR USING FSEC

Obtaining milligram quantities of pure, monodisperse protein is a general prerequisite for protein crystallography studies. As large-scale membrane protein expression and purification can be very costly and time consuming, methods to efficiently and rapidly screen genetic constructs and buffer compositions at smaller scales are necessary to optimize protein expression levels and biochemical behavior. For this project, we employed fluorescence-detection size exclusion chromatography (FSEC) [46], where the target protein is expressed as a fusion with green fluorescent protein (GFP) (or other fluorescent proteins) and crude lysate containing just nanogram quantities of the fusion protein is loaded onto a gel filtration column. The resulting chromatogram provides a quick assessment of the identity (elution time), biochemical behavior (number, sharpness, and symmetry of chromatographic peaks), and expression level (peak amplitude) of the target protein. FSEC experiments can routinely be conducted using adherent Sf9 or HEK cells in a six-well plate (Figure 14.2a) to inform larger scale expression, which can be carried out in polycarbonate Erlenmeyer flasks (Figure 14.2b). A concurrent advantage of expressing the target protein as a GFP fusion, even at the large scale, is that expression levels can be monitored in real time by epifluorescence microscopy (Figure 14.2c).

Initially, we used FSEC to screen C-terminal GFP fusions of ~20 TRPV5/6 orthologues for crystallographic studies. To facilitate affinity purification and proteolytic removal of the GFP tag, a strep tag (WSHPQFEK) or 8x His tag was added to the C-terminus and thrombin site cleavage site (LVPRGS) was inserted between the target protein and GFP. While a majority of orthologues showed poor chromatographic behavior, rat TRPV6 (rTRPV6), which displayed a single sharp, dominant FSEC peak, was selected as a promising candidate for further study. As our initial efforts to crystallize wildtype rTRPV6 protein were unsuccessful, we generated truncation mutations and screened them by FSEC. A small-scale FSEC screen of C-terminal truncations is shown in Figure 14.2d. Several of these truncations improved the expression level of rTRPV6. In parallel, we used FSEC to screen a panel of detergents used for the extraction (solubilization) and purification of the target protein (Figure 14.2e). We also used FSEC to optimize parameters for protein expression, including incubation time (Figure 14.2f) and the concentration of histone deacetylase inhibitor sodium butyrate (NaBu) added to improve the expression level (Figure 14.2g). Exploiting FSEC at the small-scale to inform large-scale purifications was critical to maximizing the efficiency of the crystallization trials.

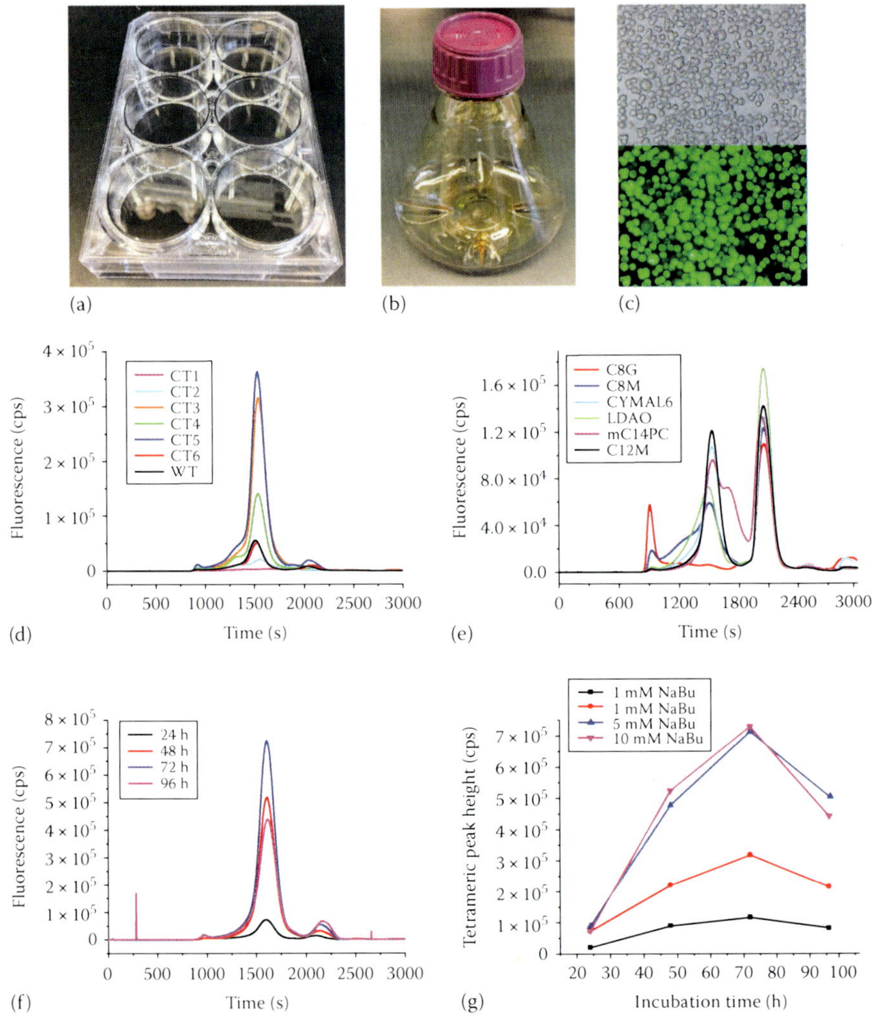

FIGURE 14.2 TRPV6 expression and FSEC screening. (a) Six-well plate used for small-scale expression of TRPV6 in adherent Sf9 or HEK cells. (b) Baffled polycarbonate Erlenmeyer flask used for TRPV6 expression in suspension-adapted HEK 293S cells. (c) HEK 293S cells transduced with baculovirus to express a TRPV6–GFP fusion construct viewed in visible light (top) and their GFP fluorescence (bottom). (d) FSEC profiles for various rat TRPV6–GFP C-terminal truncation mutants. Note the drastic differences in peak amplitude depending on the construct. (e) FSEC screen of crudely solubilized rat TRPV6–GFP in various detergents. (f) FSEC profiles for HEK 293S cells in suspension expressing a TRPV6–GFP fusion construct. Samples were taken at various time points after transduction. In this experiment, 10 mM NaBu was added 8 h after transduction. (g) FSEC tetrameric peak height for a TRPV6–GFP fusion construct depending on NaBu concentration and incubation time after transduction.

14.3 LARGE-SCALE PURIFICATION AND CRYSTALLIZATION

Large-scale purifications (800 mL–6 L) were conducted by the baculovirus infection of Sf9 cells or baculovirus-mediated transduction of HEK293 cells [47]. We found that the BacMam system provided ~2x greater yields (~1 mg/L of cells depending on the construct) than Bac-to-Bac (~0.5 mg/L). A flow chart of the process of going from a sequenced plasmid containing the desired target construct to a purified protein ready for crystallization trials, which takes approximately 2–3 weeks, is shown in Figure 14.3a. Throughout the project, we kept the same purification protocol overall but varied the expressed protein construct and the detergent (and lipid) used for solubilization.

For large-scale expression, suspension-adapted HEK 293S cells lacking *N*-acetylglucosaminyltransferase I were grown in Freestyle 293 media (Life Technologies) supplemented with 2% FBS at 37°C in the presence of 5% CO_2. The culture was transduced with P2 baculovirus once cells reached a density of 2.5–3.5 × 10^6 per mL. After 8–12 h, 10 mM NaBu was added and the temperature was changed to 30°C. Cells were harvested 48–72 h posttransduction and resuspended in a buffer containing 150 mM NaCl, 20 mM Tris–HCl pH 8.0, 1 mM β-mercaptoethanol (βME), 0.8 μM aprotinin, 2 μg/mL leupeptin, 2 mM pepstatin A, and 1 mM phenylmethysulfonyl fluoride (PMSF). The cells were disrupted using a Misonix Sonicator (12 × 15 s, power level 7), and the resulting homogenate was clarified by spinning in a Sorvall centrifuge at 7500 rpm for 15 min. Crude membranes were collected by ultracentrifugation for 1 h in a Beckman Ti45 rotor at 40,000 rpm. The membranes were mechanically homogenized and subsequently solubilized for 2–4 h in a buffer containing 150 mM NaCl, 20 mM Tris–HCl pH 8.0, 1 mM βME, 20 mM *n*-dodecyl-β-D-maltopyranoside (DDM), 0.8 μM aprotinin, 2 μg/mL leupeptin, 2 mM pepstatin A, and 1 mM PMSF. After insoluble material was removed by ultracentrifugation, streptavidin-linked resin was added to the supernatant and rotated for 4–16 h. Resin was washed with 10 column volumes of wash buffer containing 150 mM NaCl, 20 mM Tris pH 8.0, 1 mM βME, and 1 mM DDM, and the protein was eluted using wash buffer supplemented with 2.5 mM D-desthiobiotin. The eluted fusion protein was concentrated to ~1.0 mg/mL and digested with thrombin at a mass ratio of 1:100 (thrombin:protein) for 1.5 h at 22°C. The digested protein was concentrated and injected into a Superose 6 column equilibrated in a buffer composed of 150 mM NaCl, 20 mM Tris–HCl pH 8.0, 1 mM βME, and 0.5 mM DDM. Ten millimolar tris(2-carboxyethyl)phosphine (TCEP) was added to fractions with elution time corresponding to the tetrameric channel, and protein was concentrated to 2.5–3.0 mg/mL using a 100 kDa MWCO concentrator. All purification steps were conducted at 4°C. Typical purifications yielded ~1 mg of purified protein per liter of transduced cells. Sodium dodecyl sulfate polyacrylamide gel electrophoresis (SDS-PAGE) (Figure 14.3b) and FSEC (Figure 14.3c) analysis revealed a final product with high chemical and conformational homogeneity.

We produced TRPV6 protein labeled with the unnatural amino acid selenomethionine for the subsequent detection of the selenium atoms to aid sequence registry during model building. In regions of the protein devoid of methionines in the natural sequence, we introduced methionine substitutions. Protocols to express selenomethionine-labeled protein in HEK cells were adapted from literature [48].

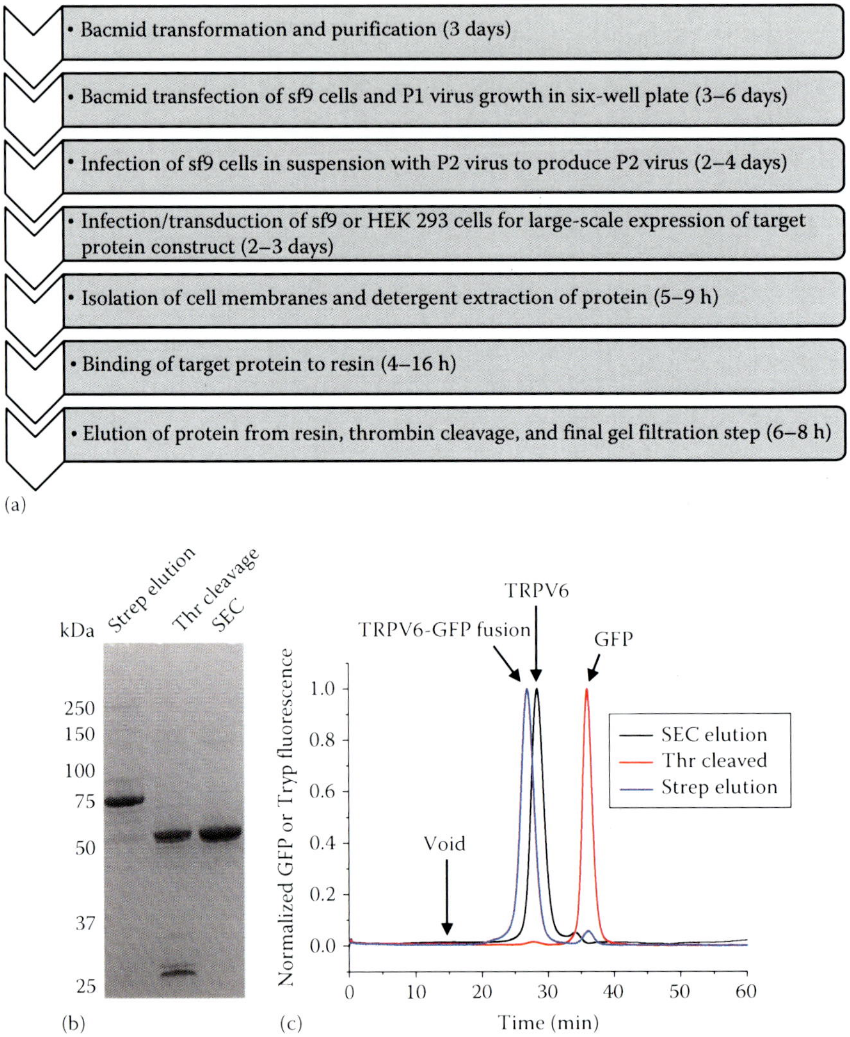

FIGURE 14.3 Purification of TRPV6. (a) Flow chart of TRPV6 expression and purification starting from bacmid isolation and ending with the protein ready for crystallization trials. (b) Example coomassie stained SDS-PAGE analysis of a TRPV6–GFP fusion construct purification. Samples were taken after the elution of fusion protein from strep resin, proteolytic cleavage of GFP by thrombin, and final separation by size exclusion chromatography (SEC). (c) FSEC profiles of the samples shown in (b). For the GFP fusion protein and post-thrombin cleavage sample, GFP fluorescence was monitored, while tryptophan (tryp) fluorescence was tracked for the final SEC-purified protein. Chromatograms were normalized to the amplitude of the major peak for comparison.

Specifically, 6–8 h after transduction with P2 baculovirus, cells were pelleted and resuspended in DMEM (Life Technologies) supplemented with 10% FBS and lacking L-methionine. After incubating methionine-depleted cells for 6 h at 37°C, 60 mg of L-selenomethionine was added per liter of cells. Thirty-six to forty-eight hours after transduction, cells were harvested and protein was purified using the same protocol as described earlier, except for the addition of 4 mM L-methionine to all purification buffers, excluding the final gel filtration buffer. This procedure yielded ~0.4 mg of selenomethionine-labeled protein per liter of transduced cells.

Purified protein was routinely subjected to high-throughput screening (~100 nL protein per drop) of crystallization facilitated by liquid dispensing robotics and automated imaging systems. Initial tests identified the best range of protein concentrations (2–3 mg/mL) and crystallization temperatures (20°C–22°C) that were maintained in a majority of subsequent crystallization screening experiments. Primary "hits" from the screening experiments were imaged for UV fluorescence (Figure 14.4a, d, and g) and scaled up (1–2 μL protein per drop) in sitting drop or hanging drop vapor diffusion trays. Overall, we found that hanging drop vapor diffusion resulted in the largest

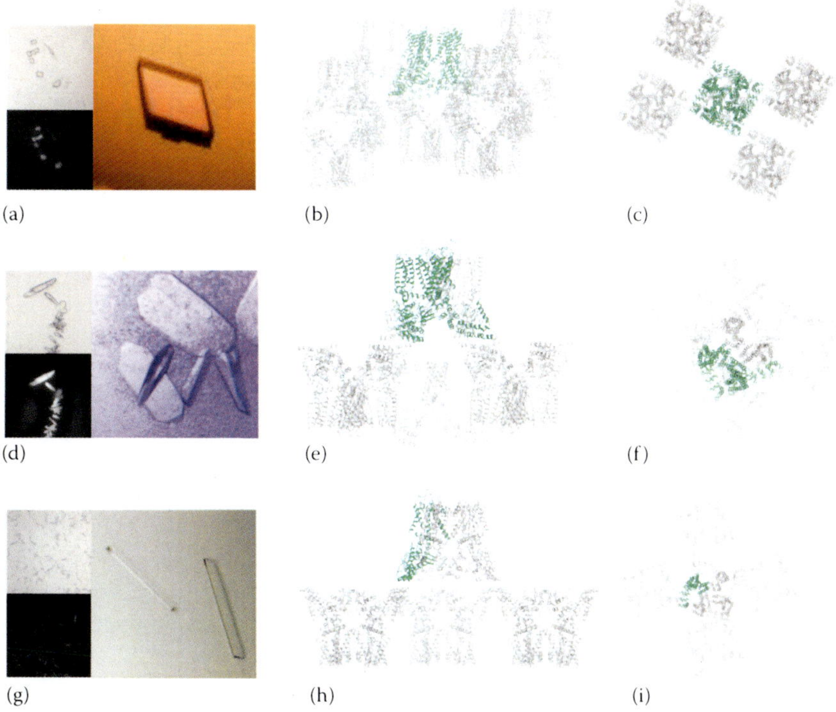

FIGURE 14.4 Crystallization of TRPV6 mutants in various space groups. (a, d, and g) Initial crystals from high-throughput screen imaged using white (top left) or UV (bottom left) light and manually optimized crystals (right) in C2 (a), $C222_1$ (d), or $P42_12$ (g) space groups. (b, c, e, f, h, and i) Orthogonal views of crystal lattice packing for C2 (b and c), $C222_1$ (e and f), or $P42_12$ (h and i) space groups. The protein contents of the asymmetric unit are shown in green.

and most reproducible crystals (Figure 14.4a, d, and g). Crystallization conditions were optimized for pH, precipitant concentration, salt concentration, salt type, and protein to mother liquor ratio. To obtain structures in the presence of various cations, protein was incubated with Ca^{2+}, Ba^{2+}, or Gd^{3+} for ~1 h prior to crystallization. Selenomethionine-labeled protein crystallized in similar conditions to native protein but yielded significantly smaller crystals. We also manually screened for additives (salts, volatiles, other small molecules, or detergents) to improve crystal size or morphology by briefly incubating the protein with the additive prior to crystallization. In addition, a variety of postcrystallization treatments to improve diffraction quality were attempted, including dehydration, chemical cross-linking of free amines, and screening various cryoprotectants [49].

14.4 COLLECTION AND PROCESSING OF DIFFRACTION DATA

X-ray diffraction data were collected at NSLS X29, APS NECAT 24 ID-C/E, and ALS 5.0.1/5.0.2 synchrotron beamlines. Prior to data collection, crystals were harvested in nylon loops matching the crystal size, cryoprotected by incubating in cryoprotectant solution (usually the mother liquor used for crystal growth with added glycerol, ethylene glycol, or low molecular weight polyethylene glycol) and plunged and stored in liquid nitrogen. Crystals were transferred to ALS-style or universal pucks for shipping and to aid remote-controlled robotics for crystal positioning and data collection at the beamline.

Crystals were initially screened for diffraction quality by visually examining diffraction patterns after exposure to x-rays at several angles and locations along the crystal. Those with satisfactory diffraction properties were selected for the collection of a complete data set. The angular coverage necessary to collect a complete diffraction data set depends on the symmetry of the crystal lattice. Thus, the strategy employed for collecting full data sets was chosen based on the crystal form. For example, for crystals in the $P42_12$ space group, the crystal must be rotated at least 90° during data collection (with 1–3 frames collected per degree, depending on the detector used and crystal mosaicity) to collect a complete data set with the maximum number of unique diffraction spots. To minimize radiation decay during data collection, crystals significantly larger than the x-ray beam cross section were translated along a vector orthogonal to the beam.

To crystallographically identify cations such as Ca^{2+}, Ba^{2+}, and Gd^{3+} (Figure 14.1e through g), the x-ray beam was tuned to specific wavelengths at the synchrotron to maximize the anomalous scattering of the cation. The wavelengths chosen were 1.75 Å for Ca^{2+} and Ba^{2+} and 1.56 Å for Gd^{3+}. To detect anomalous scattering, precise differences between each diffraction spot and its centrosymmetrically related partner (Friedel pair) must be calculated. As such, care must be taken to minimize radiation damage in the time between collections of Friedel pairs. We thus employed the "inverse beam" data collection approach, in which diffraction from small (~10°) wedges related centrosymmetrically (separated by 180°) are collected consecutively. The same approach was used to identify selenium atoms in crystals with selenomethionine-labeled protein, as well as sulfur atoms in crystals with native protein.

The experimentally collected spot intensities in diffraction data were indexed, integrated, scaled, and merged together using processing programs such as XDS [50], HKL2000 [51], and IMOSFLM [52]. The scaled intensities were converted into the structure factor amplitudes by the CTRUNCATE program in CCP4 suite [53]. While the x-ray detector provides the structure factor amplitudes, it cannot collect data to provide the corresponding phase information, which is necessary to convert the structure factors into electron density. Thus, to obtain the structure, crystallographers must solve the "phase problem" in one of several ways [54]. When homologous (~25% identity or greater) structures of the protein or its domain(s) are available, the phases can be obtained by the method of molecular replacement (MR). In MR, the homologous structure(s), or "search model(s)," are rotated and translated in a crystal lattice according to various algorithms [55] and structure factors are back calculated and compared to experimental data. If the MR program can find a search model orientation that adequately matches the experimental diffraction, electron density maps can be calculated from the obtained phases. The resulting model is then refined against the map to improve the phases and the accuracy of the structure.

We used the Phaser [56] program to obtain phases and an initial structural solution by MR with the structure of mouse TRPV6 ankyrin domain (PDB code 2RFA; consisting of residues 44–225) as a search model. The resulting structural model was iteratively refined in REFMAC [57] or PHENIX [58] and built in COOT [59]. The TM domain was built using the TRPV1 cryo-EM reconstruction (PDB code: 3J5P) [35] as a guide. The structural model improvement was monitored by the gradual decrease in R-factors. At the final stages, the model quality was assessed by low values of the R-factors and good stereochemistry (low values for Ramachandran outliers, bond angle, and bond length deviations).

14.5 PROTEIN ENGINEERING TO IMPROVE CRYSTAL PACKING

Obtaining well-diffracting crystals of membrane proteins requires the optimization of many parameters, including the target genetic construct, expression system, detergents/lipids used for protein solubilization, buffer composition, crystallization method (vapor diffusion, microbatch, lipid crystalline phase, etc.), crystallization condition, and postcrystallization treatments. Perhaps the most important determinant for the crystallization of membrane proteins is the composition of its amino acid sequence; genetic engineering of the crystallizing construct has been critical for making various classes of membrane proteins amenable to crystallization. For example, the tethering of fusion partners such as T4 lysozyme has aided the crystallization of many different G-protein coupled receptors (GPCRs) [60], while the screening of mutants to improve thermostability has been employed for various targets including ionotropic glutamate receptor [61], serotonin transporter [62,63], and GPCRs [64–66].

Initial efforts to improve crystallization of TRPV6 by construct engineering included N- and C-terminal truncations, deletions of hypervariable regions such as the extracellular loop connecting TM segments S1 and S2, mutations of exposed cysteines to prevent nonspecific aggregation, mutations of a conserved N-linked glycosylation site [67], and mutations of high entropy residues such as lysine and glutamic acid [68]. After the optimization of purification conditions and crystallization of these constructs as described earlier, the diffracting power of the crystals was limited to approximately 6 Å resolution.

While the data at this resolution were not useful for building an accurate structural model of TRPV6, we were able to obtain MR solutions (see preceding text) that accurately placed the protein's ankyrin domains in a crystal lattice, thus providing information about secondary structure elements or residues involved in crystal lattice contacts. Based on this information, we generated mutants (substitutions, deletions, and insertion of fusion partners) aimed at strengthening crystal contacts or favoring new ones. Each of these mutants was tested for expression and biochemical behavior by FSEC prior to purification and crystallization trials. While a majority of the >100 mutants informed by this MR solution-based strategy either had no appreciable effect or completely ablated crystallization, several constructs resulted in new crystal forms that have different crystal packing contacts (Figure 14.4b, c, e, f, h, and i). The construct used to build the final structural model contained three substitutions in the ankyrin domain (I62Y, L92N, and M96Q), in addition to a single substitution in the TM domain and a C-terminal truncation. Of these three ankyrin domain mutations, single substitution I62Y completely favored crystallization in the $P42_12$ (Figure 14.4g through i) space group over $C222_1$ (Figure 14.4d through f), despite identical purification protocols and similar crystallization conditions. The best crystals in the $P42_12$ space group diffracted to a much higher resolution limit (~3.3 Å) than in $C222_1$ (~4.0 Å), allowing us to build a significantly more accurate and complete structural model. Overall, these results are a striking example of how subtle changes in the protein construct can have enormous effects on the success of a membrane protein crystallization project.

14.6 COMPARISON OF TRPV6 AND ORAI STRUCTURES

Rat TRPV6 [44] and *Drosophila* Orai [45] are the only eukaryotic Ca^{2+}-selective ion channels with crystal structures available. The recently published 4.2 Å cryo-EM structure of the $Ca_v1.1$ complex [69] and crystal structures of engineered Ca^{2+}-selective prokaryotic channels [70] are beyond the scope of this discussion. Orai proteins serve as the pore-forming subunits of the CRAC channel, which is activated by interaction with the intracellular calcium sensor, the stromal interaction molecule (STIM) [71,72]. TRPV6 and Orai share similar biophysical properties, including high calcium selectivity, permeability to monovalent cations in the absence of external divalents, channel block by trivalent cations, strong inward rectification, and Ca^{2+}/calmodulin-dependent inactivation [72,73]. Due to these similarities, in 2001 TRPV6 was proposed to also compose the pore of the CRAC channel [4]. However, distinct features of TRPV6 and CRAC channels, including higher Cs^+ permeability in TRPV6, activation of CRAC channel but not TRPV6 by ionomycin-induced Ca^{2+} store depletion, and voltage-dependent Mg^{2+} block of TRPV6, but not CRAC channels, subsequently rebuffed this idea [74].

While TRPV6 and Orai differ completely in sequence, fold, and oligomeric state (Figure 14.5a and e), close comparison of their structures reveals similarities that underlie resemblances in their biophysical properties. The most striking similarities lie in the extracellular vestibule and selectivity filters. Both TRPV6 and Orai have highly

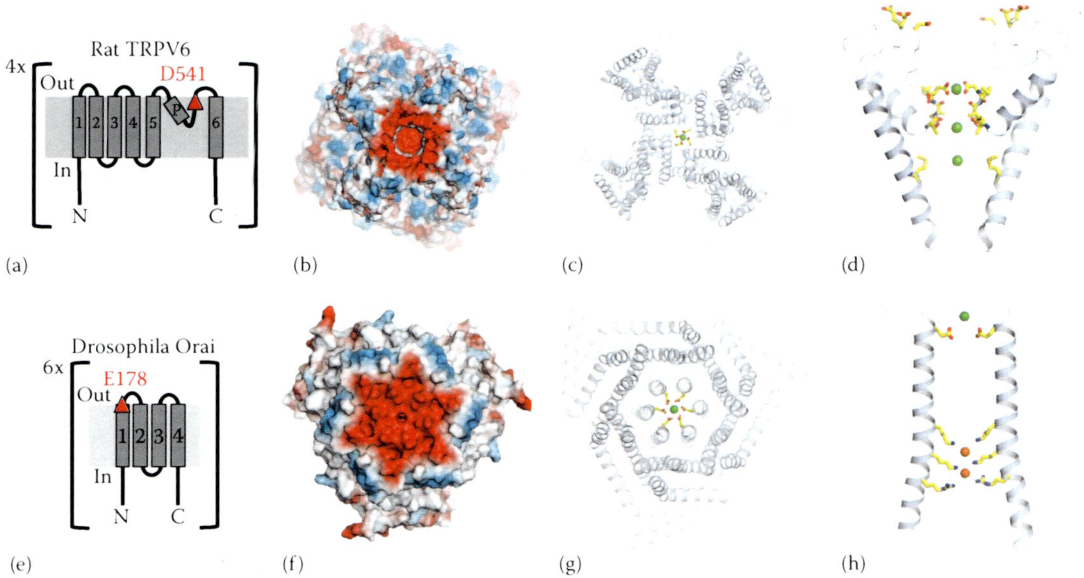

FIGURE 14.5 Structural comparison of rat TRPV6 and *Drosophila* Orai. (a and e) Membrane topology diagram of TRPV6 (a) and Orai (e). TRPV6 assembles as a tetramer with each subunit containing six TM helices (S1–S6) and a re-entrant pore loop between S5 and S6, while Orai assembles as a hexamer, with each subunit containing four TM helices. Acidic residues (D541 in rat TRPV6 and E178 in *Drosophila* Orai) shown to be important for the permeation properties of these channels are shown as red triangles and labeled. (b and f) Bird's-eye views of TRPV6 (b) and Orai (f) in surface representation, colored by electrostatic potential. Red represents negative electronegative potential, while blue represents positive. Note the highly electronegative patches for both channels surrounding the central pore axis. (c and g) Bird's-eye views of TRPV6 (b) and Orai (f) in cartoon representation, with calcium ions shown as green spheres and calcium-coordinating residues (D541 in rat TRPV6 and E178 in *Drosophila* Orai) shown as yellow sticks. (d and h) Side views of the TRPV6 (d) and Orai (h) permeation pathways. For each channel, only two diametrically opposed subunits are shown for clarity. Residues important for ion binding are shown in stick, with calcium and iron ions shown as green and magenta spheres, respectively. The iron binding sites in Orai also constitute binding sites for anions thought to stabilize a closed pore configuration.

electronegative extracellular vestibules that may serve to attract divalent metal cations (normally Ca^{2+} and Mg^{2+}) to the pore (Figure 14.5b and f). In TRPV6, acidic residues from the extracellular pore loops connecting the pore helix and selectivity filter to S5 and S6 contribute to the electronegativity (Figure 14.5c and d), while in Orai, aspartates in the M1–M2 loop are involved (Figure 14.5g and h). Interestingly, in co-crystals of TRPV6 with Ba^{2+} or Gd^{3+}, we observed peaks in the anomalous difference Fourier maps in the vicinity of acidic side chains in the extracellular vestibule (Figure 14.1f and g), indicating that these residues might constitute "recruitment sites" for divalent and trivalent cations.

Early mutagenesis studies of TRPV6 [25,75] and Orai [76–78] highlighted a single aspartate or glutamate residue (D541 in rat TRPV6, E178 in *Drosophila* Orai, D542 in human TRPV6, and E106 in human Orai1) as determinants of their permeation properties, and it was proposed for both channels that high Ca^{2+} selectivity is conferred by the coordination of Ca^{2+} by these side chains. The crystal structures reveal that these residues reside at analogous locations; their side chains protrude toward the central pore axis to produce constrictions at the pore mouth. Further, co-crystallization or soaking experiments showed that these residues comprise binding sites for Ca^{2+}, Ba^{2+}, and Gd^{3+} in both Orai and TRPV6. Interatomic distances between the cation and the carboxylate oxygen suggest that in both cases, the acidic side chain directly coordinates the (at least) partially dehydrated cation. Notably, TRPV6 contains two additional Ca^{2+}-binding sites along the permeation pathway (Figure 14.5d), while the aforementioned E178 site seems to be the only cation-binding site in Orai. On the other hand, basic residues in the lower region of the Orai pore appear to form a binding site for anion(s) that plug this pore (Figure 14.5h). Thus, apart from the binding of Ca^{2+} and other cations at acidic residues in the pore mouth, TRPV6 and Orai have distinct permeation mechanisms. Additional structures of these channels in activated states (TRPV6 requires PI(4,5)P_2 for activation [31] while Orai is opened by STIM [71,72]) will be required to obtain a more complete structural understanding of how these channels contribute to calcium entry in non-excitable cells.

ACKNOWLEDGMENTS

We thank E.C. Twomey for comments on the manuscript. This work was supported by the NIH grants R01 NS083660 (A.I.S) and T32 GM008281 (K.S.), by the Pew Scholar Award in Biomedical Sciences, the Schaefer Research Scholar Award, the Klingenstein Fellowship Award in the Neurosciences, and the Irma T. Hirschl Career Scientist Award (A.I.S.).

REFERENCES

1. Berridge, M.J. 2016. The inositol trisphosphate/calcium signaling pathway in health and disease. *Physiological Reviews* 96:1261–1296.
2. Clapham, D.E. 2003. TRP channels as cellular sensors. *Nature* 426:517–524.
3. Vennekens, R., Hoenderop, J.G.J., Prenen, J., Stuiver, M., Willems, P.H.G.M., Droogmans, G., Nilius, B., and Bindels, R.J.M. 2000. Permeation and gating properties of the novel epithelial Ca^{2+} channel. *Journal of Biological Chemistry* 275:3963–3969.
4. Yue, L., Peng, J.B., Hediger, M.A., and Clapham, D.E. 2001. CaT1 manifests the pore properties of the calcium-release-activated calcium channel. *Nature* 410:705–709.

5. Hoenderop, J.G., van der Kemp, A.W., Hartog, A., van de Graaf, S.F., van Os, C.H., Willems, P.H., and Bindels, R.J. 1999. Molecular identification of the apical Ca^{2+} channel in 1,25-dihydroxyvitamin D3-responsive epithelia. *The Journal of Biological Chemistry* 274:8375–8378.
6. Peng, J.B., Chen, X.Z., Berger, U.V., Vassilev, P.M., Tsukaguchi, H., Brown, E.M., and Hediger, M.A. 1999. Molecular cloning and characterization of a channel-like transporter mediating intestinal calcium absorption. *The Journal of Biological Chemistry* 274:22739–22746.
7. Hoenderop, J.G., van Leeuwen, J.P., van der Eerden, B.C., Kersten, F.F., van der Kemp, A.W., Merillat, A.M., Waarsing, J.H. et al. 2003. Renal Ca^{2+} wasting, hyperabsorption, and reduced bone thickness in mice lacking TRPV5. *The Journal of Clinical Investigation* 112:1906–1914.
8. Bianco, S.D., Peng, J.B., Takanaga, H., Suzuki, Y., Crescenzi, A., Kos, C.H., Zhuang, L. et al. 2007. Marked disturbance of calcium homeostasis in mice with targeted disruption of the Trpv6 calcium channel gene. *Journal of Bone and Mineral Research* 22:274–285.
9. Pike, J.W., Zella, L.A., Meyer, M.B., Fretz, J.A., and Kim, S. 2007. Molecular actions of 1,25-dihydroxyvitamin D3 on genes involved in calcium homeostasis. *Journal of Bone and Mineral Research* 22(Suppl. 2):V16–V19.
10. Meyer, M.B., Watanuki, M., Kim, S., Shevde, N.K., and Pike, J.W. 2006. The human transient receptor potential vanilloid type 6 distal promoter contains multiple vitamin D receptor binding sites that mediate activation by 1,25-dihydroxyvitamin D3 in intestinal cells. *Molecular Endocrinology* 20:1447–1461.
11. Song, Y., Peng, X., Porta, A., Takanaga, H., Peng, J.B., Hediger, M.A., Fleet, J.C., and Christakos, S. 2003. Calcium transporter 1 and epithelial calcium channel messenger ribonucleic acid are differentially regulated by 1,25 dihydroxyvitamin D3 in the intestine and kidney of mice. *Endocrinology* 144:3885–3894.
12. Balesaria, S., Sangha, S., and Walters, J.R. 2009. Human duodenum responses to vitamin D metabolites of TRPV6 and other genes involved in calcium absorption. *American Journal of Physiology—Gastrointestinal and Liver Physiology* 297:G1193–G1197.
13. Lehen'kyi, V., Raphael, M., and Prevarskaya, N. 2012. The role of the TRPV6 channel in cancer. *The Journal of Physiology* 590:1369–1376.
14. Peng, J.B., Zhuang, L., Berger, U.V., Adam, R.M., Williams, B.J., Brown, E.M., Hediger, M.A., and Freeman, M.R. 2001. CaT1 expression correlates with tumor grade in prostate cancer. *Biochemical and Biophysical Research Communications* 282:729–734.
15. Wissenbach, U., Niemeyer, B.A., Fixemer, T., Schneidewind, A., Trost, C., Cavalie, A., Reus, K., Meese, E., Bonkhoff, H., and Flockerzi, V. 2001. Expression of CaT-like, a novel calcium-selective channel, correlates with the malignancy of prostate cancer. *The Journal of Biological Chemistry* 276:19461–19468.
16. Zhuang, L.Y., Peng, J.B., Tou, L.Q., Takanaga, H., Adam, R.M., Hediger, M.A., and Freeman, M.R. 2002. Calcium-selective ion channel, CaT1, is apically localized in gastrointestinal tract epithelia and is aberrantly expressed in human malignancies. *Laboratory Investigation* 82:1755–1764.
17. Fixemer, T., Wissenbach, U., Flockerzi, V., and Bonkhoff, H. 2003. Expression of the Ca^{2+}-selective cation channel TRPV6 in human prostate cancer: A novel prognostic marker for tumor progression. *Oncogene* 22:7858–7861.
18. Bolanz, K.A., Hediger, M.A., and Landowski, C.P. 2008. The role of TRPV6 in breast carcinogenesis. *Molecular Cancer Therapeutics* 7:271–279.
19. Raphael, M., Lehen'kyi, V., Vandenberghe, M., Beck, B., Khalimonchyk, S., Vanden Abeele, F., Farsetti, L. et al. 2014. TRPV6 calcium channel translocates to the plasma membrane via Orai1-mediated mechanism and controls cancer cell survival. *Proceedings of the National Academy of Sciences of the United States of America* 111:E3870–E3879.

20. Peters, A.A., Simpson, P.T., Bassett, J.J., Lee, J.M., Da Silva, L., Reid, L.E., Song, S. et al. 2012. Calcium channel TRPV6 as a potential therapeutic target in estrogen receptor-negative breast cancer. *Molecular Cancer Therapeutics* 11:2158–2168.
21. Bowen, C.V., DeBay, D., Ewart, H.S., Gallant, P., Gormley, S., Ilenchuk, T.T., Iqbal, U. et al. 2013. In vivo detection of human TRPV6-rich tumors with anti-cancer peptides derived from soricidin. *PLoS One* 8(3):e58866.
22. Soricimed. 2015. Soricimed Completes Phase 1 Trial of SOR-C13 in Advanced Solid Tumour Cancers. http://www.soricimed.com, accessed August 2016.
23. Long, S.B., Campbell, E.B., and Mackinnon, R. 2005. Crystal structure of a mammalian voltage-dependent Shaker family K+ channel. *Science* 309:897–903.
24. Nilius, B., Vennekens, R., Prenen, J., Hoenderop, J.G., Droogmans, G., and Bindels, R.J. 2001. The single pore residue Asp542 determines Ca^{2+} permeation and Mg^{2+} block of the epithelial Ca^{2+} channel. *The Journal of Biological Chemistry* 276:1020–1025.
25. Voets, T., Janssens, A., Prenen, J., Droogmans, G., and Nilius, B. 2003. Mg^{2+}-dependent gating and strong inward rectification of the cation channel TRPV6. *The Journal of General Physiology* 121:245–260.
26. Phelps, C.B., Huang, R.J., Lishko, P.V., Wang, R.R., and Gaudet, R. 2008. Structural analyses of the ankyrin repeat domain of TRPV6 and related TRPV ion channels. *Biochemistry* 47:2476–2484.
27. Erler, I., Hirnet, D., Wissenbach, U., Flockerzi, V., and Niemeyer, B.A. 2004. Ca^{2+}-selective transient receptor potential V channel architecture and function require a specific ankyrin repeat. *The Journal of Biological Chemistry* 279:34456–34463.
28. Chang, Q., Gyftogianni, E., van de Graaf, S.F., Hoefs, S., Weidema, F.A., Bindels, R.J., and Hoenderop, J.G. 2004. Molecular determinants in TRPV5 channel assembly. *The Journal of Biological Chemistry* 279:54304–54311.
29. Nilius, B., Weidema, F., Prenen, J., Hoenderop, J.G., Vennekens, R., Hoefs, S., Droogmans, G., and Bindels, R.J. 2003. The carboxyl terminus of the epithelial Ca^{2+} channel ECaC1 is involved in Ca^{2+}-dependent inactivation. *Pflügers Archiv* 445:584–588.
30. Derler, I., Hofbauer, M., Kahr, H., Fritsch, R., Muik, M., Kepplinger, K., Hack, M.E. et al. 2006. Dynamic but not constitutive association of calmodulin with rat TRPV6 channels enables fine tuning of Ca^{2+}-dependent inactivation. *The Journal of Physiology* 577:31–44.
31. Cao, C., Zakharian, E., Borbiro, I., and Rohacs, T. 2013. Interplay between calmodulin and phosphatidylinositol 4,5-bisphosphate in Ca^{2+}-induced inactivation of transient receptor potential vanilloid 6 channels. *The Journal of Biological Chemistry* 288:5278–5290.
32. Carpenter, E.P., Beis, K., Cameron, A.D., and Iwata, S. 2008. Overcoming the challenges of membrane protein crystallography. *Current Opinion in Structural Biology* 18:581–586.
33. Cheng, Y. 2015. Single-particle Cryo-EM at crystallographic resolution. *Cell* 161:450–457.
34. Gao, Y., Cao, E., Julius, D., and Cheng, Y. 2016. TRPV1 structures in nanodiscs reveal mechanisms of ligand and lipid action. *Nature* 534:347–351.
35. Liao, M., Cao, E., Julius, D., and Cheng, Y. 2013. Structure of the TRPV1 ion channel determined by electron cryo-microscopy. *Nature* 504:107–112.
36. Cao, E.H., Liao, M.F., Cheng, Y.F., and Julius, D. 2013. TRPV1 structures in distinct conformations reveal activation mechanisms. *Nature* 504:113–118.
37. Zubcevic, L., Herzik, M.A., Chung, B.C., Liu, Z.R., Lander, G.C., and Lee, S.Y. 2016. Cryo-electron microscopy structure of the TRPV2 ion channel. *Nature Structural & Molecular Biology* 23:180–186.
38. Huynh, K.W., Cohen, M.R., Jiang, J., Samanta, A., Lodowski, D.T., Zhou, Z.H., and Moiseenkova-Bell, V.Y. 2016. Structure of the full-length TRPV2 channel by cryo-EM. *Nature Communications* 7:11130.

39. Paulsen, C.E., Armache, J.P., Gao, Y., Cheng, Y., and Julius, D. 2015. Structure of the TRPA1 ion channel suggests regulatory mechanisms. *Nature* 520:511–517.
40. Brunger, A.T. 1992. Free R value: A novel statistical quantity for assessing the accuracy of crystal structures. *Nature* 355:472–475.
41. Brown, A., Long, F., Nicholls, R.A., Toots, J., Emsley, P., and Murshudov, G. 2015. Tools for macromolecular model building and refinement into electron cryo-microscopy reconstructions. *Acta Crystallographica, Section D: Structural Biology* 71:136–153.
42. Scheres, S.H.W. and Chen, S.X. 2012. Prevention of overfitting in cryo-EM structure determination. *Nature Methods* 9:853–854.
43. Rosenthal, P.B. and Rubinstein, J.L. 2015. Validating maps from single particle electron cryomicroscopy. *Current Opinion in Structural Biology* 34:135–144.
44. Saotome, K., Singh, A.K., Yelshanskaya, M.V., and Sobolevsky, A.I. 2016. Crystal structure of the epithelial calcium channel TRPV6. *Nature* 534:506–511.
45. Hou, X.W., Pedi, L., Diver, M.M., and Long, S.B. 2012. Crystal structure of the calcium release-activated calcium channel Orai. *Science* 338:1308–1313.
46. Kawate, T. and Gouaux, E. 2006. Fluorescence-detection size-exclusion chromatography for precrystallization screening of integral membrane proteins. *Structure* 14:673–681.
47. Goehring, A., Lee, C.H., Wang, K.H., Michel, J.C., Claxton, D.P., Baconguis, I., Althoff, T., Fischer, S., Garcia, K.C., and Gouaux, E. 2014. Screening and large-scale expression of membrane proteins in mammalian cells for structural studies. *Nature Protocols* 9:2574–2585.
48. Barton, W.A., Tzvetkova-Robev, D., Erdjument-Bromage, H., Tempst, P., and Nikolov, D.B. 2006. Highly efficient selenomethionine labeling of recombinant proteins produced in mammalian cells. *Protein Science* 15:2008–2013.
49. Heras, B. and Martin, J.L. 2005. Post-crystallization treatments for improving diffraction quality of protein crystals. *Acta Crystallographica, Section D: Biological Crystallography* 61:1173–1180.
50. Kabsch, W. 2010. Xds. *Acta Crystallographica, Section D: Biological Crystallography* 66:125–132.
51. Otwinowski, Z. and Minor, W. 1997. Processing of X-ray diffraction data collected in oscillation mode. *Methods in Enzymology* 276:307–326.
52. Battye, T.G., Kontogiannis, L., Johnson, O., Powell, H.R., and Leslie, A.G. 2011. iMOSFLM: A new graphical interface for diffraction-image processing with MOSFLM. *Acta Crystallographica, Section D: Biological Crystallography* 67:271–281.
53. CCP4 Project, N. 1994. The *CCP4* suite: Programs for protein crystallography. *Acta Crystallographica* D50:760–763.
54. Taylor, G.L. 2010. Introduction to phasing. *Acta Crystallographica, Section D: Biological Crystallography* 66:325–338.
55. Scapin, G. 2013. Molecular replacement then and now. *Acta Crystallographica, Section D: Biological Crystallography* 69:2266–2275.
56. McCoy, A.J. 2007. Solving structures of protein complexes by molecular replacement with Phaser. *Acta Crystallographica, Section D: Biological Crystallography* 63:32–41.
57. Murshudov, G.N., Skubak, P., Lebedev, A.A., Pannu, N.S., Steiner, R.A., Nicholls, R.A., Winn, M.D., Long, F., and Vagin, A.A. 2011. REFMAC5 for the refinement of macromolecular crystal structures. *Acta Crystallographica, Section D: Biological Crystallography* 67:355–367.
58. Afonine, P.V., Grosse-Kunstleve, R.W., Echols, N., Headd, J.J., Moriarty, N.W., Mustyakimov, M., Terwilliger, T.C., Urzhumtsev, A., Zwart, P.H., and Adams, P.D. 2012. Towards automated crystallographic structure refinement with phenix.refine. *Acta Crystallographica, Section D: Biological Crystallography* 68:352–367.
59. Emsley, P. and Cowtan, K. 2004. Coot: Model-building tools for molecular graphics. *Acta Crystallographica, Section D: Biological Crystallography* 60:2126–2132.

60. Chun, E., Thompson, A.A., Liu, W., Roth, C.B., Griffith, M.T., Katritch, V., Kunken, J. et al. 2012. Fusion partner toolchest for the stabilization and crystallization of G protein-coupled receptors. *Structure* 20:967–976.
61. Durr, K.L., Chen, L., Stein, R.A., De Zorzi, R., Folea, I.M., Walz, T., McHaourab, H.S., and Gouaux, E. 2014. Structure and dynamics of AMPA receptor GluA2 in resting, pre-open, and desensitized states. *Cell* 158:778–792.
62. Green, E.M., Coleman, J.A., and Gouaux, E. 2015. Thermostabilization of the human serotonin transporter in an antidepressant-bound conformation. *PLoS One* 10:e0145688.
63. Coleman, J.A., Green, E.M., and Gouaux, E. 2016. X-ray structures and mechanism of the human serotonin transporter. *Nature* 532:334–339.
64. Shibata, Y., White, J.F., Serrano-Vega, M.J., Magnani, F., Aloia, A.L., Grisshammer, R., and Tate, C.G. 2009. Thermostabilization of the neurotensin receptor NTS1. *Journal of Molecular Biology* 390:262–277.
65. White, J.F., Noinaj, N., Shibata, Y., Love, J., Kloss, B., Xu, F., Gvozdenovic-Jeremic, J. et al. 2012. Structure of the agonist-bound neurotensin receptor. *Nature* 490:508–513.
66. Dore, A.S., Okrasa, K., Patel, J.C., Serrano-Vega, M., Bennett, K., Cooke, R.M., Errey, J.C. et al. 2014. Structure of class C GPCR metabotropic glutamate receptor 5 transmembrane domain. *Nature* 511:557–562.
67. Lu, P., Boros, S., Chang, Q., Bindels, R.J., and Hoenderop, J.G. 2008. The beta-glucuronidase klotho exclusively activates the epithelial Ca^{2+} channels TRPV5 and TRPV6. *Nephrology, Dialysis, Transplantation* 23:3397–3402.
68. Cooper, D.R., Boczek, T., Grelewska, K., Pinkowska, M., Sikorska, M., Zawadzki, M., and Derewenda, Z. 2007. Protein crystallization by surface entropy reduction: Optimization of the SER strategy. *Acta Crystallographica, Section D: Biological Crystallography* 63:636–645.
69. Wu, J., Yan, Z., Li, Z., Yan, C., Lu, S., Dong, M., and Yan, N. 2015. Structure of the voltage-gated calcium channel Cav1.1 complex. *Science* 350:aad2395.
70. Tang, L., El-Din, T.M.G., Payandeh, J., Martinez, G.Q., Heard, T.M., Scheuer, T., Zheng, N., and Catterall, W.A. 2014. Structural basis for Ca^{2+} selectivity of a voltage-gated calcium channel. *Nature* 505:56–61.
71. Soboloff, J., Rothberg, B.S., Madesh, M., and Gill, D.L. 2012. STIM proteins: Dynamic calcium signal transducers. *Nature Reviews. Molecular Cell Biology* 13:549–565.
72. Derler, I., Madl, J., Schutz, G., and Romanin, C. 2012. Structure, regulation and biophysics of I-CRAC, STIM/Orai1. *Calcium Signaling* 740:383–410.
73. Owsianik, G., Talavera, K., Voets, T., and Nilius, B. 2006. Permeation and selectivity of TRP channels. *Annual Review of Physiology* 68:685–717.
74. Voets, T., Prenen, J., Fleig, A., Vennekens, R., Watanabe, H., Hoenderop, J.G.J., Bindels, R.J.M., Droogmans, G., Penner, R., and Nilius, B. 2001. CaT1 and the calcium release-activated calcium channel manifest distinct pore properties. *Journal of Biological Chemistry* 276:47767–47770.
75. Voets, T., Janssens, A., Droogmans, G., and Nilius, B. 2004. Outer pore architecture of a Ca^{2+}-selective TRP channel. *Journal of Biological Chemistry* 279:15223–15230.
76. Yeromin, A.V., Zhang, S.Y.L., Jiang, W.H., Yu, Y., Safrina, O., and Cahalan, M.D. 2006. Molecular identification of the CRAC channel by altered ion selectivity in a mutant of Orai. *Nature* 443:226–229.
77. Prakriya, M., Feske, S., Gwack, Y., Srikanth, S., Rao, A., and Hogan, P.G. 2006. Orai1 is an essential pore subunit of the CRAC channel. *Nature* 443:230–233.
78. Yamashita, M., Navarro-Borelly, L., McNally, B.A., and Prakriya, M. 2007. Orai1 mutations alter ion permeation and Ca^{2+}-dependent fast inactivation of CRAC channels: Evidence for coupling of permeation and Gating. *Journal of General Physiology* 130:525–540.

15 Identifying TRP Channel Subunit Stoichiometry Using Combined Single Channel Single Molecule Determinations (SC-SMD)

Laura G. Ceballos, Alexander Asanov, and Luis Vaca

CONTENTS

15.1 INTRODUCTION

The assembly of transient receptor potential (TRP) proteins into multimers and the existence of heteromeric TRP pores of defined subunit composition were recognized early on in TRP channel research [1–3]. This was suggested by dominant negative effects of nonfunctional TRP fragments or loss-of-function mutations, as well as by the generation of other properties by the coexpression of different subunits. Thus, electrophysiological experiments using defined expression of pore proteins provide information on the stoichiometry within ion channel complexes. Experimental approaches to analyze stoichiometry include both the independent coexpression of potential subunits and the coexpression of combinations of subunits. Solid understanding of the functional properties of individual subunits is a prerequisite for such strategies. One classical strategy to confirm interactions between TRP channel proteins is to test whether loss-of-function mutations of a particular species (e.g., proteins that lack a functional pore structure) are able to prevent currents through the potential heteromerization partner. The dominant negative suppression of channel function and transfer of mutant properties to a heteromeric channel complex allow the determination of subunit stoichiometry and testing of certain concepts relating to pore properties and stoichiometry. The use of mutant channels fused to fluorescent proteins is helpful to test and confirm proper expression and targeting of the proteins. This allows for subunit stoichiometry determination by Förster resonance energy transfer (FRET) [4].

Recent advances in imaging and the use of very sensitive cameras in combination with surface-selective procedures, such as total internal reflection fluorescence microscopy (TIRFM), have facilitated the observation of single molecules [5,6]. There are several types of single molecule determinations: in this report we focus on identifying single molecules based on photobleaching steps. Single molecule detection (SMD) studies, however, provide no information about the functionality of the protein observed. On the other hand, patch clamp (PC) is a powerful technique that allows the observation in real time of single channel kinetics [7]. Unfortunately, PC electrophysiology cannot provide any information about the molecular identity of the channels studied.

In an attempt to overcome the limitations imposed by SMD or PC alone, we developed a novel method, which combines both into an integrated procedure to simultaneously detect channel stoichiometry and assess single-channel gating [8]. We have named this new method single channel single molecule detection (SC-SMD) system. As an example of the power of this new method, we have recently obtained stoichiometric information on six members of the transient receptor potential canonical (TRPC) family of cation channels, a task that using biochemistry and crystallographic studies would have taken several years, with SC-SMD was accomplished in several weeks [8].

15.1.1 Methods for the Determination of Channel Stoichiometry

It is challenging to obtain structural data from ion channels with conventional techniques such as x-ray crystallography [9], which are extremely difficult to conduct with membrane-embedded proteins (such as channels and receptors) (see Chapter 14).

Another method used to study structure of membrane proteins is circular dichroism spectroscopy [10]. However, this method is difficult to implement to determine complex stoichiometry and lacks any information about the functional state of the protein studied [11]. Several researchers have, despite these difficulties, managed to demonstrate crystal structures of ion channels, for example, MacKinnon and colleagues who presented the crystal structure of the KcsA potassium channel in 1998 [12]. The correlation of structure and function of ion channels is commonly achieved with electrophysiological techniques while using genetically modified versions of the protein, for example, site-directed mutagenesis of specific amino acids in the protein. A major disadvantage of site-directed mutagenesis of ion channels is that the methodological procedures for generating and analyzing such mutant ion channels are time consuming and laborious. Also, when designing the mutation, some preexisting knowledge or hypothesis regarding the function of the chosen residue is required. Thus, new and efficient methodological approaches are needed for the systematic evaluation of structure–function relationships in membrane proteins [11].

Studying the sequence determinants and structural features that control functionality in ion channels not only expands our understanding of the structures of these channels but also has implications for the development of therapeutic strategies for diseases associated with ion channel dysfunction. Novel approaches are desired to improve studies of structure–function aspects in ion channels [11].

15.1.2 Single Molecule Detection (SMD)

Single-molecule detection (SMD) with fluorescence microscopy is a widely used technique for biomolecule structure and function characterization. Modern light microscopes equipped with high numerical aperture objectives and sensitive CCD cameras can image fluorescent protein tags with reasonable time resolution (milliseconds) at the single molecule level [13].

The requirement for several bound chromophores to achieve SMD was relaxed as new methodologies gradually emerged to increase signal-to-noise ratios by lowering background intensity, increasing emitted light collection efficiency, and enhancing chromophore stability against photobleaching [13].

Recent advances in imaging and the use of very sensitive cameras in combination with surface-selective procedures, such as TIRFM, have facilitated the observation of single molecules [5,6]. SMD studies can be used to identify the stoichiometry of protein multimers [14,16,17].

TIRF is another near-field (also called evanescent field) technique created when excitation laser light is incident on the glass side of a glass/aqueous interface at angles greater than the critical angle for total internal reflection. Although incident light is completely reflected, an evanescent field is generated in the liquid medium, which decays exponentially with distance from the interface and excites fluorophores within ~100 nm or less of the surface [13]. Axelrod and coworkers combined TIRF with microscopy to take advantage of the small excitation volume created by the evanescent field and the high efficiency of microscopic optics to collect emitted light [14,15].

There are several forms in which TIRF can be achieved; most commercially available systems use trough-objective TIRFM (o-TIRFM). This form of TIRFM is the least efficient and the one that deviates the most from theoretical TIRF. Because emission and excitation light travel through the same optical channel (the objective), large amounts of stray light are generated and critical angles are rarely achieved. Another approach to achieve TIRFM includes the use of prisms (p-TIRFM). This form of TIRFM produces the cleanest illumination with the least generation of stray light. Importantly, in p-TIRFM emission and excitation, channels are separated.

Our group recently developed a novel form of TIRFM. It uses the patch pipette as a light guide to direct the evanescent wave to the tip of the pipette, underneath which the channel of interest is located. This form of TIRFM generates very little stray light, results in very high signal-to-noise ratios, and concentrates all the excitation light in a 1–2 μm, resulting in very high optical powers at the tip of the pipette. The high optical power attained at the tip facilitates the use of excitation light with relatively low power and prevents photobleaching of adjacent molecules in neighboring cells or even in the same cell.

15.1.2.1 Limitations of SMD Studies

For a satisfactory signal-to-background ratio, it is crucial to choose a dye with good fluorescence properties. Photostability is of major importance for the suitability of a dye, since it leads to irreversible loss of fluorescence, which limits the statistical accuracy of detection. Stoichiometry is determined by analyzing the photobleaching steps; therefore, we want those steps to occur in a controlled manner and not randomly during the experiment, hence the relevance of using photostable dyes, because of the probability of a dye to survive a certain number of excitation cycles before being photobleached [18].

Most single-molecule measurements are performed using laser scanning confocal microscopy because excellent light gathering can be achieved using a high numerical aperture objective in combination with photo detectors and photomultipliers [19]. Unfortunately, the excitation light in confocal microscopy penetrates hundreds of microns beyond the surface of the specimen, increasing the signal-to-noise ratio. Thus, many recent SMD studies have preferred more surface-selective methods, such as TIRFM. Even though SMD can provide information about the stoichiometry of the channel of interest, it does not provide any information about the functional (or dysfunctional) state of the channel being studied. It is very possible, for example, that the fusion of fluorescent proteins (such as GFP) to channel subunits of interest may affect gating kinetics or even ionic selectivity.

In order to overcome the limitations of SMD, we have developed a novel method, which combines SMD with single-channel electrophysiology studies; we have named this method the combined single channel single molecule determinations (SC-SMD). In the next sections we will focus on the reagents, procedures, and description of a typical SC-SMD experiment, including cell culture and transfection of the plasmids of interest, patch clamp pipette preparation and mounting of the SC-SMD device.

15.2 MATERIALS

15.2.1 Cell Culture and Transient Transfection

1. Human Embryonic Kidney 293 cells (HEK293)
2. Purified plasmid DNA containing the gene of interest fused to a fluorescent protein cDNA (e.g., green fluorescent protein, GFP)

3. Culture medium for HEK293 cells: Dulbecco's modified Eagle's medium (DMEM) with glutamine, 10% (v/v) heat-inactivated fetal bovine serum (FBS) and penicillin/streptomycin
4. Phosphate-buffered saline, pH 7.20–7.60
5. Trypsin-EDTA (0.2% trypsin, 1 mM EDTA pH 8.0), sterile
6. Opti-MEM® I Reduced Serum Medium pH 7.3 ± 0.1 (Gibco)
7. Polyethylenimine (PEI) solution 1 µg/µL (Sigma-Adrich)
8. Humidified incubator preset to 37°C, 5% CO_2
9. pH Meter
10. Epifluorescence microscope for monitoring cell transfection

All the procedures described for cell culture and cell transfection should take place inside a laminar flow cabinet, in order to maintain sterile conditions.

15.2.2 Patch Clamp Electrophysiology

1. Patch clamp pipettes are produced with 8161 or 8250 glass (Corning). 8250 glass is preferred due to lower autofluorescence in the 400–600 nm range. The high lead content in 8161 contributes to higher fluorescence background when recording the green fluorescent protein (GFP) with peak emission in 525 nm. PC-12 pipette puller (Narishige).
2. Electric oven for drying capillaries.
3. Long flexible needles (World Precision Instruments) MF34G-5.
4. 200A patch clamp amplifier (Axon).
5. Micromanipulators (Narishige).
6. Electron multiplied CCD camera.
7. Inverted microscope.
8. High numerical aperture objective (typically over 1.4).

Reagents/buffers needed for electrophysiology

Bath solution pH 7.4	
NaCl	140 mM
KCl	5.4 mM
HEPES	10 mM
$CaCl_2$	2 mM
$MgCl_2$	1 mM
In deionized water	
Pipette solution pH 7.2	
Cs-Methanesulfonate	110 mM
CsCl	25 mM
HEPES	30 mM
$MgCl_2$	5 mM
EGTA	3 mM
In deionized water	

15.3 METHODS

All the procedures described for cell culture and cell transfection should take place inside a laminar flow cabinet in order to maintain sterile conditions.

15.3.1 Phosphate-Buffered Saline

Dissolve the tablet in 200 mL of purified water and sterilize in an autoclave oven at 121°C, 15 lb, for 15 min. It is not necessary to adjust pH (pH 7.20–7.60).

15.3.2 Trypsin-EDTA (0.2% Trypsin, 0.5 M EDTA pH 8.0)

100 mL of 1 mM EDTA solution, pH 8.0 in water.

To prepare 100 mL aqueous solution of 0.5 M EDTA, weigh 14.61 g EDTA, molecular weight 292.24 (Sigma-Aldrich). Transfer to a precipitate glass. Add 80 mL deionized water. Use a magnetic stirrer for the dissolving process. While stirring adjust the solution pH to 8.0 with 10 N NaOH solution (see *Note 1*). Transfer the solution to a 100 mL volumetric flask and adjust the volume to 100 mL. In a bottle, sterilize the solution by autoclaving for 15 min at 121°C–124°C.

After autoclaving, cool off the 0.5 M EDTA pH 8.0 solution to room temperature. Weigh 0.2 g of trypsin and dissolve in this solution. Sterilize using a 0.22 μm polyvinylidene difluoride (PVDF) membrane filter inside a laminar flow cabinet. Aliquot and store at −20°C.

15.3.3 Polyethylenimine (PEI) Solution 1 μg/μL

Polyethylenimine (PEI), 25 kDa, linear, powder (Sigma–Aldrich).

For a stock solution: Dissolve PEI in endotoxin-free deionized water that has been previously heated to 80°C. Cool down to room temperature and neutralize to pH 7.0. Sterilize using a 0.22 μm PVDF filter. Aliquot into 1.5 mL centrifuge tubes and store at −20°C.

15.3.4 Trypsin-EDTA Treatment for Lifting of HEK-293 Cells

Treatment with trypsin-EDTA is important for disruption and lifting of HEK-293 cell monolayers at confluence for routine maintenance and when preparing cells for assay.

Before beginning the treatment, allow 0.2% trypsin-EDTA solution and DMEM to warm to room temperature.

1. If cells are at the needed confluence for the assay, remove the DMEM culture medium from the cell culture dish (see *Note 2*).
2. Eliminate residual serum by rinsing the cell monolayer with PBS (see *Note 3*).
3. Dispense dropwise with a micropipette 0.2% trypsin-EDTA pH 7.4 solution into the cell culture dish until it completely covers the cell monolayer. Place at 37°C in an incubator for 2 min (see *Note 4*).

TABLE 15.1
Recommended Quantities for Cell Culture

Dish	Surface Area (mm²)	Seeding Density	Cells at Confluency	Supplemented DMEM (mL)	Trypsin (μL)
35	962	0.3×10^6	1.2×10^6	2	300
60	2827	0.8×10^6	3.2×10^6	3	500
100	7854	2.2×10^6	8.8×10^6	10	700

4. After incubation, trypsin-treated cells were rinsed with a double amount of DMEM (Gibco) to total added trypsin volume. Detach cells from the dish by pipetting gently up and down. Dislodge any remaining adherent cells.
5. Collect the total amount of detached cells in a centrifuge tube.
6. Centrifuge collected cells at 800 rpm for 4 min at room temperature.
7. After centrifugation, the supernatant should be carefully removed with a pipette.
8. Resuspend the cell pellet obtained in 1 mL of DMEM cell culture medium (see *Note 5*).
9. Count cells using a hemocytometer and determine the microliters of cells required according to the mm² of the dish.
10. Resuspend again the cell suspension by gentle pipetting and add cells needed to the dish containing the milliliters of DMEM specified in Table 15.1.
11. Incubate at 37°C in a 5% CO_2/95% humidity-controlled incubator.
12. After 24 h incubation, check cell morphology under microscope.
13. Cell culture should be split upon confluence (see *Note 6*).

The recommended quantities of trypsin and culture medium according to the dish surface in mm² are listed in Table 15.1.

15.3.5 Transfecting Cells with Plasmids

Our method of choice for plasmid transfection is polyethylenimine (PEI), because it is easy to use, it binds to and precipitates DNA efficiently, and is very affordable [20,21].

Ratios of plasmid DNA to PEI are important for obtaining high percentages of transfected cells. In this protocol we used a DNA:PEI ratio of 1:3. Nevertheless, this is going to depend on the plasmid, therefore the optimal ratio selected for transfection should be determined experimentally.

The following procedure is for cell culture dishes of 35 mm². It is important to use a 24 h cell culture of 70%–80% confluence. According to Tom et al. [21], cell density at transfection should range from 8×10^5 to 1.2×10^6 cells/mL.

Before transfection, allow Opti-MEM to warm to room temperature and thaw the DNA and PEI-containing solutions (see *Note 7*).

1. The following quantities are for each cell culture dish to be transfected: separately on two different 1.5 mL microcentrifuge tubes (tube 1 and tube 2) dispense 80 μL of Opti-MEM® I Reduced Serum Medium pH 7.3 ± 0.1 (Gibco™).
2. Add 3.5 μg of the plasmid DNA to the Opti-MEM contained on tube 1. Vortex two times, 5 s each (see *Note 8*).
3. Add 10.5 μg of the 1 μg/μL PEI solution to the Opti-MEM contained in tube 2. Vortex two times, 5 s each (see *Note 8*).
4. Pour the total content of PEI solution (tube 2) into DNA solution (tube 1), mix by pipetting gently up and down. It is important to maintain this order of addition (see *Note 8*).
5. Incubate the mixture for 30 min at room temperature (see *Note 9*).
6. While incubation time is running, remove medium by aspiration with a micropipette. Immediately wash carefully the monolayer with 1 mL of PBS to remove all traces of antibiotic and FBS and add 600 mL of Opti-MEM®.
7. After 30 min incubation, remove 100 μL of Opti-MEM and add dropwise the mixture of PEI–plasmids to the cell culture.
8. Gently swirling, homogenize carefully the mixture on the total surface of the cell culture dish.
9. Incubate at 37°C in a 5% CO_2/95% humidity-controlled incubator (see *Note 10*).
10. After 16–24 h incubation, replace media with supplemented DMEM, observe under fluorescence microscopy the resulting cell transfection.
11. At this point, transfection has taken place. Cells can be used 48 h post transfection (see *Note 11*). Perform electrophysiological measurements if expression is sufficient according to fluorescence intensities.

15.3.6 Pipette Preparation

The capillary glass is washed with 70% ethanol and rinsed three times with distilled water. Glass is dried in an oven at 100°C for 5 h until fully dry. Capillaries are pulled using the Narishige puller. Pulling settings vary depending on the melting temperature of the glass used (8161 or 8250). The correct settings should be standardized measuring the pipette resistance using the pipette and bath solutions. The goal is to obtain patch pipettes with resistance ranging 8–12 MΩ. A typical 10 MΩ pipette has a tip of about 1 μm in diameter, which is adequate for single channel recoding and SMD. It is very important not to pull pipettes with very long tip; this will affect the geometry of the glass walls and the reflections inside the pipette walls, producing more stray light, which will impact negatively the imaging procedures. Pipettes can be stored until use in a closed chamber to avoid dust. Dust will also increase stray light and reduce the likelihood of gigaohm seal formation. Patch pipettes can be used for about a week after manufacture.

15.3.6.1 Checking Pipette Autofluorescence

Every individual pipette should be carefully checked for autofluorescence levels. High background should be avoided for obtaining robust SMD experiments. Typically, we mount each pipette in the SC-SMD system and verify background fluorescence levels prior to using them in an experiment. Because fluorescence levels depend on the power of the excitation source and the sensitivity of the imaging device, specific numbers cannot be provided, but in general, acceptable autofluorescence levels should be 20% or less of the highest GFP signal obtained.

15.3.6.2 Filling the Patch Pipette with Buffer

We use 1 mL syringes to fill patch pipettes with solution. To avoid air bubbles during filling we use long flexible needles (World Precision Instruments). After backfilling the pipette, gentle tapping is used to reduce air bubble formation. Air bubbles are the main cause of interrupted current that prevents electrical continuity in the patch clamp system. Bubbles will also increase stray light and the number of reflections, resulting in imperfect light coupling with the pipette tip.

15.3.6.3 Mounting the SC-SMD System

The SC-SMD system consists of a custom-made patch pipette holder adapted to receive the fiber optic coming from the laser used as an excitation light source (Figure 15.1a). The fiber optic is equipped with a connector at the end to plug into the laser. The pipette enters the holder via the inlet and an O-ring to ensure a tight seal between the holder and the pipette (Figure 15.1c). The silver wire is contained within the holder and provides electric continuity between the pipette solution and the patch clamp amplifier headstage (Figure 15.1c). In this regard, the pipette holder is similar to other holders used for patch clamp pipettes. It is important to note that the silver wire produces minimal or even negligible stray light. The pipette is first filed with the recording solution and then placed inside the holder from the SC-SMD system. Figure 15.1d illustrates a photograph of the holder with a pipette inside with the laser off (left panel) and on (right panel). An optical trap (black O-ring) is used to reduce stray light arising from undesirable reflection angles throughout the pipette wall. The SC-SMD system is designed to concentrate all the optical power coming from the laser (excitation light) into the tip of the patch pipette. Thus, patch pipette is used as a light guide.

The BNC connector from the SC-SMD system is designed to fit the connector found in most patch clamp amplifier headstages (Figure 15.2a and b). The connector at the end of the fiber optic fits the laser (Figure 15.2c and e).

15.3.6.4 Checking Pipette Resistance

Patch clamp pipette is mounted on the SC-SMD system, which is attached to the patch clamp amplifier headstage (Figure 15.2). Most patch clamp systems provide readout for patch pipette resistance directly from the amplifier. We use Axon Instruments 200A patch clamp amplifier, which displays patch pipette resistance directly in the readout.

15.3.6.5 Obtaining a Gigaohm Seal

The first step is to identify a particular cell for patch clamping. Under the epifluorescence microscope, cells expressing the channel or protein of interest are easily identified by fluorescence. Cell selection should not take more than a minute to avoid GFP photobleaching. We usually go for the cell that looks healthy, is firmly attached to the bottom of the petri dish, and has good fluorescence levels, ensuring the robust

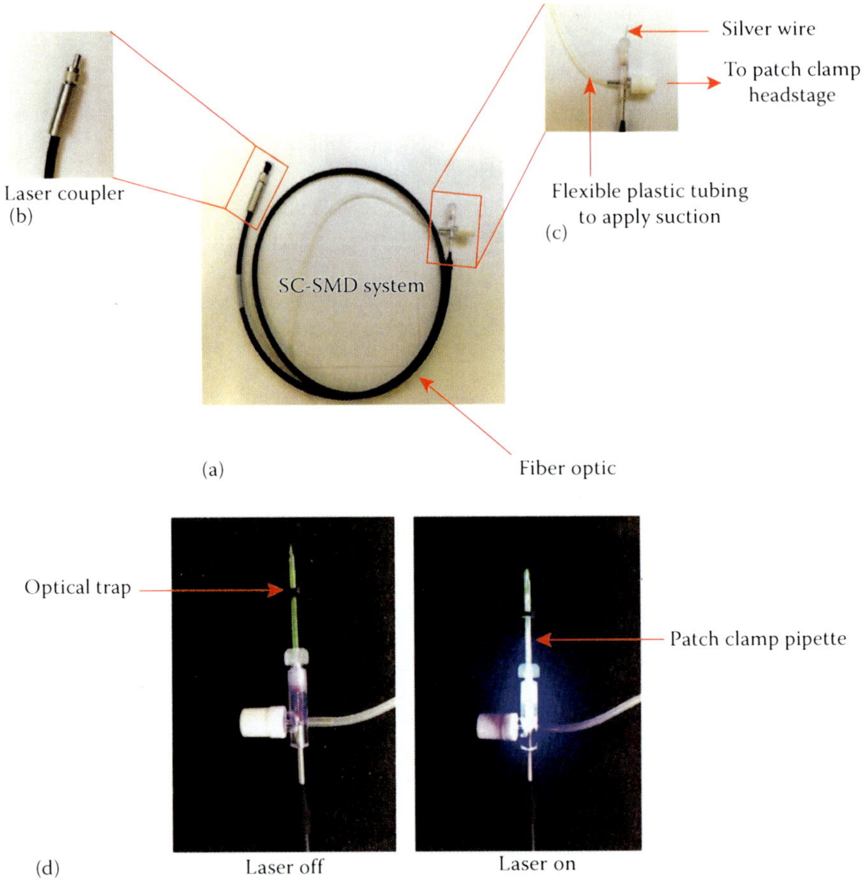

FIGURE 15.1 The SC-SMD system. (a) Photograph illustrating system components. A customized pipette holder serves as a light guide to couple the fiber optic to the patch pipette. A BNC connector couples the system to the headstage of the patch clamp amplifier. (b) Enlarged picture of the connector coupling the fiber optic to the laser. (c) Enlarged illustration of the customized pipette holder. (d) The pipette holder is shown with a patch pipette attached. The left panel shows the configuration with the laser off and right panel with the laser on.

expression of the protein of interest. However, we never select the brightest cell possible because low levels of protein expression are required to increase the chances of obtaining a single channel in the patch pipette. Once the cell is identified, the micromanipulator is used to slowly position the patch pipette tip against the cell surface while monitoring resistance using a square voltage pulse. Once resistance increases, pipette movement is stopped and gentle suction is applied using the plastic tubing in the SC-SMD system. When resistance is increased to reach 5 GΩ or more, suction stops and we are ready to initiate SMD and single channel measurements in the cell-attached mode. Alternatively, the patch of the membrane can be excised from the cell by gently pulling up the patch pipette. In this configuration we can study channels in the inside-out configuration.

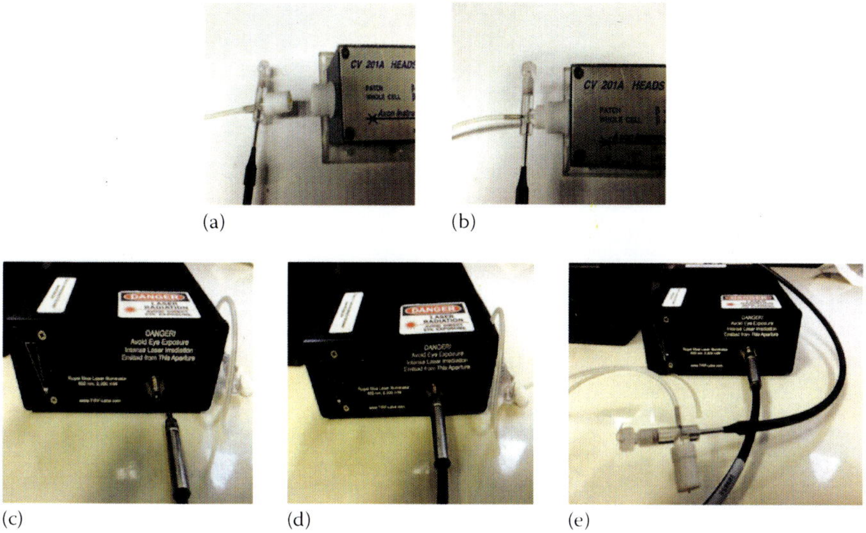

FIGURE 15.2 Attaching the SC-SMD system to the laser and headstage. (a) Photograph illustrating the customized pipette holder aligned with the headstage from the patch clamp amplifier. (b) Pipette holder attached to the headstage with a BNC plug. (c) Fiber optic coupler aligned to the laser connector. (d) Fiber optic coupler attached to the laser connector. (e) The SC-SMD system coupled to the laser.

15.3.6.6 Simultaneous Recording of Single-Channel Activity and Single Molecule Determinations

Before proceeding with single-molecule determinations, one must identify that a single functional channel is present in the patch of membrane. This verification should be conducted in the dark to reduce the possibility of photobleaching the specimen.

Typically, in the single channel configuration of patch clamp, individual channels are identified as square current transitions between the zero current level and the current conducted by the particular channel. Multiple current steps indicate the presence of more than one channel. If this is the case, we typically discard the pipette and begin again the procedure from Section 15.3.6.1.

Some channels are more active at a particular voltage; thus, the holding potential (the voltage at which the channel is maintained) should be set accordingly. Other channels may require an activator to induce the channel to open; if that is the case, the activator should be included to force the channel to open and identify in this way the number of channels present in the pipette. In the case of TRPC channels, we used as activator the SOAR fragment from STIM1 [22]. To activate the channel, a perfusion pipette was used to apply the peptide in solution to the inner leaf of the patch of membrane. This system delivers only a few microliters of the pipette solution, avoiding the necessity to bathe the entire petri dish with the expensive peptide. The idea behind these procedures is to induce a high open probability in order to identify single from multiple channels. Having multiple channels will result in higher than expected subunit compositions, thus making difficult determining channel stoichiometry. This is the main reason why pipettes containing more than one channel should be discarded.

Once it has been determined that only a single channel is present in the patch pipette (measuring single channel currents, as described earlier), we can proceed with SMD recordings. The first step is to move the patch clamp pipette away from the bottom of the petri dish; this is to reduce any background fluorescence signal arising at this location, mainly from the plastic at the bottom of the petri dish. After moving the patch pipette, the focus is adjusted using the micrometric knob on the microscope to focus on the pipette tip. To initiate photobleaching experiments, the laser power is increased to 10% (this value depends on the predefined laser power required for SMD; each laser is different and should be determined experimentally) and imaging is initiated. Images are stored on the computer hard drive for offline analysis. Simultaneously, single channel currents are stored on a separate computer for single channel analysis. To synchronize single channel patch clamp recordings with SMD, we have connected the laser and trigger from the camera to the TTL ports of the patch clamp amplifier (most patch clamp amplifiers have TTL triggers); in this way we can initiate all instruments simultaneously using the patch clamp software. In this configuration both camera and laser are slaved to the patch clamp amplifier.

15.3.6.7 SMD Analysis

The first thing we need to understand when conducting single molecule analysis is that single molecules cannot be visualized directly, simply because they are smaller than the limit of diffraction. In order to assess the number of fluorescent molecules in a protein or channel of interest, photobleaching is induced and the number of quantum steps in which fluorescence emission decays are measured. Each quantum step represents the photobleaching of a single fluorophore.

The amount of light emitted by a single fluorescent protein is very small, a few photons. Thus, it becomes essential for the success of SMD experiments to minimize background fluorescence. This is the reason why most single molecule experiments are conducted using TIRF microscopy. TIRF is a surface-selective method, which excites molecules located only a few nanometers away from the bottom of the dish or coverslip. Surface-selective excitation arises from the fact that the evanescent wave (EW) produced by TIRF decays exponentially with the distance from the source, thus illuminating only molecules in closed proximity to the surface where the EW is generated (typically the EW does not travel more than 100 nm).

In the case of our SC-SMD system, we use a patch pipette as a light guide to conduct the evanescent wave along the pipette wall (Figures 15.1 and 15.2). In this way, no complex optics or prisms are required to generate TIRF. Another advantage of our system is that all the power of the excitation light is concentrated in a very small area (tip of the patch pipette). Even when using relatively modest excitation light sources, a very high optical power is attained at the tip of the patch pipette.

Many modern fluorescent proteins have been genetically engineered to resist photobleaching; thus, they are very useful to evaluate photobleaching-induced decay steps because the probability of two or more fluorescent proteins photobleaching at the same time is small, especially if we are recording events with millisecond resolution. The photobleaching of fluorescent proteins is a stochastic phenomenon, thus, depending on the intensity of the excitation light, complete photobleaching of the proteins of interest may take place rather fast. Ideally, the photobleaching of individual fluorescent proteins occurs in a time interval sufficient for our recording camera to capture it. Thus, it is essential to use CCD electron multiplied cameras, which are very sensitive and typically have low intrinsic noise.

The adequate selection of excitation light intensity is essential for a successful SMD experiment. Unfortunately, the only way to determine the appropriate excitation intensity is through trial and error. The goal is to completely photobleach the sample in 1–2 min. Using very low light intensity may result in photobleaching times of many minutes or even hours; this is not ideal because completing a single experiment will require a lot of time. On the other hand, using very strong excitation light will result in photobleaching of the entire sample in a few milliseconds, making it impossible for our recording device (camera) to catch individual photobleaching events. In our experiments, we try to accomplish complete photobleaching in 2–3 min.

In an ideal experiment, no photobleaching has occurred until we start recording our SMD experiment. However, in real-life experiments that may not be the case. Therefore, several experiments must be conducted in order to generate statistical data from which to identify the correct stoichiometry of the protein of interest. In some instances one or two of four fluorescent proteins (in a tetrameric channel) may have been photobleached before we initiate imaging. However, after analyzing the pooled data, it becomes evident that the majority of the events point to the correct stoichiometry (in this example, four subunits).

The steps involved in the analysis of SMD experiments are as follows:

1. Use the raw data for analysis; do not transform the images into a compressed format, such as GIF, JPG, compressed TIF, etc. Doing so will result in data loss and poor resolution of the SMD experiment.
2. The next step is to distinguish signal from noise. Because of the low emission resulting from single molecule experiments, typically such recordings are very noisy (see *Note 10*). There are several types of single molecule determinations; in this report we focus on identifying single molecules based on photobleaching steps. The photobleaching of a single molecule should result in a quantum step. The intensity of such step would depend on the recording conditions, equipment used, properties of the glass (patch pipette), excitation efficiency, and emission yield. Thus, photobleaching steps (even from two seemingly identical fluorescent proteins) will not be equal; uneven steps are the rule. Such variations arise from the quantum yield of each fluorescent protein, affected by the surrounding microenvironment. Regardless of the magnitude of each photobleaching step, all steps would reflect the photobleaching of an individual fluorescent protein. Thus, the initial step to separate signal from noise is to identify what is known as the centroid, the pixels that received the light emitted by the single molecule. Typically, the centroid will be the area with the brightest pixels, because it contains the signal and noise, whereas surrounding pixels would only contain noise (Figure 15.3a). The same area of the centroid should

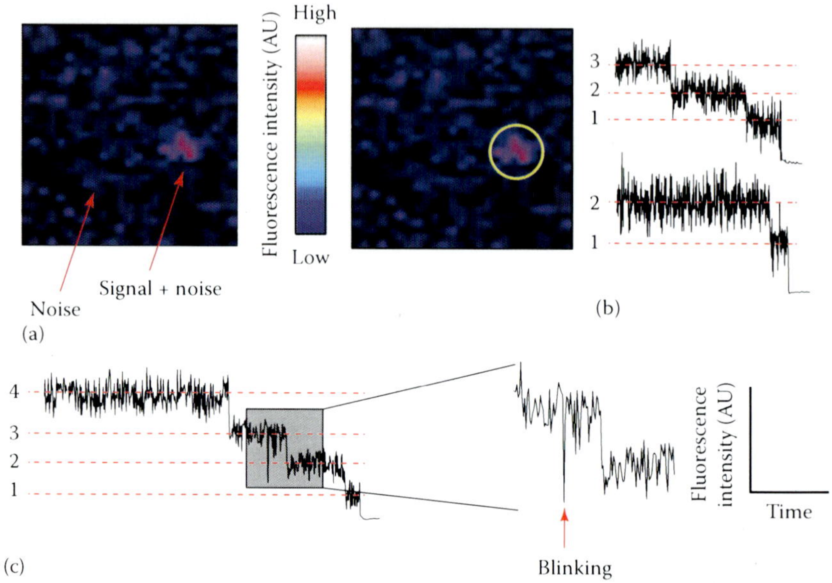

FIGURE 15.3 Analysis of an SC-SMD experiment. (a) Typical raw images obtained from the SC-SMD system. Red arrows point to areas with noise only and noise + signal. Images presented in pseudocolor to enhance the signal (the calibration bar at the center shows the pseudocolor scale used). The image on the right panel shows the centroid in yellow. (b) Measurements of fluorescence intensity over time illustrating two examples: the top panel shows an experiment with three molecules (three steps) and the lower panel an experiment with two molecules. Red dotted lines show the individual steps. (c) Measurements of fluorescence intensity over time illustrating the blinking of a fluorescent protein. The panel at the right shows a zoom area (gray rectangle) and the red arrow points to the blinking step.

be used for all images acquired in the experiment. Changing the area may result in overestimating or underestimating the steps.

3. Once the centroid has been identified (Figure 15.3a, right panel, yellow circle), we can proceed to plot the variations in intensity over time (Figure 15.3b). In some cases, before proceeding to plot intensity variations over time, it is helpful to perform a background subtraction. In photobleaching experiments, fluorescence intensity should always decrease with time (never increase), although blinking is a property of most fluorophores and thus intensity recovery to a previous step is often observed (Figure 15.3c). From the number of quantum steps in fluorescence decrements, we can identify the number of subunits in the protein of interest. There are several programs that help in the analysis of SMD data; many are available free of charge. The software that we use frequently is the Single Molecule Analysis Research Tool (SMART) [23]. SMART not only facilitates SMD analysis but also helps in organizing large amounts of data collected from this type of experiment.

4. Blinking is a process by which fluorescence is turned off for a few milliseconds to come back later on. Most fluorophores studied to this date blink. Blinking can be identified when fluorescence results in a down step that returns to the previous level rapidly (Figure 15.3c, right panel). The blinking phenomenon results from the spontaneous and fast switching between on and off of the emission from a single molecule. This phenomenon can complicate the determination of stoichiometry but new probes are being developed that blink less than currently available fluorophores.

15.4 CONCLUSIONS

There are a handful of methods developed to determine the stoichiometry of protein complexes. Unfortunately many are easier to apply in the case of soluble proteins and do not provide any information about the functionality of the proteins studied. We have developed a novel method, which combines single molecule determinations with single channel electrophysiology (SC-SMD). With SC-SMD we have recently demonstrated that TRPC channels (TRPC1–TRPC6) are functional tetramers [8]. Furthermore, we have identified the stoichiometry of calmodulin–TRPC channel complexes [8]. We showed that TRPC1, TRPC4, and TRPC5 are activated by STIM1, while TRPC3 and TRPC6 are not [8]. All these single molecule experiments and stoichiometry determinations could be concluded within a few weeks. Time courses of the association or dissociation of the channel modulators (SOAR or calmodulin) could also be identified using SC-SMD [8].

Even though we have established that all TRPC family members are functional *homo*tetramers, our method could be used to identify the formation of *hetero*tetramers as well. In fact, SC-SMD can be used to identify higher stroichiometries, pentamers or hexamers. However, the accurate determinations of stoichiometry become more complicated as the subunit numbers increase. Counting more than 7–10 molecules becomes extremely cumbersome and accuracy declines. Other methods can be utilized to complement SC-SMD data [24]. Using conventional x-ray crystallography studies, identifying the stoichiometry of these complexes would have taken years. With SC-SMD, the stoichiometries were resolved in a few weeks. Furthermore, SC-SMD facilitates the study of association and dissociation constants among members of the protein complex of interest. SC-SMD provides very high signal-to-noise ratios because excitation only occurs at the tip of the pipette [8]. Moreover, all the optical power of the excitation light is concentrated in a few microns, thus providing very high optical power at the tip of the pipette. With SC-SMD, one can potentially use several fluorescent proteins simultaneously, facilitating the study of protein interactions and even FRET between the members of the complex.

15.5 NOTES

1. EDTA dissolves completely only when pH of the solution is ~8.0.
2. Remove carefully, avoiding the loosening of the monolayer.
3. Serum affects the activity of trypsin. Wash carefully avoiding the loosening of the monolayer.

4. Do not incubate for more than 2 min. Trypsin causes cell membrane damage and the time of exposure should be kept to a minimum.
5. 1 mL of DMEM per cell culture dish.
6. It is recommended to split cells every 48 h, but this will depend on the confluency required. Forty-eight hours is the optimal recommended time because HEK293 have the "correct" morphology and good growth. Nevertheless, it is important to frequently monitor the morphology, confluence, and adherence of the mammalian cells.
7. Avoid repeated PEI freeze thaw cycles. Freezing at −20°C in small aliquots is the optimal storage condition and minimizes damage due to freezing and thawing. Aliquots should be frozen and thawed once, with any remainder kept at 4°C.
8. It is important after vortexing to ensure there is no Opti-MEM on the tube wall. If that occurs, centrifuge for 15 s at 1000 rpm in order to force all medium to the bottom of the tube.
9. This incubation time is important for PEI–DNA complex formation.
10. After transfection with plasmids of fluorescent proteins fused to the genes of interest, cells must be kept in the dark to reduce the risk of photobleaching of individual fluorescent proteins.
11. The optimal time for transfection will vary depending on the DNA species. It should be empirically determined to obtain the best expression of the protein.
12. It is important to evaluate autofluorescence for every pipette; we have found large variations, especially when using 8161 capillary glass.

REFERENCES

1. Lee KP, Yuan JP, So I, Worley PF, Muallem S. STIM1-dependent and STIM1-independent function of transient receptor potential canonical (TRPC) channels tunes their store-operated mode. *J Biol Chem* 2010;285(49):38666–38673.
2. de Souza LB, Ambudkar IS. Trafficking mechanisms and regulation of TRPC channels. *Cell Calcium* 2014;56:43–50.
3. Salgado A, Ordaz B, Sampieri A, Zepeda A, Glazebrook P, Kunze D, Vaca L. Regulation of the cellular localization and function of human transient receptor potential channel 1 by other members of the TRPC family. *Cell Calcium* 2008;43(4):375–387.
4. Hoppe A, Christensen K, Swanson JA. Fluorescence resonance energy transfer-based stoichiometry in living cells. *Biophys J* 2002;83(6):3652–3664.
5. Poulter NS, Pitkeathly WTE, Smith PJ, Rappoport JZ. The physical basis of total internal reflection fluorescence (TIRF) microscopy and its cellular applications. *Methods Mol Biol* 2015;1251:1–23.
6. Latty SL, Felce JH, Weimann L, Lee SF, Davis SJ, Klenerman D. Referenced single-molecule measurements differentiate between GPCR oligomerization states. *Biophys J* 2015;109(9):1798–1806.
7. Penner R. A practical guide to patch clamping. In: *Single-Channel Recording*, Sakmann B and Neher E, eds., Plenum Publishing Corporation, New York, pp. 3–30, 1995.
8. Asanov A, Sampieri A, Moreno C, Pacheco J, Salgado A, Sherry R et al. Combined single channel and single molecule detection identifies subunit composition of STIM1-activated transient receptor potential canonical (TRPC) channels. *Cell Calcium* 2015;57(1):1–13.

9. Sharon M. How far can we go with structural mass spectrometry of protein complexes? *J Am Soc Mass Spectrom* 2010;21(4):487–500.
10. Miles AJ, Wallace BA. Circular dichroism spectroscopy of membrane proteins. *Chem Soc Rev* 2016;45(18):4859–4872.
11. Trkulja CL, Jansson ET, Jardemark K, Orwar O. Probing structure and function of ion channels using limited proteolysis and microfluidics. *J Am Chem Soc* 2014;136(42):14875–14882.
12. Doyle DA, Morais Cabral J, Pfuetzner RA, Kuo A, Gulbis JM, Cohen SL et al. The structure of the potassium channel: Molecular basis of K^+ conduction and selectivity. *Science* 1998;280(5360):69–77.
13. Burghardt TP, Ajtai K. Single-molecule fluorescence characterization in native environment. *Biophys Rev* 2010;2(4):159–167.
14. Murcia MJ, Garg S, Naumann CA. Single-molecule fluorescence microscopy to determine phospholipid lateral diffusion. *Methods Mol Biol* 2007;400:277–294.
15. Axelrod D. Cell-substrate contacts illuminated by total internal reflection fluorescence. *J Cell Biol* 1981(1):141–145.
16. Axelrod D, Thompson NL, Burghardt TP. Total internal inflection fluorescent microscopy. *J Microsc* January 1983;129(Pt 1):19–28.
17. Thompson NL, Burghardt TP, Axelrod D. Measuring surface dynamics of biomolecules by total internal reflection fluorescence with photobleaching recovery or correlation spectroscopy. *Biophys J* 1981;33(3):435–454.
18. Eggeling C, Widengren J, Rigler R, Seidel CAM. Photobleaching of fluorescent dyes under conditions used for single-molecule detection: Evidence of two-step photolysis. *Anal Chem* 1998;70(13):2651–2659.
19. Emory JM, Peng Z, Young B, Hupert ML, Rousselet A, Patterson D et al. Design and development of a field-deployable single-molecule detector (SMD) for the analysis of molecular markers. *Analyst* 2012;137(1):87–97.
20. Cervera L, Gutiérrez-Granados S, Martínez M, Blanco J, Gòdia F, Segura MM. Generation of HIV-1 Gag VLPs by transient transfection of HEK 293 suspension cell cultures using an optimized animal-derived component free medium. *J Biotechnol* 2013;166(4):152–165.
21. Tom R, Bisson L, Durocher Y. Transfection of HEK293-EBNA1 cells in suspension with linear PEI for production of recombinant proteins. *Cold Spring Harb Protoc* 2008;3(3):1–5.
22. Yuan JP, Zeng W, Dorwart MR, Choi YJ, Worley PF, Muallem S. SOAR and the polybasic STIM1 domains gate and regulate Orai channels. *Nat Cell Biol* 2009;11(3):337–343.
23. Greenfeld M, Pavlichin DS, Mabuchi H, Herschlag D. Single molecule analysis research tool (SMART): An integrated approach for analyzing single molecule data. *PLoS One* February 20, 2012;7(2):e30024.
24. Ulbrich MH, Isacoff EY. Subunit counting in membrane-bound proteins. *Nat Methods* 2007;4(4):319–321.

16 Pharmacology of Store-Operated Calcium Entry Channels

Gary S. Bird and James W. Putney, Jr.

CONTENTS

16.1 INTRODUCTION

In general, calcium signaling in non-excitable cells is primarily initiated by the activation of surface membrane receptors coupled to phospholipase C (PLC) and stimulates a calcium signaling process that is complex both spatially and temporally, involving the interplay of calcium channels and calcium pumps [1]. Receptor activation of PLC leads to a breakdown of phosphatidylinositol 4,5-bisphosphate in the plasma membrane and production of diacylglycerol and inositol 1,4,5-trisphosphate (IP_3) [2]. Fundamentally, receptor activation results in a biphasic process of calcium mobilization composed of the release of intracellular calcium ions from an intracellular organelle, which is coupled to and activates the entry of calcium ions across the plasma membrane of the cell. This second phase of calcium entry is known as store-operated calcium entry (SOCE).

Our ability to identify and define underlying calcium signaling processes and mechanisms is greatly facilitated and influenced by two chief experimental approaches to monitor and characterize the mobilization and movement of Ca^{2+} ions. Fluorescence-based techniques using calcium-sensitive ion probes provide the ability to measure calcium signals with high temporal and spatial resolution, and simultaneously in multiple cells. However, the measured "fluorescent calcium signal" is the result of multiple processes involving calcium pumps and calcium channels that contribute to a steady-state flux of Ca^{2+} ions. To identify and discern specific calcium signaling pathways using this technique, it has been very important to employ pharmacological manipulations that help define, or rule out, specific mechanisms. As will be described later, these approaches can help define the underlying calcium entry process as SOCE or non-SOCE, and the potential involvement of Orai and canonical transient receptor potential (TRPC) family proteins.

The other major and complementary technique for defining calcium signaling processes involves the electrophysiological measurement of ion movement. Importantly, this technique can define and distinguish the biophysical properties of underlying calcium channel activities. In concert with pharmacological manipulations, this technique can be used to identify PLC-activated SOCE ion currents either as I_{CRAC} or I_{SOC} and distinguish this from the PLC-activated and non-SOCE ion current I_{ARC} (see Chapters 1 and 11).

Much is known about the regulation of intracellular Ca^{2+} stores by IP_3 [3] and the nature of the SOCE process [4–6]. Discoveries within the past decade have helped identify the molecular players underlying PLC-coupled Ca^{2+} entry, the Ca^{2+} sensors STIM1 and STIM2, and the SOCE channel subunit proteins Orai1, Orai2, and Orai3 [7]. Indeed, one can describe three types of channels, I_{CRAC}, I_{SOC}, and I_{ARC}. I_{CRAC} represents the most extensively characterized store-operated channel and is composed of the pore-forming subunit Orai1, Orai2, or Orai3. I_{SOC} is characterized as a less Ca^{2+}-selective SOCE channel compared to I_{CRAC} that, in addition to the Orai subunit, combines in an incompletely understood manner with TRPC family members (see Chapter 10). As mentioned earlier, there is also a non-store-operated current, I_{ARC}, which is gated by arachidonic acid and involves Orai1, Orai3, and STIM1. Since I_{ARC} and I_{CRAC} are composed of Orai subunits, they share some similar properties, yet it is possible to clearly distinguish these calcium entry pathways by both biophysical and pharmacological techniques. I_{CRAC} is a small, strongly inwardly rectifying current activated by Ca^{2+} store depletion and inhibited by the drug 2-APB (discussed later). I_{ARC} is a similarly small and strongly inwardly rectifying current, activated by a ligand (not by store depletion), has a different pH sensitivity, exhibits reduced or lacks fast Ca^{2+}-dependent inactivation (CDI), does not rapidly depotentiate, and is not inhibited by 2-APB. In addition, Orai1 was recently discovered to be expressed as two isoforms due to alternative translation initiation, Orai1α (long) and Orai1β (short) [8]. Channels composed of either Orai1α and Orai1β can associate with STIM1 and form CRAC or SOC channels. However, only Orai1α, and not Orai1β, undergoes CDI, and only Orai1α appears to form channels underlying I_{ARC} [9] (see Chapter 11).

Today, our ability to pharmacologically dissect and manipulate the SOCE calcium signaling pathway remains a readily accessible way to understand receptor-regulated calcium signaling in a wide variety of biological systems. This is particularly useful in systems where molecular biological strategies are difficult to employ. However, it always remains a challenge to ensure these pharmacological approaches provide some degree of specificity and control.

16.2 PLC ACTIVATION AND STORE-OPERATED CALCIUM CHANNELS

As mentioned earlier, PLC activation results in a biphasic process of calcium mobilization. The first phase of calcium release is often attributable to IP_3, which acts by binding to a specific receptor on the endoplasmic reticulum (ER) [10]. The second phase of calcium entry is most commonly attributed to SOCE. This process is not regulated by direct actions of IP_3 on the plasma membrane but by a process of retrograde signaling, whereby the depletion of an intracellular calcium storage organelle produces a signal for calcium ion entry across the plasma membrane [4,11] (see Chapter 3). This biphasic signaling process is best illustrated under conditions where the receptors are maximally activated.

However, under more physiological conditions of receptor activation, the resulting calcium signals are complex both spatially and temporally. Rather that the simple, bimodal response observed with high agonist concentrations, activation of PLC-coupled receptors with lower, more physiological concentrations of agonists results in a complex, repetitive cycling of $[Ca^{2+}]_i$, known as $[Ca^{2+}]_i$ oscillations [12,13]. These calcium oscillations depend upon complex mechanisms of regenerative intracellular signaling events, either at the level of PLC activity or the IP_3 receptor calcium release channel. These oscillatory calcium events arise from a process that depends on an interrelationship between calcium release and calcium entry, and Ca entry is necessary to sustain this process for extended periods of receptor activation (discussed also in Chapter 5).

There is some discussion that the Ca entry process triggered under more physiological conditions of PLC activation may differ from that activated under maximal receptor activation. The suggestion being that PLC activation can regulate a separate Ca^{2+} entry pathway in addition to SOCE [14]. To discern the underlying route for receptor-mediated Ca^{2+} entry, it has been critical to employ molecular and pharmacological interventions. This provides a means to identify SOCE and distinguish it from pathways that do not involve SOCE [13,15].

16.3 PHARMACOLOGICAL ACTIVATION OF STORE-OPERATED CHANNELS

The single initial signal for the activation of SOCE and I_{CRAC} is the depletion of intracellular Ca^{2+} stores located in the ER. PLC-coupled receptors initiate this process through the production of IP_3. However, this process represents an uncontrolled approach for manipulating SOCE since (1) the magnitude and kinetics of PLC activation may vary and (2) PLC may activate pathways that are unrelated or interfere with SOCE.

A preferable strategy is to employ pharmacological approaches that target the depletion of ER Ca^{2+} pools directly to activate SOCE, independent of activating PLC-coupled receptors. In general, there are several ways in which this can be achieved: (1) blockade of SERCA pumps, (2) use of a Ca^{2+} ionophore, (3) direct activation of the IP_3 receptor, and (4) "passive depletion" of ER Ca^{2+} pools (usually by patch-clamp technique; see Chapter 1).

16.3.1 SERCA Pump Inhibition

The ER serves as a critical Ca^{2+} buffer with SERCA Ca^{2+} ATPase pumps that can rapidly sequester Ca^{2+} ions from the cell cytoplasm. This activity serves to prevent untoward changes in $[Ca^{2+}]_i$ and replenish intracellular Ca^{2+} stores following PLC activation. Even in unstimulated cells considered "at rest," Ca^{2+} ions are continually cycling across the ER membrane with the actions of SERCA pumps sequestering Ca^{2+} balanced against a poorly defined "Ca^{2+} leak" process out of the ER. However, by inhibiting this SERCA Ca^{2+}ATPase activity, the prevailing "Ca^{2+} leak" process will result in depletion of ER Ca^{2+} stores and full activation of SOCE [16] (see Chapter 1).

There is a selection of membrane permeant SERCA inhibitors available that provide a noninvasive technique for manipulating ER Ca^{2+} pools in intact cells. These include thapsigargin, cyclopiazonic acid (CPA), and tBHQ [17]. Historically, it was the discovery of thapsigargin that first provided the clearest demonstration and important validation of the SOCE pathway. Importantly, treatment of cells with these inhibitors made it possible to deplete the IP_3-sensitive Ca^{2+} stores and activate SOCE without formation of any inositol phosphates associated with agonist activation [18,19]. Of the three SERCA inhibitors, only the more water-soluble CPA can be readily washed out of cells. This property of CPA provides the ability to partially deplete intracellular Ca^{2+} stores and thus partially activate SOCE [20,21]. An example of the use of thapsigargin to determine the effects of a pharmacological inhibitor on the size of the intracellular Ca^{2+} pool and the magnitude of SOCE is illustrated in Figure 16.1.

16.3.2 Ca^{2+} Ionophores

The use of Ca^{2+} ionophores provides an alternative strategy for depleting intracellular Ca^{2+} from the ER and activating SOCE. A23187 [22] and ionomycin [23] are lipid-soluble carboxylic acid antibiotics that transport divalent but not monovalent cations. In general, Ca^{2+} ionophores have proven useful in transporting Ca^{2+} ions across a variety of membranes and manipulating intracellular Ca^{2+} pools in intact cells. Experimentally, ionomycin performs better and is more selective than A23187 in transporting Ca^{2+} ions [23].

However, the actions of ionomycin on manipulating Ca^{2+} movements are complex and concentration dependent. At concentrations ~10 μM, ionomycin increases the permeability of Ca^{2+} ions across all cell membranes. However, at concentrations

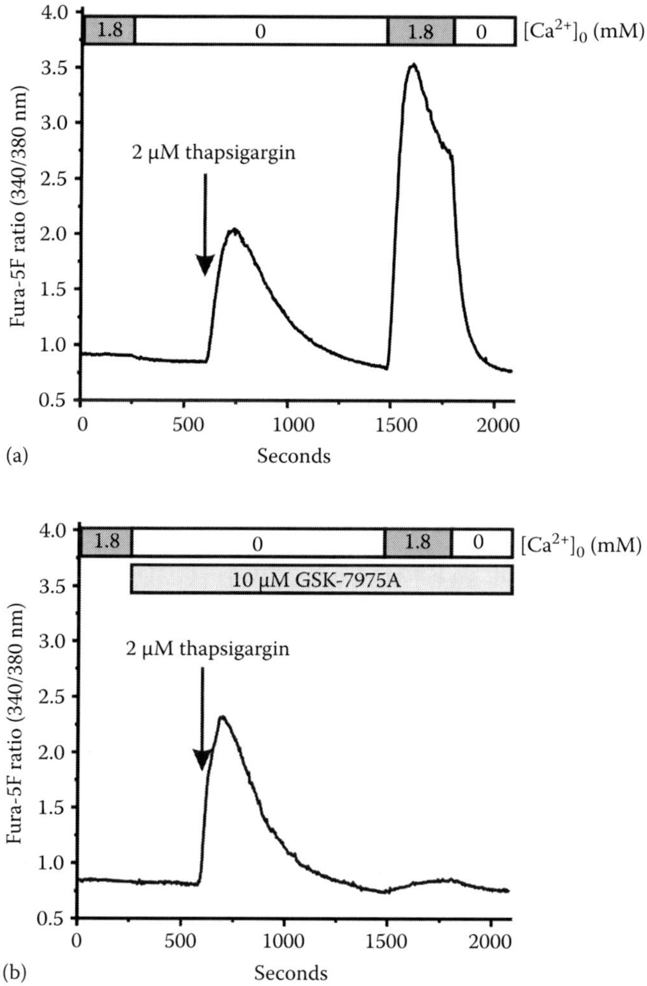

FIGURE 16.1 Effect of a SOCE inhibitor on the thapsigargin-induced biphasic calcium signaling using the calcium readdition protocol in fura-5F-loaded HEK 293 cells. HEK 293 attached to glass cover slips were loaded with the calcium indicator fura-5F, and cytoplasmic $[Ca^{2+}]_i$ was measured at the single-cell level as described by Bird and Putney [13]. The cells are bathed in a HEPES-buffered salt solution (HBSS) containing 1.8 mM extracellular Ca^{2+}. (a) Cells were subject to the calcium readdition protocol and treated with the SERCA Ca^{2+} ATPase inhibitor, thapsigargin (2 μM). Specifically, extracellular Ca^{2+} was removed from the bathing medium and thapsigargin was added. After 15 min, during which the ER Ca^{2+} stores were fully depleted, extracellular Ca^{2+} was restored to the medium and SOCE measured. This protocol illustrates the biphasic nature of the calcium response. (b) Cells undergo the same calcium readdition protocol and treatment with thapsigargin. In this case, the cells were also treated with the SOCE inhibitor GSK 7975A (10 μM). As the data illustrate, the initial phase of thapsigargin-induced Ca^{2+} release, indicating intracellular Ca^{2+} store content, was normal, but the second phase of SOCE Ca^{2+} entry was blocked. Each data trace represents the average signal from 25 to 30 cells.

$<1\ \mu M$, ionomycin appears to selectively partition into intracellular membranes and release intracellular Ca^{2+} stores without greatly increasing the permeability of the plasma membrane to extracellular calcium [24]. At low concentrations, ionomycin can enhance Ca^{2+} influx by stimulating store-regulated cation entry and not by a direct action at the plasma membrane [25]. Under these conditions, this effect of low concentrations of ionomycin can be used to activate SOCE (see Chapter 5).

16.3.3 Activation of IP_3 Receptors

Monitoring currents associated with SOCE is achieved using the whole-cell patch-clamp technique that, in this mode, allows the intracellular milieu to be modified directly with the internal patch pipette solution. Thus, rather than relying on the external application of SERCA pump inhibitors to activate SOCE channels, the intracellular Ca^{2+} store depletion can be achieved by addition of metabolizable or nonmetabolizable analogs of IP_3 (in conjunction with EGTA or BAPTA) [26] directly to the internal pipette solution (see Chapter 1).

16.3.4 Passive Depletion of ER Ca^{2+} Pools

As mentioned earlier, the use of the whole-cell patch-clamp technique allows the intracellular milieu to be modified directly with the internal patch pipette solution. By simply breaking into cells with a patch pipette solution containing a high concentration of the Ca^{2+} chelators BAPTA or EGTA, the intracellular ER Ca^{2+} stores are gradually emptied. This "passive depletion" of Ca^{2+} results in I_{CRAC} activation that is equivalent to that activated by IP_3 or thapsigargin; however, this passive process is slow to develop fully (in the order of minutes). This is a technique that can be used in conjunction with IP_3 to facilitate a rapid and maximal activation of I_{CRAC} [27]. The slow onset of I_{CRAC} with passive depletion can be of experimental advantage as it provides sufficient time to obtain control, baseline current measurements before significant store depletion is achieved (discussed in Chapter 1).

16.3.5 Membrane Potential

By definition, activation of Ca^{2+} entry in non-excitable cells results from IP_3-mediated Ca^{2+} store depletion rather than through voltage activation [1]. However, SOCE is an electrogenic process and can be influenced by the concentration gradient of Ca^{2+} ions and membrane potential across the plasma membrane. Thus, less calcium enters upon plasma membrane depolarization, and hyperpolarization will promote calcium influx [28] (see Figure 1.1). Thus, it is important to control the effects of membrane potential when considering the specificity of a pharmacological agent to modulate a Ca^{2+} entry signal. This may especially be important with pharmacological manipulations that inhibit Ca^{2+} entry where toxic insults to a cell might damage the plasma membrane and cause depolarization. Should this be a concern, the effects of pharmacological agents can be studied using the patch-clamp technique where fluxes of calcium ions across the membrane are measured and membrane potential controlled (i.e., voltage clamp).

16.4 PHARMACOLOGICAL INHIBITION OF STORE-OPERATED CHANNELS

While pharmacological activation of SOCE with SERCA pump inhibitors such as thapsigargin was crucial for developing our understanding of SOCE in intact cells, our ability to manipulate SOCE channels directly and inhibit them with any degree of specificity has historically been rather limited. However, the progress made in identifying the molecular players underlying the SOCE pathway has provided an opportunity to screen for and develop drugs that can modulate channel activity or act directly at the pore of the channel. In this section, we summarize a range of pharmacological agents that target SOCE channel inhibitors and are commercially available.

16.4.1 Lanthanides

The lanthanides La^{3+} and Gd^{3+} have been the most widely employed tools for blocking SOCE [29]. Interestingly, lanthanides were initially used to block both Ca^{2+} entry and Ca^{2+} efflux via the plasma membrane Ca^{2+}ATPase (PMCA) [30]. These effects, on entry and efflux, however, can be dissociated due to differential sensitivities of the processes to lanthanides. At low concentrations, lanthanides can block SOCE (<1 μM) [31]. With identification of Orai, studies suggest that lanthanides act by blocking access of Ca^{2+} ions to the selectivity filter and pore, a likely target being acidic residues in loop I–II [32].

At concentrations above 100 μM, lanthanides begin to block PMCA activity [13,30] and appear to block completely at concentrations at or above 1 mM. The cytoplasm then appears isolated or "insulated" from the extracellular space. Under these conditions, this "gadolinium insulation" is blocking both the entry and efflux of Ca^{2+} ions and presents an opportunity to investigate complex intracellular calcium signaling events independently of contributions made by constituents in the extracellular space [13].

16.4.2 2-APB (2-Aminoethyldiphenyl Borate)

2-APB has broadly been used as an inhibitor for SOCE and I_{CRAC}. This compound was originally described as a noncompetitive inhibitor of the IP_3 receptor. The effects of 2-APB are far more complex, while it has been found to modulate many other ion channels [7,33–36]. The effects of 2-APB on SOCE and I_{CRAC} appear to be extracellular and act independently of IP_3 receptor inhibition [37,38]. However, by careful application, 2-APB can still be useful in distinguishing between SOCE and other channels, particularly in overexpression systems and interrogating the actions of STIM1 and the role of Orai isoforms in the activation of SOCE and I_{CRAC}.

2-APB has a dose-dependent bimodal effect on SOCE. At low concentrations (<5 μM), SOCE is enhanced, then transient enhancement and inhibition occurs at doses >20 μM [34,39,40]. At high doses, 2-APB was observed to inhibit STIM1 puncta formation at the plasma membrane, suggesting that current inhibition involved disruption of coupling to and activation of Orai channels [34,41]. However, this simple mechanism does not explain the full spectrum of 2-APB effects. For example, when overexpressing STIM1 and Orai1 together in HEK cells, 2-APB is less effective in inhibiting STIM1 puncta formation yet retains its potency for inhibiting SOCE and I_{CRAC} [34].

2-APB differs in its actions against the three Orai isoforms; as discussed earlier, it has a biphasic effect on Orai1 channels (and most native CRAC channels), a weaker effect against Orai2 channels, whereas it directly and potently activates Orai3 channels [34,40,41].

Two dimeric derivatives of 2-APB, DPB162-AE and DPB163-AE, were identified in a screen of chemical analogs with more potency in inhibiting SOCE and I_{CRAC} and with considerable specificity for action on STIM1 and Orai1-dependent SOCE [42,43]. While DPB163-AE retained a bimodal effect on SOCE with dose-dependent potentiation and inhibition, DPB162-AE was shown to only inhibit SOCE [44].

16.4.3 ML-9

ML-9, an inhibitor of myosin light chain kinase (MLCK), was discovered to inhibit SOCE and thus suggested a possible role for MLCK in the mechanism of SOCE activation [45]. The mechanism of action of ML-9 remains unclear but appears to be centered on disrupting the coupling of STIM1 and Orai1. ML-9 was found to disperse thapsigargin-induced STIM1 puncta, an event that appeared to precede the loss of SOCE. However, these effects of ML-9 on SOCE were independent of MLCK. An alternative MLCK inhibitor, wortmannin (20 μM), had no effect on SOCE, and the effects of ML-9 were unaffected by the knockdown of MLCK. ML-9 has limited application in cells overexpressing STIM1 and/or Orai1: overexpression of STIM1 reduces the effect of ML-9 on SOCE, and co-overexpression of STIM1 with Orai1 renders ML-9 ineffective.

16.4.4 BTP2 (YM-58483)

BTP2 is a member of a family of *bis*(trifluoromethyl)pyrazoles that have been shown to inhibit thapsigargin-induced SOCE and I_{CRAC} activation [46–48]. Experimentally, BTP2 proves a more potent and irreversible inhibitor of SOCE if cells are exposed to the drug for many hours prior to cell activation. While this is inconvenient from a practical standpoint, it does suggest that the mechanism of action of BTP2 is indirect and not likely a channel pore blockade. The specificity of BTP2 for directly inhibiting SOCE has also been challenged on the basis that this compound can activate TRPM4 [47]. It is suggested that activation of TRPM4, a Na-permeable channel, can depolarize the plasma membrane potential. As discussed earlier, membrane depolarization will reduce the driving force for Ca^{2+} entry and thus indirectly also SOCE.

16.4.5 Synta 66

Synta 66 appears to be a potent and highly selective inhibitor of SOCE and I_{CRAC} activation [49,50]. Synta 66 does share some of the properties of the structurally related BTP2: it is a slow and irreversible inhibitor, requiring cells to be exposed to the compound for long periods before calcium signaling is activated. As with BTP2, this would suggest that Synta 66 is not targeting the channel pore. However, Synta 66 has no effect on STIM1 puncta formation [51] and likely acts downstream of STIM1 oligomerization.

16.4.6 GSK-7975A and GSK-5503A

GSK-7975A and GSK-5503A are two pyrazole derivatives that appear to be highly selective inhibitors of SOCE and I_{CRAC} activation ([52] and Figure 16.1). Somewhat like Synta 66 and BTP2, cells have to be preincubated with these compounds in order for them to be fully effective in blocking Ca^{2+} entry. However, in this case, cells need only be exposed for a period of minutes rather than hours. Focusing on GSK-7975A, Derler et al. [52] investigated possible mechanisms of inhibition. They concluded that GSK-7975A was acting downstream of STIM1 oligomerization and STIM1/Orai1 interaction and its inhibitory effect is likely due to interference with the ion permeation through the Orai pore. This study also screened the effects of GSK-7975A against a panel of 16 ion channels concluding that this compound retained a high degree of selectivity. A weak inhibitory effect was observed for an L-type (Ca_v) Ca^{2+} channel and a more robust inhibition for TRPV6 channels.

16.4.7 RO2959

RO2959 is a potent SOCE and I_{CRAC} inhibitor that blocked an IP_3-dependent current in RBL-2H3 mast cells, in CHO cells stably expressing human Orai1 and STIM1 and in human $CD4^+$ T lymphocytes [53]. As with BTP2, Synta 66, and the GSK compounds, cells need to be treated with RO2959 at least 30 min before cell activation. The mechanism of RO2959 inhibition has not been addressed, although the required preincubation period would suggest it is not a pore blocker.

16.4.8 AnCoA4

AnCoA4 is a SOCE inhibitor discovered in a small-molecule microarray screen targeting the SOCE pathway. Instead of using a cell-based screening assay, Sadaghiani et al. [54] identified peptides encompassing Orai1 and STIM1 domains that are important for the gating of CRAC channels, purified, and immobilized them in microarrays. This was used to screen a library of molecules and identify those that interacted with these STIM1/Orai1 domains. In follow-up functional tests, AnCoA4 was found to inhibit thapsigargin-induced SOCE and CRAC channels at concentrations in the low micromolar range. The inhibitory effect of AnCoA4 was more potent when added before cell activation (~5 μM) when compared to the concentrations required for inhibition after cell activation (20 μM). This suggests that AnCoA4 is more effective when presented to cells before STIM1 starts to interact with Orai1.

16.4.9 SKF96365 and Other Imidazoles

A number of related imidazole compounds are inhibitors of SOCE and include SKF96365, econazole, miconazole, and clotrimazole. The use of these compounds is problematic and certainly not a first choice. While their mechanism of action remains elusive, a chief concern is their lack of specificity for SOCE. Indeed, these compounds have been shown to block voltage-gated Ca^{2+} channels [55,56], potassium channels [57], and TRP family members [58,59].

16.4.10 Diethylstilbestrol

Diethylstilbestrol (DES) is a synthetic estrogen agonist that has been described as an inhibitor of thapsigargin-induced SOCE and I_{CRAC} [60]. The effect of DES is rapid, requiring no pretreatment, is reversible and exerts its effect rapidly via the extracellular side of the plasma membrane. DES also had no effect on STIM1 clustering and translocation [61]. These characteristics suggest DES may inhibit SOCE at the pore. In terms of selectivity, Zakharov and colleagues demonstrated that DES had no effect on monovalent cation currents that are mediated by TRPM7 channels [60]. However, in earlier studies in smooth muscle cells, it was found that DES inhibited nonselective cation currents and K^+ currents [62]. On a cautionary note, a study in A7r5 smooth muscle cells [63] demonstrated that the observed effects of DES on inhibiting the calcium entry process were transient. However, this was not due to a transient inhibition of SOCE itself but rather an off-target effect of DES to promote other Ca^{2+} transport mechanisms.

16.5 CONCLUDING REMARKS

In this short chapter, we have attempted to outline some of the general aspects of pharmacological and experimental manipulation of SOCE, as well as briefly summarizing the effects of reagents that are currently available to the experimentalist. As researchers continue to search for potent and selective agonists and antagonists, the next phase of SOCE pharmacology may well include significant clinical aspects of this widely encountered and physiologically important pathway.

ACKNOWLEDGMENT

Work from the authors' laboratory was supported by the Intramural Research Program of the National Institutes of Health, National Institute of Environmental Health Sciences.

REFERENCES

1. Bird, G.S., Aziz, O., Lievremont, J.P., Wedel, B.J., Trebak, M., Vazquez, G., and Putney, J.W. 2004. Mechanisms of phospholipase C-regulated calcium entry. *Curr. Mol. Med.* **4**:291–301.
2. Berridge, M.J. and Irvine, R.F. 1984. Inositol trisphosphate, a novel second messenger in cellular signal transduction. *Nature* **312**:315–321.
3. Berridge, M.J. 1993. Inositol trisphosphate and calcium signalling. *Nature* **361**:315–325.
4. Putney, J.W. 1986. A model for receptor-regulated calcium entry. *Cell Calcium* **7**:1–12.
5. Putney, J.W. 2010. Pharmacology of store-operated calcium channels. *Mol. Interv.* **10**:209–218.

6. Putney, J.W. 2011. The physiological function of store-operated calcium entry. *Neurochem. Res.* **36**:1157–1165.
7. Prakriya, M. and Lewis, R.S. 2015. Store-operated calcium channels. *Physiol. Rev.* **95**:1383–1436.
8. Fukushima, M., Tomita, T., Janoshazi, A., and Putney, J.W. 2012. Alternative translation initiation gives rise to two isoforms of orai1 with distinct plasma membrane mobilities. *J. Cell Sci.* **125**:4354–4361.
9. Desai, P.N., Zhang, X., Wu, S., Janoshazi, A., Bolimuntha, S., Putney, J.W., and Trebak, M. 2015. Multiple types of calcium channels arising from alternative translation initiation of the Orai1 message. *Sci. Signal.* **8**:ra74.
10. Berridge, M.J. 1986. Inositol phosphates as second messengers. In *Phosphoinositides and Receptor Mechanisms*, J.W. Putney (ed.), Alan R. Liss, Inc., New York, pp. 25–46.
11. Parekh, A.B. and Putney, J.W. 2005. Store-operated calcium channels. *Physiol. Rev.* **85**:757–810.
12. Woods, N.M., Cuthbertson, K.S., and Cobbold, P.H. 1986. Repetitive transient rises in cytoplasmic free calcium in hormone-stimulated hepatocytes. *Nature* **319**:600–602.
13. Bird, G.S. and Putney, J.W. 2005. Capacitative calcium entry supports calcium oscillations in human embryonic kidney cells. *J. Physiol.* **562**:697–706.
14. Shuttleworth, T.J., Thompson, J.L., and Mignen, O. 2004. ARC channels: A novel pathway for receptor-activated calcium entry. *Physiology (Bethesda)* **19**:355–361.
15. Wedel, B., Boyles, R.R., Putney, J.W., and Bird, G.S. 2007. Role of the store-operated calcium entry proteins, Stim1 and Orai1, in muscarinic-cholinergic receptor stimulated calcium oscillations in human embryonic kidney cells. *J. Physiol.* **579**:679–689.
16. Camello, C., Lomax, R., Petersen, O.H., and Tepikin, A. 2002. Calcium leak from intracellular stores—The enigma of calcium signalling. *Cell Calcium* **32**:355–361.
17. Putney, J.W. 2001. Pharmacology of capacitative calcium entry. *Mol. Interv.* **1**:84–94.
18. Takemura, H., Hughes, A.R., Thastrup, O., and Putney, J.W. 1989. Activation of calcium entry by the tumor promoter, thapsigargin, in parotid acinar cells. Evidence that an intracellular calcium pool, and not an inositol phosphate, regulates calcium fluxes at the plasma membrane. *J. Biol. Chem.* **264**:12266–12271.
19. Jackson, T.R., Patterson, S.I., Thastrup, O., and Hanley, M.R. 1988. A novel tumour promoter, thapsigargin, transiently increases cytoplasmic free Ca^{2+} without generation of inositol phosphates in NG115-401L neuronal cells. *Biochem. J.* **253**:81–86.
20. Sedova, M., Klishin, A., Hüser, J., and Blatter, L.A. 2000. Capacitative Ca^{2+} entry is graded with degree of intracellular Ca^{2+} store depletion in bovine vascular endothelial cells. *J. Physiol. (Lond.)* **523**:549–559.
21. Luik, R.M., Wang, B., Prakriya, M., Wu, M.M., and Lewis, R.S. 2008. Oligomerization of STIM1 couples ER calcium depletion to CRAC channel activation. *Nature* **454**:538–542.
22. Reed, P.W. and Lardy, H.A. 1972. A23187: A divalent cation ionophore. *J. Biol. Chem.* **247**:6970–6977.
23. Liu, C.-M. and Herman, T.E. 1979. Characterization of ionomycin as a calcium ionophore. *J. Biol. Chem.* **253**:5892–5894.
24. Morgan, A.J. and Jacob, R. 1994. Ionomycin enhances Ca^{2+} influx by stimulating store-regulated cation entry and not by a direct action at the plasma membrane. *Biochem. J.* **300**:665–672.
25. Albert, P.R. and Tashjian, A.H., Jr. 1984. Relationship of thyrotropin-releasing hormone-induced spike and plateau phases in cytosolic free Ca2+ concentrations to hormone secretion. Selective blockade using ionomycin and nifedipine. *J. Biol. Chem.* **259**:15350–15363.
26. Hoth, M. and Penner, R. 1992. Depletion of intracellular calcium stores activates a calcium current in mast cells. *Nature* **355**:353–355.

27. Parekh, A.B. and Penner, R. 1997. Store depletion and calcium influx. *Physiol. Rev.* **77**:901–930.
28. Kamouchi, M., Droogmans, G., and Nilius, B. 1999. Membrane potential as a modulator of the free intracellular Ca^{2+} concentration in agonist-activated endothelial cells. *Gen. Physiol. Biophys.* **18**:199–208.
29. Bird, G.S., DeHaven, W.I., Smyth, J.T., and Putney, J.W., Jr. 2008. Methods for studying store-operated calcium entry. *Methods* **46**:204–212.
30. Van Breemen, C., Farinas, B., Gerba, P., and McNaughton, E.D. 1972. Excitation-contraction coupling in rabbit aorta studied by the lanthanum method for measuring cellular calcium influx. *Circ. Res.* **30**:44–54.
31. Broad, L.M., Cannon, T.R., and Taylor, C.W. 1999. A non-capacitative pathway activated by arachidonic acid is the major Ca^{2+} entry mechanism in rat A7r5 smooth muscle cells stimulated with low concentrations of vasopressin. *J. Physiol. (Lond.)* **517**:121–134.
32. McNally, B.A., Yamashita, M., Engh, A., and Prakriya, M. 2009. Structural determinants of ion permeation in CRAC channels. *Proc. Natl. Acad. Sci. USA* **106**:22516–22521.
33. Zhang, S.L., Kozak, J.A., Jiang, W., Yeromin, A.V., Chen, J., Yu, Y., Penna, A., Shen, W., Chi, V., and Cahalan, M.D. 2008. Store-dependent and -independent modes regulating Ca^{2+} release-activated Ca^{2+} channel activity of human Orai 1 and Orai 3. *J. Biol. Chem.* **283**:17662–17671.
34. DeHaven, W.I., Smyth, J.T., Boyles, R.R., Bird, G.S., and Putney, J.W., Jr. 2008. Complex actions of 2-aminoethyldiphenyl borate on store-operated calcium entry. *J. Biol. Chem.* **283**:19265–19273.
35. Prakriya, M. and Lewis, R.S. 2002. Separation and characterization of currents through store-operated CRAC channels and Mg^{2+}-inhibited cation (MIC) channels. *J. Gen. Physiol.* **119**:487–507.
36. Chokshi, R., Fruasaha, P., and Kozak, J.A. 2012. 2-Aminoethyl diphenyl borinate (2-APB) inhibits TRPM7 channels through an intracellular acidification mechanism. *Channels (Austin)* **6**:362–369.
37. Bootman, M.D., Collins, T.J., Mackenzie, L., Roderick, H.J., Berridge, M.J., and Peppiatt, C.M. 2002. 2-Aminoethoxydiphenyl borate (2-APB) is a reliable blocker of store-operated Ca^{2+} entry but an inconsistent inhibitor of InsP3-induced Ca^{2+} release. *FASEB J.* **16**:1145–1150.
38. Lievremont, J.P., Bird, G.S., and Putney, J.W. 2005. Mechanism of inhibition of TRPC cation channels by 2-aminoethoxydiphenylborane. *Mol. Pharmacol.* **68**:758–762.
39. Prakriya, M. and Lewis, R.S. 2001. Potentiation and inhibition of Ca^{2+} release-activated Ca^{2+} channels by 2-aminoethyldiphenyl borate (2-APB) occurs independently of IP3 receptors. *J. Physiol. (Lond.)* **536**:3–19.
40. Lis, A., Peinelt, C., Beck, A., Parvez, S., Monteilh-Zoller, M., Fleig, A., and Penner, R. 2007. CRACM1, CRACM2, and CRACM3 are store-operated Ca^{2+} channels with distinct functional properties. *Curr. Biol.* **17**:794–800.
41. Peinelt, C., Lis, A., Beck, A., Fleig, A., and Penner, R. 2008. 2-APB directly facilitates and indirectly inhibits STIM1-dependent gating of CRAC channels. *J. Physiol.* **586**:3061–3073.
42. Zhou, H., Iwasaki, H., Nakamura, T., Nakamura, K., Maruyama, T., Hamano, S., Ozaki, S., Mizutani, A., and Mikoshiba, K. 2007. 2-Aminoethyl diphenylborinate analogues: Selective inhibition for store-operated Ca^{2+} entry. *Biochem. Biophys. Res. Commun.* **352**:277–282.
43. Hendron, E., Wang, X., Zhou, Y., Cai, X., Goto, J., Mikoshiba, K., Baba, Y., Kurosaki, T., Wang, Y., and Gill, D.L. 2014. Potent functional uncoupling between STIM1 and Orai 1 by dimeric 2-aminodiphenyl borinate analogs. *Cell Calcium* **56**:482–492.
44. Goto, J., Suzuki, A.Z., Ozaki, S., Matsumoto, N., Nakamura, T., Ebisui, E., Fleig, A., Penner, R., and Mikoshiba, K. 2010. Two novel 2-aminoethyl diphenylborinate (2-APB)

analogues differentially activate and inhibit store-operated Ca^{2+} entry via STIM proteins. *Cell Calcium* **47**:1–10.
45. Smyth, J.T., DeHaven, W.I., Bird, G.S., and Putney, J.W., Jr. 2008. Ca^{2+}-store-dependent and -independent reversal of Stim 1 localization and function. *J. Cell Sci.* **121**:762–772.
46. Ishikawa, J., Ohga, K., Yoshino, T., Takezawa, R., Ichikawa, A., Kubota, H., and Yamada, T. 2003. A pyrazole derivative, YM-58483, potently inhibits store-operated sustained Ca^{2+} influx and IL-2 production in T lymphocytes. *J. Immunol.* **170**:4441–4449.
47. Takezawa, R., Cheng, H., Beck, A., Ishikawa, J., Launay, P., Kubota, H., Kinet, J.P., Fleig, A., Yamada, T., and Penner, R. 2006. A pyrazole derivative potently inhibits lymphocyte Ca^{2+} influx and cytokine production by facilitating transient receptor potential melastatin 4 channel activity. *Mol. Pharmacol.* **69**:1413–1420.
48. Zitt, C., Strauss, B., Schwarz, E.C., Spaeth, N., Rast, G., Hatzelmann, A., and Hoth, M. 2004. Potent inhibition of Ca^{2+} release-activated Ca^{2+} channels and T-lymphocyte activation by the pyrazole derivative BTP2. *J. Biol. Chem.* **279**:12427–12437.
49. Ng, S.W., Di Capite, J., Singaravelu, K., and Parekh, A.B. 2008. Sustained activation of the tyrosine kinase Syk by antigen in mast cells requires local Ca^{2+} influx through Ca^{2+} release-activated Ca^{2+} channels. *J. Biol. Chem.* **283**:31348–31355.
50. Di, S.A., Rovedatti, L., Kaur, R., Spencer, J.P., Brown, J.T., Morisset, V.D., Biancheri, P. et al. 2009. Targeting gut T cell Ca^{2+} release-activated Ca^{2+} channels inhibits T cell cytokine production and T-box transcription factor T-bet in inflammatory bowel disease. *J. Immunol.* **183**:3454–3462.
51. Li, J., McKeown, L., Ojelabi, O., Stacey, M., Foster, R., O'Regan, D., Porter, K.E., and Beech, D.J. 2011. Nanomolar potency and selectivity of a Ca^{2+} release-activated Ca^{2+} channel inhibitor against store-operated Ca^{2+} entry and migration of vascular smooth muscle cells. *Br. J. Pharmacol.* **164**:382–393.
52. Derler, I., Schindl, R., Fritsch, R., Heftberger, P., Riedl, M.C., Begg, M., House, D., and Romanin, C. 2013. The action of selective CRAC channel blockers is affected by the Orai pore geometry. *Cell Calcium* **53**:139–151.
53. Chen, G., Panicker, S., Lau, K.Y., Apparsundaram, S., Patel, V.A., Chen, S.L., Soto, R. et al. 2013. Characterization of a novel CRAC inhibitor that potently blocks human T cell activation and effector functions. *Mol. Immunol.* **54**:355–367.
54. Sadaghiani, A.M., Lee, S.M., Odegaard, J.I., Leveson-Gower, D.B., McPherson, O.M., Novick, P., Kim, M.R. et al. 2014. Identification of Orai 1 channel inhibitors by using minimal functional domains to screen small molecule microarrays. *Chem. Biol.* **21**:1278–1292.
55. Merritt, J.E., Armstrong, W.P., Benham, C.D., Hallam, T.J., Jacob, R., Jaxa-Chamiec, A., Leigh, B.K., McCarthy, S.A., Moores, K.E., and Rink, T.J. 1990. SK & F 96365, a novel inhibitor of receptor-mediated calcium entry. *Biochem. J.* **271**:515–522.
56. Singh, A., Hildebrand, M.E., Garcia, E., and Snutch, T.P. 2010. The transient receptor potential channel antagonist SKF96365 is a potent blocker of low-voltage-activated T-type calcium channels. *Br. J. Pharmacol.* **160**:1464–1475.
57. Schwarz, G., Droogmans, G., and Nilius, B. 1994. Multiple effects of SK & F 96365 on ionic currents and intracellular calcium in human endothelial cells. *Cell Calcium* **15**:45–54.
58. Zhu, X., Jiang, M., and Birnbaumer, L. 1998. Receptor-activated Ca^{2+} influx via human Trp3 stably expressed in human embryonic kidney (HEK)293 cells. Evidence for a non-capacitative calcium entry. *J. Biol. Chem.* **273**:133–142.
59. Boulay, G., Zhu, X., Peyton, M., Jiang, M., Hurst, R., Stefani, E., and Birnbaumer, L. 1997. Cloning and expression of a novel mammalian homolog of Drosophila Transient Receptor Potential (Trp) involved in calcium entry secondary to activation of receptors coupled by the G_q class of G protein. *J. Biol. Chem.* **272**:29672–29680.

60. Zakharov, S.I., Smani, T., Dobrydneva, Y., Monje, F., Fichandler, C., Blackmore, P.F., and Bolotina, V.M. 2004. Diethylstilbestrol is a potent inhibitor of store-operated channels and capacitative Ca^{2+} influx. *Mol. Pharmacol.* **66**:702–707.
61. Zeng, B., Chen, G.L., and Xu, S.Z. 2012. Store-independent pathways for cytosolic STIM1 clustering in the regulation of store-operated Ca^{2+} influx. *Biochem. Pharmacol.* **84**:1024–1035.
62. Nakajima, T., Kitazawa, T., Hamada, E., Hazama, H., Omata, M., and Kurachi, Y. 1995. 17beta-Estradiol inhibits the voltage-dependent L-type Ca^{2+} currents in aortic smooth muscle cells. *Eur. J. Pharmacol.* **294**:625–635.
63. Brueggemann, L.I., Markun, D.R., Henderson, K.K., Cribbs, L.L., and Byron, K.L. 2006. Pharmacological and electrophysiological characterization of store-operated currents and capacitative Ca^{2+} entry in vascular smooth muscle cells. *J. Pharmacol. Exp. Ther.* **317**:488–499.

Index